Strategy, Leadership & Innovation

The International CTO Programme

David Ivell

DEDICATION

For my wife, best friend and greatest supporter

ISBN: 978-1-916989-05-4

www.pocketbookcompany.com

THE INTERNATIONAL CTO PROGRAMME

At the Marlow Business School, through The International CTO Programme, we pride ourselves on providing accreditation opportunities tailored to recognise and leverage the rich experience of technology leaders. Our accreditation programme acknowledges the invaluable expertise gained from previous technology leadership roles, offering a pathway for recognition and advancement in the field. Additionally, our meticulously designed online courses in strategy, leadership, and innovation cater specifically to Chief Technology Officers (CTOs) and senior technology professionals, providing them with the tools and insights necessary to navigate the complexities of modern business landscapes. Through expert instruction from industry leaders, participants engage with best practices in technology leadership, gaining a comprehensive understanding of current industry trends, emerging technology and leadership.

Our courses are crafted to offer practical, actionable insights that empower CTOs and senior technology professionals to drive innovation and spearhead strategic initiatives within their organisations. Whether you are seeking to enhance your leadership acumen or stay abreast of the latest technological advancements, The International CTO Programme equips you with the knowledge and skills needed to excel in today's dynamic business environment and accelerate your career.

www.marlowbusinessschool.com

ABOUT THE AUTHOR

David Ivell has held senior IT and Digital positions as Chief Technology Officer, Chief Information Officer, Chief Innovation Officer, Chief Digital Officer, Chief Product and Technology Officer and Consulting Principal across a wide spectrum of organisations. These have included IBM, Credit Suisse, The British Film Institute, The Prince's Trust, Royal Botanic Garden's Kew and The Empowering Group. His experience spans multinational corporations, development labs, startups, and accelerators, covering diverse industry sectors such as Technology, Banking, Insurance, Energy, Media, Edtech, Environment, Not-for-Profit, Government, and Engineering.

David has earned professional recognition, being named as one of the world's top 100 CIOs and consistently features in the UK's CIO top 100. He has been honoured as CDO of the Year, received the Technology and Data for Good Award, coached the Digital Team of the Year, and secured the Social Impact Digital Product of the Year. Additionally, he has served as a regular judge for prestigious events like the AI & Machine Learning Awards, Tech Leaders Awards, Digital Leaders Awards, and Women in IT Awards. As a regular conference speaker David has presented at leading industry events including the opening address for London Tech Week.

David also founded the technology for good group, Energise Resources, in 2018, now with over 300 leading technology professionals around the world donating their time to social and environmental challenges.

Currently a CPTO and Non Exec Director within the Edtech sector and Principal Lecturer at the Marlow Business School.

LIBRARY

Book 1 – Creating a Technology Strategy

In this book, we delve into the art of crafting and executing a robust technology, innovation, or digital strategy that becomes a cornerstone within the broader business objectives. We discuss the essential principles and practical insights needed to seamlessly integrate technology initiatives into the core fabric of organisational success.

Book 2 – Technology Leadership

This book offers a thorough exploration of technology leadership and challenges confronting CTOs today. It covers navigating leadership styles, building strategic relationships, fostering innovation, driving digital transformation, embracing ethics, continuous learning, managing remote/hybrid teams, diversity, inclusion, and navigating mergers and acquisitions.

Book 3 – Driving Innovation

Innovation in technology is the catalyst for industry progress, transforming ideas into user-centric solutions. CTOs play a pivotal role in cultivating a culture of experimentation and learning, driving companies to stay ahead by embracing innovation throughout the product lifecycle. This book sets the stage for a detailed exploration of strategies and best practices for accelerating innovation through the hands of the CTO.

CONTENTS

Book 1 - Strategy

Book 2 – Leadership and Innovation

Book 3 – Innovation

Appendix

Book 1 – Technology Strategy

In this book, we delve into the art of crafting and executing a robust technology, innovation, or digital strategy that becomes a cornerstone within the broader business objectives. We discuss the essential principles and practical insights needed to seamlessly integrate technology initiatives into the core fabric of organisational success.

Chapter 1

Introduction to Strategic Vision and Planning

The role of a Chief Technology Officer (CTO) is not merely confined to the realm of coding and technical expertise; it extends far beyond into the strategic domain. A successful CTO must possess the ability to craft and execute a comprehensive technology, innovation or digital strategy that doesn't just align with the overall business objectives but forms a cornerstone within it.

I have designed strategies for organisations across a wide range of industries, Banking, Energy, Ed-Tech, Not for Profit, Engineering and in organisations from the size of IBM and Credit Suisse right down to the other end of the scale, with start-ups and accelerators. What I have learnt is that no matter the organisation size, or industry, or how well organised the team you are a part of, that moment when you sit down with a piece of paper to start to create that strategy, the task seems like climbing a mountain. To climb a mountain you need a map! These books should give that first piece of structure, or map you need to start designing that step change for your organisation. I won't pretend that this will be everything you need as everyone's style is different and the culture of your organisation will direct your approach, but starting in this chapter, we delve into the crucial aspects of strategic vision and planning that form the bedrock of a CTO's responsibilities.

I have broken down this chapter into four components; Understanding the Business Landscape, Defining Technical Objectives, Creating a Technology Roadmap and Strategy Creation.

1. Understanding the Business Landscape

Before crafting a technology strategy, it's imperative to have a good understanding of the business landscape. A CTO should be well-versed in the industry trends, market dynamics, and the competitive landscape. Conducting regular market analysis, staying abreast of technological advancements, and engaging in cross-functional collaboration are key activities to foster this understanding.

1. Industry Trends & Dynamics
2. Staying Abreast of the Evolving Business Landscape
3. Navigating Regulatory Compliance
4. Globalisation & Geopolitical Influences
5. Embracing Emerging Technologies
6. Customer Centric Technology Trends
7. Innovation Through Partnerships
8. Future Proofing Strategies
9. Competitive Analysis
10. Market Research & Customer Insights

2. Defining Technological Objectives

With a clear grasp of the business landscape, the CTO can then define the technological objectives that will drive the company forward. These objectives should be aligned with the overall business strategy and focus on leveraging technology to gain a competitive edge. Whether it's enhancing operational efficiency, improving customer experience, or exploring new revenue streams, the CTO must set clear and measurable goals.

1. Crafting the Technological Vision
2. Visionary Technology Leadership
3. Strategic Alignment with Business Goals
4. SMART Goal Setting
5. Stakeholder Engagement & Alignment
6. Elevating Innovation
7. Metrics for Success
8. Cultivating a Culture of Ownership

3. Creating a Technology Roadmap

A strategic vision requires a roadmap to guide its implementation and outlines the steps to achieve the defined objectives. This roadmap should encompass short-term and long-term goals, taking into consideration the scalability and adaptability of the chosen technologies. Regularly revisiting and updating the roadmap is essential to ensure alignment with the evolving business landscape.

1. Defining Strategic Milestones
2. Prioritising Initiatives and Dependencies
3. Aligning with Technological Trends
4. Resource Allocation and Budgeting
5. Iterative Development and Agile Principles
6. Communicating the Roadmap Effectively
7. Monitoring and Key Performance Indicators (KPIs)
8. Specialised Tools for creating Roadmaps
9. Technology Roadmap Outline Structure

4. Strategy Creation

In this chapter, we draw the distinction between technology and digital strategies, whilst unravelling the nuanced relationship between the two. Investigating how we prepare a blueprint for digital transformation with a roadmap for seamless integration. We also give some hints to the art of presenting this strategy to the C-Suite and other key stakeholders, ensuring we land the blend of technological vision and business objectives. The chapter concludes with some high level summaries of strategies everyone will recognise that are concise enough to show how a strategy can be transformational.

1. Technology V Digital Strategy
2. The Strategy Presentation
3. Measuring Success
4. Examples of Transformational Strategies

A CTO's role in shaping and executing a technology strategy is pivotal for the success of any modern organisation. By cultivating a deep understanding of the business landscape, setting clear technological objectives, creating a roadmap, balancing innovation and stability, engaging stakeholders, and continuously monitoring and adapting, a CTO can steer the technological direction of the company towards sustained growth and success. Let's delve a little deeper into each of these.

Chapter 2.

Understanding the Business Landscape

In the CTO role, I've learnt that it's not merely advantageous but absolutely crucial to possess an in-depth understanding of the business landscape that the business sits in. Especially in my early career, I hadn't always given enough attention to the business landscape and been surprised when things outside my projects or outside my control caused challenges.

This chapter delves into the complexities of the business environment and how a CTO can leverage this understanding to shape a robust technology strategy.

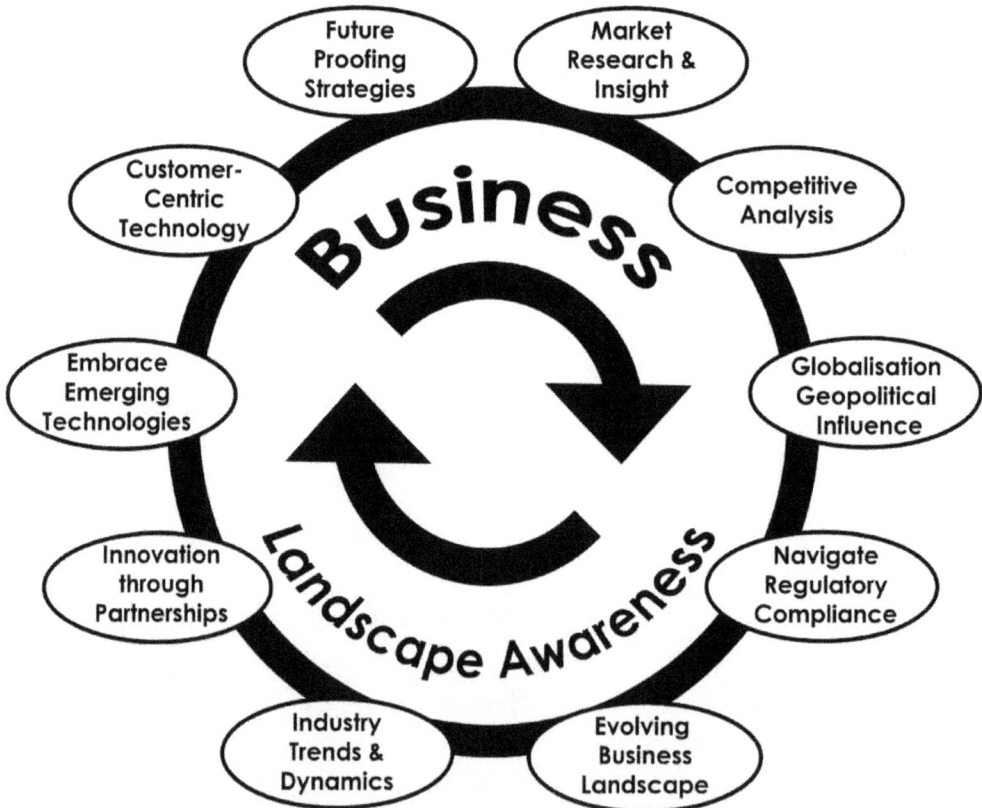

2.1 Industry Trends and Dynamics

Understanding the business landscape hinges on staying informed about industry trends and changes. A CTO should be observant and mindful of emerging technologies, market shifts, and disruptive innovations. This involves keeping an eye on industry publications, attending conferences, and participating in forums. We all know that finding time for this is always easier said than done but by having a deeper awareness of the industry's current state, a CTO can identify opportunities for technological progress and anticipate challenges. From navigating the evolution of emerging technologies to understanding market dynamics, we CTO's can guide the organisation through the complexities of innovation with insight, foresight, and adaptability.

Recognising that industry trends are interconnected elements within broader market dynamics, we require a comprehensive analysis of our market, competitors, and their behaviours. This involves collecting economic indicators, as well at technological ones, while gaining insights into industry positioning, potential disruptors, and areas of opportunity. This strategic awareness forms the foundations for informed decision-making and proactive responses to shifts in the market.

To achieve this, we must engage in forecasting long-term trends, investing in scalable technologies, and fostering a mindset that embraces change. Implementing future-proofing strategies ensures that the organisation isn't merely a passive observer but an active participant in shaping the technological future.

2.2 Staying Abreast of the Evolving Business Landscape:

Continuous learning and adaptability are key in navigating the dynamic business landscape. Let's delve into practical examples of how CTOs can effectively gather and analyse information from diverse sources:

Industry Reports and Analysis:

Industry reports play a pivotal role in the field of technology and innovation by providing comprehensive insights and analysis. These reports offer a panoramic view of the current technological landscape, outlining trends, challenges, and opportunities. Analysts delve into emerging technologies, market dynamics, and competitive landscapes, aiding businesses in making informed decisions. Companies often rely on these reports to assess the feasibility of adopting specific technologies, understand the competitive environment, and strategise for future innovation. Additionally, these reports serve as valuable resources for policymakers, researchers, and investors, fostering a deeper understanding of the technological ecosystem and its implications.

In-Depth Analysis and Forecasting: Leading firms specialising in technology and innovation produce reports that go beyond surface-level observations, offering in-depth analysis and forecasting. These reports delve into the intricacies of technological advancements, providing a nuanced understanding of disruptive technologies and their potential impact on industries. Moreover, they often include market forecasts, enabling businesses to anticipate trends, plan R&D initiatives, and align their innovation strategies with projected market demands. The comprehensive nature of these reports aids stakeholders in mitigating risks, seizing opportunities, and should enable them to position themselves strategically.

Reports to put on your reading list.

1. **Gartner Technology Predictions:** *Gartner is renowned for its comprehensive technology predictions and analyses. Their annual reports cover a wide spectrum of technologies, offering insights into emerging trends, market dynamics, and strategic recommendations. From artificial intelligence to cybersecurity, Gartner's reports are considered authoritative guides for technology decision-makers.*

2. **Forrester's Technology Adoption Reports:** *Forrester Research provides in-depth technology adoption reports that help businesses understand how various technologies are being adopted across industries. These reports offer insights into market trends, potential challenges, and opportunities for innovation. Forrester's analysis is widely used by companies seeking a holistic view of the technology landscape.*

3. **IDC's Worldwide IT Industry Forecasts:** *International Data Corporation (IDC) releases comprehensive forecasts for the global IT industry. Covering hardware, software, services, and emerging technologies, IDC's reports are instrumental in strategic planning. They provide a forward-looking perspective on technology spending, enabling businesses to align their innovation initiatives with projected market trends.*

4. **TechCrunch's Annual State of Tech Report:** *TechCrunch, a leading technology media platform, releases an annual State of Tech report that encapsulates the major trends and disruptions in the tech industry. This report is particularly valuable for startups, investors, and tech enthusiasts looking for a concise overview of the key happenings and shifts in the technology sector.*

5. **McKinsey's Technology Insights:** *McKinsey & Company is globally recognised for*

its strategic insights, and their technology reports are highly regarded. McKinsey's technology insights cover a wide range of topics, including digital transformation, innovation strategy, and the impact of technology on business models. These reports are trusted resources for executives navigating the complexities of technological change.

6. ***Tech Nation's UK Tech on the Global Stage:*** *Tech Nation, a UK-based organisation, releases annual reports that delve into the state of the UK's tech ecosystem. These reports offer a deep dive into sectors like fintech, artificial intelligence, and cybersecurity, providing a detailed analysis of key players, investment trends, and global competitiveness.*

7. ***IDC's UK IT Market Outlook:*** *For businesses seeking a specific focus on the UK IT market, IDC produces reports that forecast technology spending and adoption trends within the country. These reports consider local factors that impact the technology landscape, helping organisations align their strategies with the evolving dynamics in the UK.*

Customer Feedback and Engagement:

Engaging with customers and deriving meaningful insights from their feedback is essential for businesses looking to enhance their products, services, and overall customer experience. One significant advantage of customer engagement is the establishment of a stronger and more loyal customer base. When businesses actively seek and value customer opinions, it fosters a sense of trust and connection. Engaging through various channels, such as social media, surveys, and direct interactions, allows customers to feel heard and appreciated, contributing to increased brand loyalty. In the end we all value being asked and we accept change better if we feel the organisation has made a change in conjunction with listening to their customers.

To effectively engage with customers, businesses employ various techniques. Social media platforms serve as powerful tools for real-time interactions. Companies actively participate in online conversations, respond (hopefully) promptly to customer queries and comments, and leverage social listening tools to monitor discussions about their brand. Additionally, conducting customer surveys, whether through email, website pop-ups, or dedicated survey platforms, provides structured insights. These surveys can address specific aspects of products or services, allowing businesses to gather targeted feedback. Furthermore, implementing feedback mechanisms directly within digital products or services enables businesses to capture in-the-moment reactions, providing valuable insights for continuous improvement.

Customer feedback is a goldmine of information that, when appropriately analysed, can drive strategic decision-making. Advancements in data analytics allow businesses to sift through large volumes of feedback data efficiently. Sentiment analysis tools can gauge the overall mood of customer comments, helping businesses understand the emotional undertones of feedback. Additionally, businesses utilise key performance indicators (KPIs) and metrics derived from customer feedback to quantify satisfaction levels and identify areas for improvement. This data-driven approach ensures that businesses make informed decisions based on the genuine sentiments and preferences of their customer base, ultimately leading to a more customer-centric and competitive business model.

Explore: Brand24 is an AI-powered social listening tool that covers the major social platforms as well as blogs, news sites, and other online sources

Networking and Collaboration:

Engaging in industry networking offers valuable opportunities for professionals to broaden their connections, stay abreast of industry trends, and discover potential collaborations. One common type of industry networking event is professional conferences and trade shows. These events bring together professionals, experts, and leaders from the technology sector, providing an environment conducive to knowledge exchange, idea sharing, and establishing meaningful connections. The trade floor, in itself, can be highly valuable but scout out the workshops and seminars within conferences which often offer more focused interactions.

UK Technology Trade Shows:

1. **Tech Show North (TSN):** Tech Show North is one of the largest technology events in the UK, bringing together professionals, innovators, and businesses to showcase the latest advancements in technology. It covers a wide range of sectors, including cybersecurity, AI, and digital innovation.

2. **London Tech Week:** London Tech Week is an annual event that celebrates the vibrant tech scene in London and the UK. It features conferences, exhibitions, and networking opportunities, attracting global tech leaders and showcasing the city's position as a tech hub.

3. **The Gadget Show Live:** Focused on consumer electronics and gadgets, The Gadget Show Live provides a platform for companies to exhibit their latest products. It attracts a diverse audience, including tech enthusiasts, industry

professionals, and the general public.

USA Technology Trade Shows:

1. **CES (Consumer Electronics Show):** Held annually in Las Vegas, CES is one of the world's largest technology trade shows. It showcases a broad spectrum of consumer electronics, from cutting-edge gadgets to emerging technologies, drawing industry leaders, innovators, and enthusiasts from around the globe.

2. **SXSW (South by Southwest):** While not exclusively a tech event, SXSW in Austin, Texas, incorporates a significant technology track. It brings together professionals from various industries, including technology, film, and music, providing a unique platform for networking, innovation, and the exploration of new ideas.

3. **TechCrunch Disrupt:** TechCrunch Disrupt is an annual technology conference that gathers startups, investors, and industry experts. Held in various locations across the United States, it features a startup competition, insightful panels, and opportunities for networking, making it a prominent event in the tech industry.

Others not to miss

1. **Computex Taipei - Taiwan:** Computex Taipei is a major international information technology exhibition held annually in Taipei, Taiwan. It focuses on showcasing the latest advancements and innovations in computer hardware, electronics, and information technology. It is one of the largest and most important events in the Asia-Pacific region for the technology industry.

2. **Mobile World Congress (MWC) - Barcelona, Spain:** Mobile World Congress is a globally recognised event that brings together professionals and companies from the mobile technology industry. It serves as a platform for showcasing new mobile devices, discussing industry trends, and networking. While it is not limited to a specific country, it is hosted annually in Barcelona, Spain, and is one of the largest gatherings in the mobile technology sector.

Meetups

Meetups also play a crucial role in the field of product management by providing a platform for professionals to connect, share insights, and stay updated on industry trends. These gatherings offer a unique opportunity for product managers to network with peers, exchange ideas, and learn from each other's experiences. Attendees often gain valuable insights into best practices, emerging technologies, and innovative approaches to product development. For instance, a product management meetup in Silicon Valley might feature seasoned product managers sharing success stories and discussing the challenges they faced while bringing products to market.

Furthermore, meetups in product management foster a sense of community and collaboration. Attendees can engage in interactive discussions, participate in workshops, and even collaborate on real-world problem-solving activities. These events enable product managers to build a strong professional network that extends beyond the meetup, creating lasting connections that can lead to potential collaborations or partnerships. An example could be a meetup focused on agile product management methodologies, where practitioners gather to share their experiences in implementing agile frameworks, exchange tips on overcoming common obstacles, and discuss the latest tools that enhance agile product development processes.

UK Meetups:

1. **London Tech Meetup: Description:** London Tech Meetup is a vibrant community that brings together tech enthusiasts, entrepreneurs, and professionals. They host regular events featuring guest speakers, startup pitches, and networking opportunities. **Website:** London Tech Meetup

2. **TechNation Meetups: Description:** TechNation is a network that supports and connects tech entrepreneurs across the UK. Their meetups cover a range of topics, from funding to emerging technologies, providing a platform for knowledge sharing and networking. **Website:** TechNation Meetups

3. **Silicon Roundabout Meetup: Description:** Focused on the tech hub in East London, Silicon Roundabout Meetup gathers entrepreneurs, developers, and investors. Events cover various aspects of technology, from coding workshops to discussions on the latest industry trends. **Website:** Silicon Roundabout Meetup

USA Meetups:

1. **SF New Tech: Description:** Based in San Francisco, SF New Tech is a meetup that showcases new and innovative technologies. It provides a platform for startups

to demo their products, share ideas, and connect with the tech community.
Website: SF New Tech

2. **NYC Tech Meetup: Description:** NYC Tech Meetup is one of the largest tech meetups globally, based in New York City. It features live demos and presentations from local startups, tech professionals, and industry experts, fostering collaboration and idea exchange. **Website:** NYC Tech Meetup

3. **Tech in Motion: Silicon Valley: Description:** Tech in Motion organises events across several major tech hubs, including Silicon Valley. Their meetups cover a broad spectrum of tech topics, from AI to cybersecurity, attracting professionals and enthusiasts eager to stay informed about industry developments. **Website:** Tech in Motion: Silicon Valley

Professional Organisations

Another avenue for industry networking is through professional associations and organisations. These groups often host regular meetings, webinars, and networking sessions tailored to the interests and needs of their members. Joining these associations provides access to a network of like-minded professionals, creating opportunities for mentorship, collaboration, and staying informed about industry advancements. Additionally, online platforms, such as LinkedIn groups or industry-specific forums, offer virtual networking spaces where professionals can engage in discussions, share resources, and connect with peers globally.

United Kingdom (UK): BCS, The Chartered Institute for IT: BCS is a professional body for the IT industry in the UK. It offers professional development, certifications, and resources to IT practitioners, promoting excellence in the field. **Website:** BCS, The Chartered Institute for IT

United States of America (USA): IEEE Computer Society: IEEE Computer Society is a leading organisation for computing professionals globally. It provides resources, publications, and events to advance technology and promote collaboration. **Website:** IEEE Computer Society

Europe: DigitalEurope: DigitalEurope is a leading trade association representing the digital technology industry in Europe. It advocates for policies that foster innovation, digital transformation, and a competitive digital market. **Website:** DigitalEurope

Japan: Information-technology Promotion Agency, Japan (IPA): IPA is a government-affiliated organisation in Japan that promotes the use of information technology. It works to enhance cybersecurity, support IT research, and contribute to the development of the technology industry. **Website:** Information-technology Promotion Agency, Japan

Australia: Australian Computer Society (ACS): ACS is the professional association for the Information and Communication Technology (ICT) sector in Australia. It provides networking opportunities, resources, and support for ICT professionals. **Website:** Australian Computer Society (ACS)

To maximise the impact of industry networking, professionals should adopt proactive approaches. Setting clear goals before attending events, whether physical or virtual, helps individuals focus on specific outcomes, such as connecting with potential collaborators or gaining insights into emerging trends. Don't just attend events, actually participate! Actively participating in discussions, asking thoughtful questions, and sharing experiences during networking events contribute to more meaningful engagement. Additionally, maintaining a strong online presence, regularly updating professional profiles (EG LinkedIn), and contributing to relevant conversations on social media platforms foster continuous networking beyond traditional events. This can really help create an impactful professional network.

Implementing a Strategic Analysis Process: A 10 Step Action Plan

Implementing a robust analysis process is essential for CTOs to stay ahead of the evolving business landscape. This 10-step guide provides detailed insights and practical examples for each stage:

1. Define Clear Objectives:

- Clearly articulate the objectives of your analysis process. If your organisation is focused on exploring emerging technologies, specify that the objective is to identify innovative solutions that align with business goals and customer needs. For instance, the objective could be to evaluate the feasibility of integrating blockchain technology into existing platforms.

2. Identify Key Information Sources:

- Pinpoint crucial sources of information for your industry. Key sources may include competitor websites for product offerings and pricing strategies, industry reports on online retail trends, and customer reviews on popular platforms. By diversifying your information sources you gain a holistic view of market dynamics.

3. Establish a Monitoring Schedule:

- Devise a regular monitoring schedule to review information sources. In a tech-driven company, this could involve checking industry reports quarterly to stay updated on technological advancements, engaging with customers through monthly surveys to gauge satisfaction and preferences, and attending relevant technology conferences annually for firsthand insights and networking opportunities.

- If you don't have the time, task each member of your team to investigate one area and report back at the beginning of monthly meetings. I used this technique at The Prince's Trust and it proved very successful.

4. Utilise Technology Tools:

- Leverage technology tools for efficient data collection. Implement social listening tools to track online conversations about your brand or industry. For example, monitoring social media discussions in the aftermath of a product launch can provide real-time insights into customer sentiments and identify areas for improvement. Checkout *Hootsuite, Brandwatch, Awario, SocialBee, Digimind*

5. Encourage Cross-Functional Collaboration:

- Foster collaboration with teams beyond technology. For example, collaborating with the marketing team can uncover insights into shifting trends. Regular cross-functional meetings can lead to a shared understanding of market demands, influencing the development of technology solutions that address emerging needs.

6. Develop Customer Feedback Mechanisms:

- Implement mechanisms to collect and analyse customer feedback. How do you get to hear the voice of your customer? Other departments may have a closer ear, but how do you engage? Try starting with a focus group to get detailed thoughts, ideas and critiques, then build into a survey to check the principles more widely, then maybe bring back a focus group to review. Have an area on your customer facing applications where they can leave their thoughts.

7. Participate in Networking Opportunities:

- Actively engage in industry events, conferences, and forums. Attending a technology summit not only exposes you to the latest advancements but also provides networking opportunities.

8. Establish KPIs for Analysis:

- Define KPIs that align with your analysis objectives. KPIs could include metrics like user adoption rates, software performance benchmarks, and customer satisfaction scores. By setting measurable indicators, you gain quantifiable insights into the success of your technological endeavors.

9. Conduct Regular SWOT Analysis:

- Perform periodic SWOT analysis to evaluate internal and external factors. For example, identifying weaknesses in existing security measures through SWOT analysis can drive targeted investments. This process ensures that the organisation is fortified against evolving cyber threats and remains resilient in the face of emerging challenges.

10. Iterate and Adapt:

- Embrace a culture of continuous improvement, but if your gut feel is that it's not working and that no number of iterative tweaks are going to fix it, don't be afraid to throw it all away and start again.

By incorporating these detailed insights, CTOs can construct a thorough and effective analysis process that not only enhances their understanding of industry dynamics but also positions them as strategic leaders driving technological excellence.

Additional Resources for Technological Trend Alignment:

1. **_TechCrunch:_** A leading technology media property, dedicated to obsessively profiling startups, reviewing new Internet products, and breaking tech news.

2. **_Wired_**: Offers in-depth coverage of current and future trends in technology, business, and culture.

3. **_Gartner's Technology Trends:_** Gartner's official source for technology trends and insights.

4. **_Trend Hunter:_** A platform that showcases innovative ideas and trends across various industries.

5. **_Statista:_** Provides statistical data on a wide range of topics, including technology trends and market forecasts.

6. **_Google Alerts:_** A free service that delivers email updates of the latest relevant Google search results based on chosen queries.

7. **_Feedly:_** A content aggregation tool that allows you to subscribe to various sources, keeping you informed about industry trends.

8. **_GitHub:_** A platform for developers to collaborate on and discover new projects, often indicating emerging trends in software development.

9. **_Tech News websites:_** Provides up-to-date news on the latest technology trends and developments.

10. **_Information Commissioner's Office (ICO):_** The official UK regulatory authority for data protection and privacy, essential for staying informed about evolving data protection laws.

2.3 Navigating the Regulatory and Compliance Landscape: A Crucial Dimension in Understanding the Business Environment

If you are like me, the moment we hear the words "regulatory" and "Compliance", it's hard to stop your eyes from starting to glaze over. It feels like a hindrance or a brake being applied to our innovation, but understanding the regulatory and compliance landscape is crucial for a CTO to grasp the broader business context. Keeping up with regulatory trends and compliance requirements is not just a legal obligation but also a strategic need. This landscape is ever-changing, with laws and standards continuously evolving. Key areas may encompass data protection laws, cybersecurity regulations, data sovereignty and broader legal frameworks that govern the industry. The CTO needs to be skilled at tracking these changes, understanding their implications, and adjusting technological strategies accordingly.

Continuous Compliance Monitoring:

Monitoring regulatory updates is important. This involves keeping a close eye on changes in data protection regulations, such as GDPR or CCPA, and understanding how they impact the organisation's handling of sensitive information. For instance, adjusting data storage practices or enhancing cybersecurity measures in response to updated regulations ensures not only legal adherence but also a fortified technological infrastructure.

Strategic Alignment with Legal Requirements:

A CTO should not view regulatory compliance in isolation but rather as an integral part of the overall business strategy. By understanding and adapting to the regulatory landscape, a CTO can ensure that technological strategies align seamlessly with legal requirements and ethical standards. For example, in the financial sector, compliance with stringent financial regulations may either lead to the implementation of secure transaction protocols and advanced fraud detection systems, or if regulatory changes are missed you will be sitting in a very uncomfortable chair in front of your CEO or investors.

Ethical Considerations:

Incorporating ethical considerations into technological strategies is increasingly important. Not just because it's the right thing to do, and it is, but additionally because you will be judged internally and externally how you handle ethical considerations. Your judgement will be considered, but it's also a sensible business strategy. This involves not only complying with existing regulations but also anticipating and preparing for emerging ethical standards within the industry. Proactive measures, such as implementing ethical AI frameworks or ensuring transparency in data usage, position the organisation as a responsible technology leader. This is covered in greater detail in future chapters.

Collaboration with Legal Experts:

Collaboration with legal experts is a strategic move. Having expert legal opinion in your strategy design will always keep the "nay sayers" at bay. How many times have you heard "you'll never be allowed to do that, or that won't work with GDPR" only to find that the other experts in the room may not be the legal experts that they think they are. More often than not, your original view or idea will be correct, and the temptation for other business function leaders to find a reason why not do something, especially if it wasn't their idea, will be extinguished up front. Establishing a working relationship with legal professionals ensures that the technology team has access to real-time legal insights. This collaboration facilitates the development of compliance protocols that are not only legally sound but also conducive to the seamless integration of technology with legal requirements.

Education and Training:

Given the complexity of regulatory landscapes, educating the technology team on legal nuances is sensible. Conducting regular training sessions ensures that the team is well-informed about compliance requirements relevant to their work and minimises the risk of inadvertent non-compliance, whilst fostering a culture of legal awareness within the organisation. Try inviting a specialist to a team meeting or point the team at a video for them to watch.

Impact on Technological Innovation:

Understanding the regulatory landscape doesn't merely limit technological innovation but can be a catalyst for innovation and can often be seen as an opportunity. Where there is change, someone is making money. For example, anticipating stricter cybersecurity regulations might prompt the development of advanced encryption algorithms or the implementation of blockchain technologies to enhance data security.

Global Considerations:

We should consider not only local regulations but also international standards. This involves staying informed about global data protection initiatives, regional cybersecurity norms, and international legal frameworks. Adapting technology strategies to comply with these global considerations positions the organisation as a responsible and forward-thinking player on the world stage but more so it enables us to spot new commercial opportunities or to streamline how our systems are designed.

This may be an area you only decide to explore when you have a project that warrants it, however regulations are changing all the time. For example the UK and Australia signed a free trade agreement in December 2021, the first new UK trade deal the UK had signed since Brexit. It included real positives for CTO's with international customers in both UK and Australia in regards Data Sovereignty and sharing data. Very few organisations have taken these changes into account and are missing opportunities. It meant that UK businesses can now avoid setting up servers and data storage in Australia and freely collect, process and transfer data between the two countries. As long as the UK's data protection laws are being followed then that will be acceptable in Australia and we can service those customers in a more modern way. It also enabled access to certain anonymised government data sets we did not have authority to access before.

Mastery of the regulatory and compliance landscape is increasingly a cornerstone of a CTO's role in understanding the business environment. By being vigilant, strategically aligned, ethically conscious, and globally aware, a CTO not only ensures legal adherence but also leverages the regulatory landscape as a guide for responsible and innovative technological leadership.

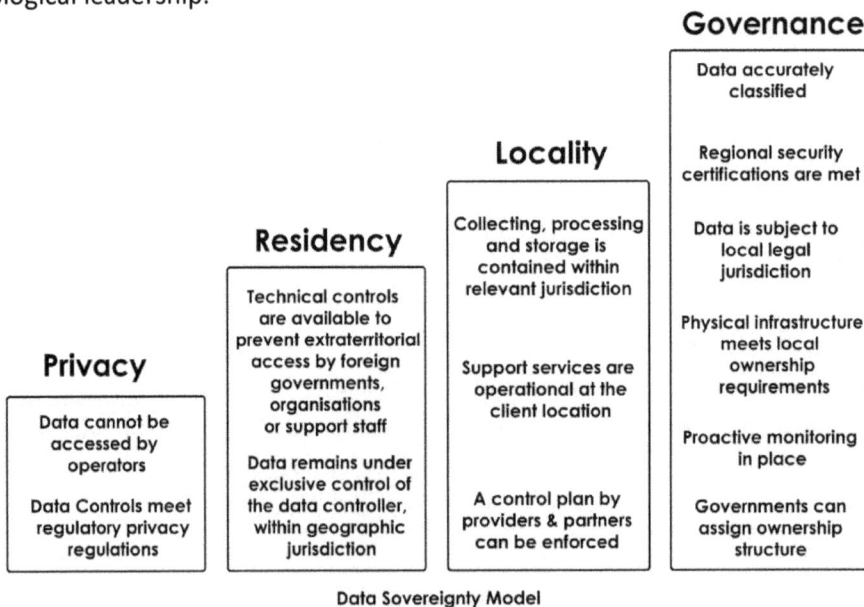

Governance

Data accurately classified

Regional security certifications are met

Locality

Data is subject to local legal jurisdiction

Residency

Collecting, processing and storage is contained within relevant jurisdiction

Physical infrastructure meets local ownership requirements

Privacy

Technical controls are available to prevent extraterritorial access by foreign governments, organisations or support staff

Support services are operational at the client location

Proactive monitoring in place

Data cannot be accessed by operators

Data remains under exclusive control of the data controller, within geographic jurisdiction

A control plan by providers & partners can be enforced

Governments can assign ownership structure

Data Controls meet regulatory privacy regulations

Data Sovereignty Model

Navigating Regulatory and Compliance: A 10-Step Action Plan

Understanding and navigating the regulatory and compliance landscape is paramount for CTOs to ensure that technological strategies align seamlessly with legal requirements and ethical standards. This 10-step guide suggests steps for mastering this complex dimension of the business environment:

1. Establish a Regulatory Intelligence System:

Deploy a robust regulatory intelligence system utilising specialised platforms and legal databases. This system should provide real-time alerts and comprehensive updates on changes to data protection laws, cybersecurity regulations, and broader legal frameworks. https://www.cube.global/ is good example.

2. Create a Compliance Calendar:

Develop a dynamic compliance calendar that not only lists regulatory deadlines but also includes preparatory milestones. For instance, if GDPR or Cyber compliance is on the horizon, the calendar should outline tasks such as data audit, policy review, and staff training to ensure readiness.

3. Conduct Regular Compliance Audits:

Conduct thorough compliance audits at regular intervals. These audits should encompass a meticulous review of existing policies, procedures, and technology implementations. EG In healthcare technology, this might involve evaluating systems against the Health Insurance Portability and Accountability Act (HIPAA) standards.

4. Establish a Cross-Functional Compliance Team:

Form a cross-functional team that includes legal experts, technology professionals, and compliance specialists. This team should meet regularly to discuss evolving regulations, interpret their implications, and strategise on aligning technological advancements with compliance requirements.

5. Educate and Train the Team:

Implement a continuous education program for the technology team. Regular training sessions should cover legal nuances, compliance intricacies, and ethical considerations. In e-commerce, for instance, the team should be well-versed in consumer protection laws and their impact on technology development.

6. Integrate Compliance into Technology Workflows:

Embed compliance checkpoints directly into technology development workflows. This integration ensures that compliance considerations are inherent in the development process. In user data protection, for example, compliance with regulations like the California Consumer Privacy Act (CCPA) or European GDPR should be a fundamental part of every feature development cycle.

7. Collaborate with External Legal Experts:

Establish a proactive relationship with external legal experts specialising in technology-related regulations. These experts should provide ongoing guidance on emerging legal landscapes. In AI development, for instance, collaboration could involve periodic consultations to navigate evolving ethical considerations.

8. Implement Ethical Frameworks:

Develop and implement comprehensive ethical frameworks that surpass minimum legal requirements. In AI development, this could involve creating guidelines addressing issues such as bias, transparency, and user consent. Ethical frameworks ensure that technology aligns with evolving ethical standards.

9. Regularly Assess Global Compliance Standards:

Stay abreast of global compliance standards, especially if operating in an international context. Adjust technology strategies to comply with these global considerations, enhancing the organisation's global reputation.

10. Foster a Proactive Compliance Culture:

Instil a proactive compliance culture within the organisation. Encourage employees to report potential compliance issues early, fostering an open communication channel.

This guide can help can navigate the regulatory and compliance landscape. This approach ensures that technological strategies not only comply with existing laws but also position the organisation as an ethical and responsible leader in the ever-evolving business environment.

...and, of course, when someone asks "where's your plan for regulatory compliance", here it is.

2.4 Embracing Globalisation and Geopolitical Influences: A Strategic Imperative

The interconnected nature of the global economy introduces a layer of complexity that directly impacts technological strategies and the overall success of an organisation, as decisions made in one corner of the world reverberate across continents.

A prime example is the semiconductor shortage in Asia, causing production halts worldwide and affecting entities ranging from European phone manufacturers to Silicon Valley AI startups. This scenario underscores the imperative for CTOs to adopt a global perspective, recognising that technological strategies and decisions are not isolated but resonate on a worldwide scale. It necessitates thinking beyond borders, anticipating disruptions, and capitalising on opportunities that may arise globally.

The dynamics of international trade and market access play a pivotal role in shaping technology landscapes. CTOs need to grasp the intricacies of trade dynamics and international policies to formulate resilient technology strategies. For instance, tariff impositions or trade agreements can alter the cost structure of technology components, influencing production costs and subsequently impacting the pricing and demand for end-user products. Understanding these geopolitical factors is critical for crafting strategies that account for potential disruptions or opportunities stemming from shifts in global trade dynamics. I have been bitten by this before in a plan for a major new employee equipment roll out when, half way through, a trade dispute blocked 500 HP laptops being delivered, causing a well planned project to flounder, with knock on to training and application launch.

More now than ever, markets transcend national boundaries and evolve under the influence of diverse cultural, economic, and political forces. Consider a video streaming service vying for global dominance; to thrive, it must comprehend regional content preferences, navigate the infrastructure disparities of each continent, and align with diverse regulatory and cultural landscapes. Operating on a global scale demands a nuanced understanding of the unique factors shaping each market, highlighting the importance of cultural sensitivity and adaptability in technology strategies.

Geopolitical shifts, such as tensions between nations or policy changes, can exert profound effects on technology markets. CTOs who can anticipate these shifts gain the ability to develop adaptive strategies that ensure resilience and sustained growth. For instance, being aware of upcoming governmental elections, understanding the potential impact of new policies, or predicting changes in tax regulations allows CTOs to proactively plan for shifts in the technology landscape. For example, in the realm of Ed-

Tech, anticipating fluctuations in Government Education Department budgets becomes crucial, enabling companies to strategise responses to potential increases or decreases that may impact the schools they collaborate with.

Opportunities in International Collaborations:

Collaborations on a global scale can spark innovation and open new avenues for technological advancements. Global collaborations present significant opportunities for innovation and technological advancement. The ability to strategically position an organisation to leverage international partnerships is crucial, particularly for forming collaborations with research institutions or technology companies in emerging markets. This approach provides access to diverse talent pools, fosters the exchange of innovative ideas, and unlocks untapped markets, contributing to multifaceted organisational development.

The diverse regulatory frameworks across borders pose significant challenges for technology strategies on a global scale. Navigating this complex terrain is crucial to ensuring compliance with diverse legal requirements and ethical standards. An example is the implementation of a data-driven technology solution, where adherence to various data protection laws, such as GDPR in Europe and HIPAA in the United States, is essential to avoid legal complications. A holistic approach that considers and accommodates these regulatory variations is necessary for fostering a technology strategy that is both globally effective and ethically sound.

Cultural Sensitivity in Technology Deployment:

Cultural sensitivity in technology deployment is a crucial factor influencing the success of global strategies. Beyond language translation, it encompasses an understanding of the cultural nuances that shape user behaviours and preferences. CTOs with an appreciation for cultural diversity can tailor technology strategies to align seamlessly with local expectations, facilitating smoother and more successful deployments. This extends beyond mere adaptation of language; it involves a comprehensive understanding of how cultural factors influence user interaction with technology.

A concrete example of cultural sensitivity in technology deployment lies in user interface design. Different cultures may have distinct preferences regarding the layout, visual elements, and navigation of user interfaces. A technology solution that resonates with one cultural group may not necessarily have the same impact elsewhere. CTOs must recognise these differences and, when necessary, implement localisation strategies to customise user interfaces for diverse global user bases. Additionally, language preferences play a pivotal role, and offering a multilingual interface can significantly enhance user experience, ensuring inclusivity and accessibility across various linguistic backgrounds.

However, it is crucial for CTOs to be mindful of potential pitfalls associated with cultural insensitivity, especially in the production of content. Dangers arise when content created for one region inadvertently causes cultural offence in another. Certain images, symbols, or terminology that may be innocuous in one cultural context could be perceived as inappropriate or offensive in another. The mismatch between the intended cultural understanding and the interpretation by users from different regions can lead to negative user experiences and, in extreme cases, harm the reputation of the technology product, service or brand. To mitigate these risks, rigorous cultural reviews and consultations with local experts are essential to ensure that the content in a knowledge library, on company websites or even on social media is culturally appropriate and sensitive to the diverse perspectives of global users. By proactively addressing these concerns, CTOs can not only enhance user satisfaction but also avoid potential cultural mishaps that may impact the acceptance and success of technology deployments on a global scale.

Example: in 2019 Gillette created content that the marketing team thought would be a winner on the back of the "me too" campaign and challenging "toxic masculinity". It included a man shaving off his traditional style moustache. However, the act of shaving a moustache can be associated with mourning or shame in some cultures. It was seen as insensitive and disrespectful, leading to outrage and accusations of cultural ignorance. Gillette apologised and immediately pulled the campaign, but they faced legal action, lost sales and market share which some analysts suggested may have been up to 10% in Middle East geographies, lost the investment in that campaign and had to create new campaigns and PR to try and retrieve brand image. A generation of brand loyalty was lost in some Middle East countries and lifetime value was at rock bottom.

A CTO's understanding of globalisation and geopolitical influences is not merely a strategic advantage; it is an essential component for navigating the complex and interconnected world of technology leadership. By embracing these global dimensions, CTOs position their organisations for agility, resilience, and innovation on the international stage.

Globalisation and Geopolitical Shifts: A 10-Step Action Plan

Globalisation and geopolitical influences demand strategic preparedness. This 10-step guide provides detailed insights into each step, offering actionable information for CTOs to navigate and prepare for changes in the global landscape:

1. Establish a Global Intelligence System:

Create a robust global intelligence system by leveraging data analytics. Explore creating a custom dashboard that aggregates real-time geopolitical news, economic indicators, and technology trends relevant to your industry. Utilise automated alerts for timely updates on critical developments. A good source is https://www.spglobal.com/en/enterprise/geopolitical-risk/ If you have an intranet then consider feeding that with curated content. Here is a test: If you don't know what is happening right now with the G20 (Governments and central banks) or G8, or the impact of Saudi Arabia joining BRICS bloc on your business, then your intelligence system needs reviewing.

2. Diversify Supply Chain Strategies:

Conduct a thorough analysis of your supply chain vulnerabilities. Identify key components and technologies sourced from specific regions. Develop a diversified sourcing strategy by qualifying alternative suppliers, assessing their geopolitical risks, and establishing contingency plans for seamless transitions during disruptions.

3. Scenario Planning Workshops:

Facilitate cross-functional scenario planning workshops. Engage representatives from technology, supply chain, legal, and strategic planning teams. Create *plausible* geopolitical scenarios, assess their impact on operations, and collaboratively develop agile contingency plans that align with technology strategies.

4. Cultural Competence Training:

Implement a comprehensive cultural competence training program. Engage cultural experts or collaborate with external training providers to create immersive learning experiences. Incorporate real-world case studies to enhance the team's ability to design technology solutions with cultural sensitivity. For example, if you want to be successful trading in New Zealand you need to understand how to communicate respectfully. It is not the same as the UK or US. Try using words such as "We celebrate the diversity and knowledge of First Nations people and acknowledge that we work, walk and talk on your lands. We thank and pay respect to your ancestors and commit to truth telling and reconciliation".

5. Regular Regulatory Compliance Audits:

Regularly update a centralised database with global regulations, ensuring that the technology team has instant access to the latest compliance requirements.

6. Strategic Global Partnerships:

Foster global partnerships through proactive networking. Attend international conferences, collaborate with technology consortia, and explore joint ventures with organisations from different regions. Develop a partnership framework that facilitates knowledge exchange, joint research, and collaborative projects.

7. Continuous Education on Geopolitical Trends:

Implement a continuous learning culture within the technology team. Organise regular webinars featuring global affairs experts, geopolitical analysts, or industry leaders. Encourage team members to participate in relevant workshops, conferences, and online courses to stay informed about geopolitical trends.

8. Technology for Global Impact Initiatives:

Formulate initiatives that align technology with global impact goals. Establish cross-functional task forces to identify areas where technology can contribute to global challenges such as climate change, healthcare disparities, or education accessibility. Develop pilot projects to test and refine these impactful solutions.

9. Agile Policy Development:

Adopt an agile policy development approach that incorporates feedback loops. Establish cross-functional policy review teams that include legal experts, compliance specialists, and technology leaders. Utilise cloud-based collaboration platforms for real-time policy editing, ensuring swift adaptation to changing geopolitical and regulatory landscapes.

10. Regular Technology Risk Assessments:

Conduct comprehensive technology risk assessments with a specific focus on geopolitical risks. Engage third-party experts to perform penetration testing, vulnerability assessments, and geopolitical risk analysis. Develop a risk register that prioritises identified risks and guides the implementation of risk mitigation strategies.

By enhancing each step with detailed implementation strategies, CTOs can ensure that their organisations are not only prepared for the challenges of globalisation and geopolitical shifts but are also strategically positioned to thrive in this ever-evolving global landscape.

2.5 Embracing Emerging Technologies:

Driving Innovation and Competitive Edge:

Remaining at the forefront of emerging technologies is not just a strategic choice but a necessity for organisations aspiring to foster a culture of innovation and secure a competitive edge. It involves not only understanding these cutting-edge technologies but also effectively harnessing them to drive continuous innovation and gain strategic advantages within respective industries.

When presenting a technology strategy, referring to some of these resources to show that our own strategy will not be left behind, and more so that we are positioning ourselves to benefit from these evolving technologies, will be seen as a positive. Showing our direction or technology stack aligns with thinking referenced in tech reports from Deloitte, Accenture, and IBM, will position the team as leaders in the tech landscape.

Creating a culture of continuous learning is instrumental in achieving leadership in emerging technologies. Platforms like Coursera, edX, and Udacity provide a plethora of courses covering the latest technologies, offering team members opportunities to upskill. Actively participating in industry conferences such as CES, TechCrunch Disrupt, and Web Summit, either in person or virtually, is equally crucial. This engagement provides invaluable exposure to the latest trends, networking opportunities, and insights from industry leaders, ensuring that organisations remain well-versed in the rapidly evolving technological landscape.

Another critical aspect is the aggregation of insights through news, podcasts, and innovation tools. Technology news aggregators like Techmeme, Hacker News, and Slashdot deliver up-to-date information on industry trends. Engaging with podcasts and webinars, such as "The Vergecast," "AI with AI," and TED Talks Technology, facilitates absorbing information on emerging technologies and industry insights. Implementing innovation management tools like Spigit/Idea Place, IdeaScale, and Brightidea streamlines the collection of ideas, manages innovation projects, and encourages collaboration within the organisation.

Navigating the ecosystem involves exploring startup ecosystems through platforms like Crunchbase, AngelList, and CB Insights to identify potential collaborators or competitors. Collaborative platforms for research and development, such as GitHub, GitLab, and Bitbucket, facilitate efficient collaboration on innovative projects. Engaging with research and market reports from industry authorities like Gartner, Forrester, and

McKinsey Insights provides in-depth views into emerging technologies and market trends. Staying ahead of regulatory changes with RegTech tools like ComplyAdvantage and Onfido ensures compliance in the tech landscape.

Community engagement and tech trend reports play a crucial role. Participating in online communities and forums such as Stack Overflow, Reddit's Technology Subreddits, and LinkedIn Groups enables organisations to engage in discussions and share experiences related to emerging technologies. Exploring thought leadership platforms like Towards Data Science and TechCrunch for industry opinions, case studies, and emerging tech insights further enriches our understanding.

The involvement of organisations specialising in predicting emerging technology trends becomes crucial. These "future scouts" employ various methods, including analysing patent filings, research papers, convening expert panels, and scouring the internet. Techtrend, a global powerhouse, stands out in this arena, maintaining a pulse on tech advancements. Their researched annual reports, packed with expert insights, have become a benchmark for businesses and policymakers. Techtrend leverages its network of researchers and analysts to identify promising technologies across fields like artificial intelligence, biotechnology, and quantum computing, providing an invaluable roadmap.

In conclusion, understanding emerging technologies is one linchpin of technological leadership, driving innovation, enabling strategic decisions, and ensuring organisations remain agile, competitive, and resilient in an ever-evolving technological landscape. Organisations that embrace this imperative lead themselves into the future and shape the technological frontier for industries at large.

Identifying Emerging Technologies: A 10-Step Action Plan

Navigating the landscape of emerging technologies requires a comprehensive approach. This expanded 10-step guide provides practical examples for each step to help CTOs systematically identify technologies that could influence their business or be utilised by competitors:

1. Establish a Technology Scouting Team:

Form a cross-functional team with individuals specialising in areas such as artificial intelligence, cybersecurity, blockchain, and other relevant fields. For example, include a data scientist, a cybersecurity expert, and a blockchain specialist.

2. Continuous Market Monitoring:

Subscribe to industry reports from reputable sources such as Gartner, Forrester, and IDC. For instance, Gartner's "Hype Cycle" reports provide insights into the maturity and adoption trends of emerging technologies. Set up custom alerts on technology news platforms like TechCrunch, Wired, and MIT Technology Review to receive real-time updates.

3. Competitor Technology Audits:

Conduct regular audits using tools like Crunchbase to track funding rounds and partnerships of competitors. For example, if a competitor in the e-commerce sector partners with a machine learning startup, it indicates a potential adoption of AI technologies for personalised customer experiences.

4. Engage in Industry Forums and Conferences:

Attend conferences like CES (Consumer Electronics Show) for a broad overview and niche conferences such as the International Conference on Machine Learning (ICML) for in-depth insights. Engage with industry forums like Stack Overflow or GitHub discussions to stay connected with developers and technologists.

5. Foster Cross-Functional Collaboration:

Initiate cross-functional workshops with marketing, sales, and product development teams. For instance, a joint workshop might uncover opportunities for integrating augmented reality (AR) into a product, enhancing both marketing and user experience.

6. Leverage External Consultants and Advisors:

Collaborate with technology advisory firms like Accenture or Deloitte. External consultants can provide insights into emerging technologies based on their diverse client experiences. Engage with renowned technology thought leaders or academic advisors who can offer tailored perspectives on the latest technological advancements.

7. Monitor Patent Filings and Research Papers:

Utilise platforms like Google Patents or the United States Patent and Trademark Office (USPTO) for patent searches. Analyse research papers on platforms like arXiv or Google Scholar. For instance, patent filings related to quantum computing technologies or breakthroughs in materials science can indicate areas of potential disruption.

8. Analyse Venture Capital and Startup Activity:

Explore databases like CB Insights for insights into venture capital investments. For example, if a significant investment is made in a startup focusing on edge computing, it suggests the rising importance of this technology in the industry.

9. Establish Technology Scanning Workshops:

Host workshops where team members actively contribute to identifying emerging technologies. Encourage brainstorming on platforms like Miro or MURAL. For instance, a collaborative session might identify the potential of integrating virtual reality (VR) in training programs.

10. Develop a Technology Radar Framework:

Create a visual technology radar using tools like ThoughtWorks Technology Radar. Classify technologies into categories such as "Adopt," "Trial," "Assess," and "Hold." For example, a technology radar might place edge computing in the "Assess" category, signalling potential future adoption based on current trends.

By incorporating these examples into the 10-step guide, CTOs can execute a robust strategy for identifying emerging technologies, ensuring that their organisations remain agile, innovative, and well-prepared for the technological shifts shaping the industry

2.6 Embracing Customer-Centric Technology Trends:

CTOs play a pivotal role in navigating the intricate interplay between technology and customer expectations.

Enhancing User Experience and Satisfaction:

Customer-centric technology trends are inherently focused on enhancing user experiences. By understanding and incorporating these trends, CTOs can ensure that technological solutions align with user expectations, leading to heightened satisfaction and loyalty.

Example: The rise of chatbots in customer service applications exemplifies a trend focused on providing instant, personalised assistance, contributing to improved user satisfaction.

Anticipating Changing Consumer Behaviours:

Customer behaviours and preferences undergo continuous evolution. Understanding customer-centric technology trends is essential for anticipating these shifts, enabling organisations to proactively adjust technological solutions to meet changing consumer expectations.

Consumer expectations are in a constant state of flux, influenced by technological advancements, cultural shifts, and external factors. To stay ahead of these changes, utilising tools and online resources for tracking and analysing these shifts is crucial. Customer Relationship Management (CRM) platforms such as Salesforce and HubSpot provide insights into individual customer interactions, enabling the creation of personalised strategies. Social listening tools like Brandwatch and Hootsuite assist in monitoring online conversations, identifying emerging trends, and gauging sentiment. Additionally, data analytics tools such as Google Analytics and Mixpanel offer quantitative insights into user behaviour, facilitating data-driven decision-making. Embracing these tools empowers organisations to proactively adapt technology solutions, ensuring alignment with the ever-evolving landscape of customer expectations.

Example: The increasing preference for mobile payments and contactless transactions highlights a shift in consumer behaviour, prompting organisations to invest in secure and convenient payment technologies.

Driving Personalisation and Customisation:

Personalisation is foundational in achieving customer-centricity, enabling tailored experiences that address the unique needs and preferences of individual users.

This transformative shift extends beyond generic one-size-fits-all solutions, allowing organisations to adapt their systems to individual preferences and needs. In the e-commerce sector, advanced algorithms analyse users' browsing and purchase history to provide personalised product recommendations, enhancing user experience and driving engagement. For instance, a notable case study is Amazon's recommendation engine, which utilises sophisticated algorithms to suggest products based on a user's past interactions and purchase history. This approach has significantly contributed to Amazon's success, fostering customer satisfaction and increasing sales through a more personalised shopping experience.

Similarly, in the realm of streaming services, industry leaders like Netflix employ sophisticated recommendation engines to curate content based on users' viewing habits, tailoring their offerings to individual preferences. The emphasis on personalisation emerges as a significant factor not only in fostering user satisfaction but also in playing a pivotal role in customer retention and loyalty.

Navigating the intricacies of personalisation and meticulous tracking of user activities requires a nuanced approach to ethical considerations. Striking the right balance between personalisation and user privacy is a delicate task that necessitates careful consideration. Ethical considerations in technology-driven personalisation have been explored in notable research pieces such as "The Age of Personalisation" by Joseph Pine and "Personalisation at Scale" by McKinsey & Company. These resources delve into the ethical implications of personalisation at scale, shedding light on how organisations can navigate this terrain responsibly, respecting user privacy while delivering tailored experiences.

Acknowledging and addressing these ethical considerations is not only a responsible approach but also a strategic one. Users are becoming increasingly conscious of how their data is utilised, making transparency and ethical practices critical elements in maintaining trust. Armed with insights from ethical research studies, organisations can navigate the complexities of personalisation confidently, ensuring that their technological strategies align not only with user expectations but also with ethical standards. This approach not only fosters positive user experiences but also contributes to the long-term sustainability and success of customer-centric technology implementations.

Adapting to Omni-Channel Experiences:

Customer interactions span multiple channels, from websites to mobile apps and social media. CTOs who understand omni-channel technology trends can seamlessly integrate these channels, providing a cohesive and synchronised experience across all touchpoints.

Identifying the requirements for an effective omni-channel strategy begins with a comprehensive understanding of customer needs, preferences, and expectations. Employing customer journey mapping techniques helps to pinpoint touchpoints across various channels, shedding light where customers interact with the brand. Surveys, feedback mechanisms, and data analytics also play a crucial role in gathering insights into customer behaviour. Additionally, understanding the context in which customers switch between channels provides valuable information to streamline their experiences. For instance, a customer looking for a new Aston Martin Car might begin a product search on a mobile app, continue research on a desktop, and make the final purchase in-store. Recognising these patterns allows CTOs to design a seamless and consistent experience across channels.

To leverage different channels effectively in an omni-channel strategy, it is essential to integrate and synchronise communication channels for a cohesive customer journey. Employing technologies such as Customer Relationship Management (CRM) systems facilitates a unified view of customer interactions, regardless of the channel. Chatbots and virtual assistants enhance real-time engagement, providing instant support and information. Social media platforms become valuable channels for customer feedback, inquiries, and brand engagement. Additionally, deploying analytics tools helps in monitoring and measuring customer interactions across channels, enabling data-driven decision-making.

Leveraging Data-Driven Insights:

Leveraging data-driven insights in the realm of customer-centric technology trends offers a multitude of benefits for organisations aiming to enhance their strategies and deliver superior customer experiences. Firstly, data-driven insights provide a deep understanding of customer behaviours, preferences, and pain points. By analysing vast datasets, organisations can uncover patterns and trends, gaining valuable insights that inform the development of targeted technological solutions tailored to meet specific customer needs.

Secondly, the integration of advanced analytics tools, such as Tableau, Power BI, Brandwatch, Clarabridge, SAS, and IBM SPSS, enables organisations to enhance their decision-making processes. These tools facilitate interactive data visualisation, sentiment analysis, and predictive analytics, empowering decision-makers to make informed choices based on quantitative and qualitative data. This not only streamlines operations but also ensures that strategies are aligned with customer expectations, fostering a more customer-centric approach.

Thirdly, the ability to anticipate future customer behaviours is a key advantage of leveraging data-driven insights. Predictive analytics tools allow organisations to forecast trends and adapt their strategies accordingly. This proactive approach enables businesses to stay ahead of evolving customer needs, pre-emptively addressing challenges and seizing opportunities. For example, in the e-commerce sector, predictive analytics can help optimise inventory management and enhance product recommendations, contributing to a more personalised and satisfying customer experience.

Moreover, the integration of various analytics tools offers a comprehensive understanding of customer interactions across different channels. This holistic view enables organisations to create a unified customer experience, ensuring consistency and coherence in their technological solutions. Whether through analysing browsing behaviour on e-commerce platforms or monitoring sentiments on social media, data-driven insights help in crafting a seamless and engaging customer journey.

Fostering Customer Trust and Transparency:

Trust is integral to customer relationships. Technology trends that prioritise transparency and data security contribute to building and maintaining customer trust. CTOs who understand and implement such trends safeguard the organisation's reputation.

Building and maintaining trust with customers is a foundational element in cultivating long-term relationships and enhancing lifetime value. Transparency and open communication play pivotal roles in fostering trust. Providing clear and honest information about products, services, and business practices establishes a foundation of reliability. Consistency across all touchpoints, whether online or in-person, reinforces a sense of dependability. Additionally, actively seeking and incorporating customer feedback demonstrates a commitment to improvement and responsiveness, further solidifying trust. Establishing trust also involves prioritising data security and privacy. Implementing robust cybersecurity measures and clearly communicating how customer data is handled instils confidence and reduces concerns about privacy breaches. Ultimately, delivering on promises, consistently exceeding expectations, and maintaining a customer-centric approach contribute significantly to building and reinforcing trust over time.

Measuring trust is a nuanced task, often requiring a combination of quantitative and qualitative metrics. Net Promoter Score (NPS), customer satisfaction surveys, and customer reviews provide quantitative insights into overall satisfaction and loyalty. Analysing customer interactions and engagement across various channels, including social media and customer support, can offer qualitative indicators of trust. Monitoring repeat business, customer retention rates, and the length of customer relationships also provides valuable data on the depth of trust within the customer base. Additionally, tracking customer referrals and recommendations can serve as a qualitative measure of the trust customers have in the brand. By employing a combination of these metrics, organisations can gain a comprehensive understanding of the level of trust they have established and identify areas for continuous improvement.

Example: Companies adopting blockchain technology for transparent supply chain tracking demonstrate a commitment to providing customers with accurate information about the origin and journey of products. For example, Walmart used IBM Food Trust to track the movement of mangoes from Mexico to its stores in the U.S. This enabled them to identify and address a contamination issue much faster than traditional methods, preventing potentially harmful products from reaching consumers. The transparency also allowed Walmart to demonstrate to its customers their commitment to food safety and ethical sourcing.

Rapid Response to Customer Feedback:

Customer feedback is a valuable source of insights. CTOs who stay attuned to customer-centric technology trends can implement agile development processes, enabling rapid responses to feedback and ensuring that technological solutions align with evolving customer needs.

The importance of rapid responses to customer feedback, inquiries, and complaints cannot be overstated in today's fast-paced business environment. Timely responses demonstrate a commitment to customer satisfaction and indicate that their concerns are taken seriously. In the age of instant communication, customers expect quick resolutions to their queries and issues. A prompt and efficient response not only addresses immediate concerns but also showcases a company's dedication to customer service. This proactive approach significantly impacts trust as customers feel heard and valued. Rapid responses convey a sense of reliability, creating a positive impression that resonates with the customer, fostering a deeper sense of trust in the brand.

The impact of swift customer responses extends beyond trust, influencing customer loyalty and the potential for future product purchases. Customers who experience prompt and effective resolutions to their feedback or complaints are more likely to develop a sense of loyalty to the brand, enhancing the likelihood of repeat business. Numerous studies and customer satisfaction surveys consistently support the assertion that customers who receive swift and efficient resolutions to their feedback or complaints are significantly more inclined to develop a strong sense of loyalty to the brand. Research has shown that a positive customer service experience, characterised by quick problem-solving and effective communication, has a direct impact on customer retention rates. When businesses demonstrate a genuine commitment to addressing customer concerns promptly, it not only mitigates dissatisfaction but also fosters a positive perception of the brand. Additionally, satisfied customers are more likely to share their positive experiences with others, contributing to positive word-of-mouth marketing and further enhancing the brand's reputation.

In today's competitive market, where customer choices abound, businesses that prioritise rapid responses not only retain existing customers but also create an environment conducive to attracting new customers who value attentive and responsive service.

Example: Social media listening tools enable organisations to promptly address customer feedback, allowing for swift adjustments and improvements to products or services.

Social media listening tools have become indispensable for businesses aiming to gain real-time insights into their online presence and customer sentiments. These tools, such as *Brand24*, *Brandwatch*, and *Mention*, sift through vast amounts of social media data, uncovering valuable information about brand mentions, customer conversations, and industry trends. By monitoring various social media channels, these tools can identify emerging discussions and sentiments surrounding a brand or product. Social media listening tools also enable businesses to track and analyse competitor activities, gauge the effectiveness of marketing campaigns, and identify potential issues or opportunities in the market. Through sentiment analysis, by categorising mentions as positive, negative, or neutral, these tools provide a nuanced understanding of customer perceptions. Overall, social media listening tools empower businesses to make data-driven decisions, enhance customer engagement, and proactively manage their online reputation in the dynamic landscape of social media.

Engaging in Conversational Interfaces:

The increasing significance of Natural Language Processing (NLP) in advancing conversational chatbots to emulate human-like interactions is transforming the landscape of customer engagement. As NLP technology evolves, there is a growing body of evidence suggesting that consumers find it increasingly challenging to distinguish between chatbots and human agents. According to a study by Oracle, 80% of businesses plan to incorporate chatbots, showcasing the pervasive adoption of this technology. The seamless integration of NLP into chatbots enables them to comprehend and respond to user queries with a level of nuance and contextual understanding that closely mirrors human conversation. As per a survey conducted by Drift, a conversational marketing platform, 64% of consumers believe 24-hour service is the best feature of chatbots, underlining the convenience and accessibility that these human-like interfaces provide. The convergence of NLP advancements and chatbot capabilities not only streamlines customer interactions but also underscores the need for businesses to continuously innovate in the pursuit of creating chatbots that are indistinguishable from human agents, ultimately enhancing user satisfaction and engagement.

The importance of understanding customer-centric technology trends extends far beyond meeting immediate user needs—it is a strategic imperative that shapes the very fabric of an organisation's success. CTOs who embrace these trends not only propel their organisations to the forefront of customer-centricity but also contribute to the ongoing evolution of technology in response to the dynamic landscape of user expectations.

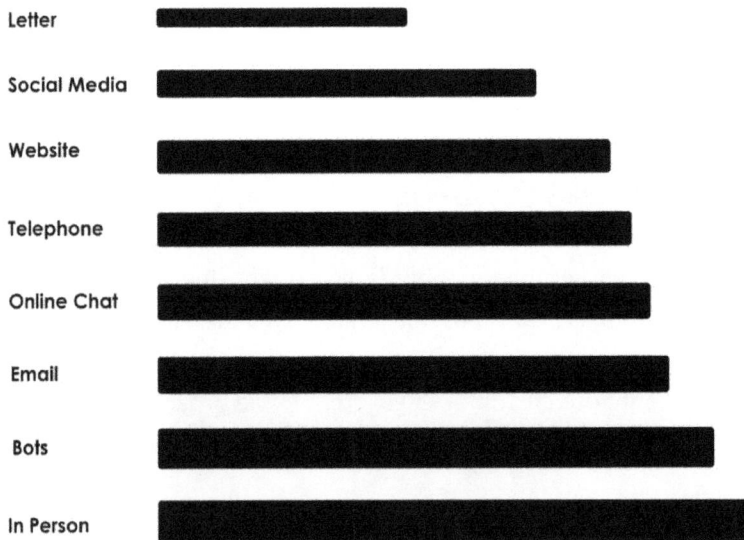

Channel Impact by Benefits - Data Marlow Business School

Customer-Centric Technology Trends: A 10-Step Action Plan

In the era of customer-centricity, staying attuned to evolving technology trends is imperative for CTOs. This 10-step guide provides a systematic approach, along with examples of data sources to identify customer-centric technology trends:

1. Establish Customer Feedback Mechanisms:

- **Implementation:** Implement robust customer feedback mechanisms, such as surveys, reviews, and social media monitoring. Analyse platforms like *Trustpilot* and *G2 Crowd* for customer sentiments and preferences.

2. Utilise Web and Mobile Analytics:

- **Implementation:** Leverage web and mobile analytics tools, such as *Google Analytics* and *Mixpanel*, to track user behaviour. Identify popular features, navigation patterns, and areas for improvement based on user engagement data.

3. Monitor Social Media Platforms:

- **Implementation:** Actively monitor social media platforms using tools like Hootsuite or Brandwatch. Analyse discussions, comments, and mentions related to your industry and products to uncover emerging trends and customer sentiments.

4. Engage in Customer Journey Mapping:

- **Implementation:** Conduct comprehensive customer journey mapping workshops. Utilise insights from tools like customer journey analytics platforms or CRM systems to identify pain points and opportunities for technology enhancements throughout the customer journey.

5. Explore Customer Support and Ticket Data:

- **Implementation:** Analyse customer support data using tools like *Zendesk* or *Freshdesk.* Identify recurring issues and areas where technology interventions can streamline support processes and improve overall customer satisfaction.

6. Leverage E-commerce and Transaction Data:

- **Implementation:** Explore e-commerce platforms and transactional data. Analyse purchase patterns, cart abandonment rates, and user preferences. Platforms like *Shopify* or *Magento* offer valuable insights into customer behaviours.

7. Implement Voice of the Customer (VoC) Platforms:

- **Implementation:** Deploy Voice of the Customer platforms like *Medallia* or *Qualtrics*. Collect and analyse direct customer feedback through surveys, interviews, and sentiment analysis to understand evolving expectations and preferences.

8. Stay Informed Through Industry Reports:

- **Implementation:** Subscribe to industry reports from research firms like *Forrester* and *Gartner*. These reports provide comprehensive insights into emerging technology trends shaping customer experiences across various industries.

9. Engage in Beta Testing and Early Adoption:

- **Implementation:** Actively participate in beta testing programs for emerging technologies. Engage with vendors offering innovative solutions and consider early adoption to gain firsthand experience of technologies that align with evolving customer needs.

10. Collaborate with User Experience (UX) Design Teams:

- **Implementation:** Foster collaboration between the technology team and UX design teams. Utilise user testing platforms like User Testing or Lookback to gather real-time feedback on technology prototypes, ensuring alignment with user expectations.

By implementing this practical 10-step guide, CTOs can proactively identify customer-centric technology trends and align their technological strategies with the evolving needs and expectations of their user base. Using a combination of direct customer feedback, analytics tools, and industry insights, CTOs can position their organisations as leaders in delivering technology solutions that truly resonate with and enhance the customer experience.

2.7 Fostering Innovation through Partnerships and Ecosystem Collaboration

The importance of understanding partnerships and ecosystem collaboration cannot be overstated. CTOs play a pivotal role in shaping the technological trajectory of their organisations through strategic collaboration.

Identifying and engaging with potential partners for cross-pollination of ideas and accessing cutting-edge technologies is a strategic initiative that can propel innovation and growth for any organisation. Firstly, the identification process involves a thorough analysis of the industry landscape. Researching key players, attending industry conferences, and actively participating in professional networks can uncover potential partners with complementary expertise or technologies. Leveraging digital platforms such as LinkedIn and industry-specific forums helps in identifying organisations that align with your innovation goals. Collaborative innovation ecosystems, like incubators and accelerators, can also serve as fertile grounds for finding partners with pioneering technologies.

Once potential partners are identified, the engagement process is critical for fostering a mutually beneficial relationship. Initiating conversations through tailored outreach, demonstrating a clear understanding of their business, and highlighting the synergies between organisations pave the way for collaboration. Face-to-face meetings, workshops, and joint events can further solidify the connection, fostering an environment conducive to cross-pollination of ideas. Establishing shared goals, open communication channels, and a framework for collaboration lays the foundation for a successful partnership. Engaging in pilot projects or joint research initiatives can also serve as a tangible demonstration of the collaborative potential, encouraging the exchange of cutting-edge technologies and innovative insights.

To maximise the impact of cross-pollination, it's crucial to create a collaborative culture that encourages continuous exchange of ideas and expertise. Utilising collaborative tools, such as project management platforms and virtual communication channels, facilitates seamless information sharing. Regular check-ins and feedback loops ensure that both parties stay aligned with the evolving landscape of cutting-edge technologies. Establishing clear expectations, defining roles and responsibilities, and nurturing a culture of mutual learning contribute to a robust partnership that not only accesses cutting-edge technologies but also propels both organisations towards sustained innovation and growth.

.

Accelerating Time-to-Market:

Collaborative ecosystems often enable organisations to accelerate their time-to-market. By leveraging the resources and expertise of partners, CTOs can streamline development processes, bringing products and solutions to market faster than if developed in isolation.

Example: Partnering with a technology consortium can expedite the development of interoperable solutions, reducing time-to-market for complex projects.

Navigating Complex Regulatory Landscapes:

Collaborating with industry partners and regulatory experts provides insights into navigating complex regulatory landscapes. CTOs can draw on the expertise of their ecosystem to ensure compliance and minimise risks associated with evolving regulations.

Example: Partnering with legal experts in a collaborative ecosystem helps the organisation stay abreast of changing data protection laws, ensuring technology strategies align with regulatory requirements.

Enhancing Resource Scalability:

Ecosystem collaboration enhances resource scalability. By forming partnerships, CTOs can tap into additional resources, whether it be talent, infrastructure, or funding, enabling the organisation to scale its technological endeavours more efficiently.

Example: Collaborating with venture capitalists provides access to funding that can fuel the development of ambitious projects and initiatives.

Facilitating Market Expansion:

Strategic partnerships facilitate market expansion. By collaborating with organisations in different regions or industries, CTOs can access new markets, diversify their customer base, and identify opportunities for growth.

Example: Partnering with a global technology firm opens avenues for market expansion, tapping into diverse customer demographics and preferences.

Driving Open Innovation Initiatives:

Ecosystem collaboration encourages open innovation initiatives. CTOs can engage in hackathons, innovation challenges, and collaborative projects that harness the collective creativity of the ecosystem, driving breakthrough technological advancements.

Example: Hosting a hackathon with startup partners fosters open innovation, leading to the development of prototypes and solutions that may not have been conceivable within the confines of the organisation.

Cultivating a Network of Expertise:

Collaborative ecosystems provide access to a network of expertise. CTOs can leverage the knowledge and skills of partners, creating a dynamic environment where the organisation can respond effectively to technological challenges and opportunities.

Example: Building relationships with industry experts through collaborative partnerships creates a knowledge-sharing network, enriching the organisation's collective intelligence.

Responding to Industry Disruptions:

Ecosystem collaboration enhances the organisation's ability to respond to industry disruptions. By having a network of partners, CTOs can quickly adapt strategies, share insights, and collectively navigate challenges posed by disruptive technologies or market shifts.

Example: Collaborating with industry peers allows for a collective response to disruptions, ensuring the organisation remains resilient in the face of technological and market uncertainties.

Paving the Way for Sustainable Growth:

Fostering partnerships and ecosystem collaboration is integral to sustainable growth. CTOs who strategically build and nurture collaborative networks pave the way for long-term success by tapping into shared resources, expertise, and market opportunities.

Example: Establishing long-term partnerships with key players in the ecosystem creates a foundation for sustained growth, as the organisation continues to evolve in tandem with technological advancements and market dynamics.

Understanding the importance of partnerships and ecosystem collaboration is not just a strategic consideration—it is a fundamental imperative for CTOs shaping the technological destiny of their organisations. By embracing collaborative ecosystems, CTOs unlock a myriad of opportunities, from accessing cutting-edge technologies to fostering a culture of innovation that propels their organisations into a future of sustainable growth and technological leadership.

Partnership Case Study: Partnership between Pfizer, a global pharmaceutical company, and BioNTech, a German biotechnology company, in the development of the COVID-19 vaccine.

Pfizer-BioNTech COVID-19 Vaccine Collaboration:

Objective:

- The primary goal of this collaboration was to develop a safe and effective COVID-19 vaccine to address the global pandemic.

Key Components:

1. **mRNA Technology:**

    - BioNTech specialises in mRNA technology, and the collaboration utilised BioNTech's expertise in this field. The mRNA technology allows the development of vaccines that instruct cells to produce specific proteins, triggering an immune response.

2. **Global Research and Development:**

    - Pfizer and BioNTech combined their research and development efforts to accelerate the vaccine's timeline. The collaboration involved scientists and researchers from both organisations working collaboratively.

3. **Clinical Trials:**

    - The development process included rigorous clinical trials, with participants from various regions worldwide. The collaboration facilitated the recruitment of diverse populations for these trials, ensuring the vaccine's effectiveness across different demographics.

4. **Manufacturing and Distribution:**

    - Pfizer's global manufacturing and distribution capabilities played a crucial role in scaling up production and distributing the vaccine on a global scale.

Outcomes:

1. **First Authorised COVID-19 Vaccine:**

   - The Pfizer-BioNTech COVID-19 vaccine, named BNT162b2, became one of the first authorised vaccines for emergency use against COVID-19. It demonstrated high efficacy in preventing illness caused by the SARS-CoV-2 virus.

2. **Global Vaccination Campaign:**

   - The successful collaboration enabled the rapid deployment of the vaccine, contributing significantly to global vaccination efforts. Millions of doses were distributed worldwide, playing a pivotal role in managing and mitigating the impact of the pandemic.

3. **Scientific Advancements:**

   - The collaboration showcased the potential of mRNA technology in vaccine development. The success of the Pfizer-BioNTech vaccine has spurred further interest and investment in mRNA-based approaches to address various diseases.

4. **Public-Private Partnership:**

   - The collaboration highlighted the effectiveness of public-private partnerships in responding to global health crises. It demonstrated how the expertise and resources of a pharmaceutical company (Pfizer) and a biotechnology company (BioNTech) could be combined to achieve a common goal.

The Pfizer-BioNTech collaboration serves as a recent and notable example of how joint efforts between organisations can lead to groundbreaking advancements, addressing critical challenges and benefiting global public health.

Partnerships and Ecosystem Collaboration: A 10 Step Action Plan

Forming strategic partnerships and engaging in ecosystem collaboration is paramount. This 10-step guide provides CTOs with detailed insights to effectively identify and cultivate partnerships:

1. Define Strategic Objectives:

- Utilise internal data sources such as organisational strategy documents, technology roadmaps, and innovation goals. ***Analyse past performance metrics to identify areas where external collaboration can fill gaps or accelerate progress.***
- **Example:** Review internal strategic documents and reports outlining key focus areas for technology development and innovation.

2. Conduct Ecosystem Mapping:

- Leverage industry reports from research firms like Gartner, Forrester, or industry-specific publications. Utilise online directories such as ***Crunchbase, AngelList***, and industry association member lists for a comprehensive ecosystem mapping.
- **Example:** Use ***Crunchbase*** to identify startups, research institutions, and technology companies relevant to your industry.

3. Attend Industry Conferences and Events:

- Access conference attendee lists for events relevant to your industry. Utilise event platforms like ***Eventbrite, Meetup***, or official conference websites for detailed participant information.
- **Example:** Explore the attendee list on the official website of a technology conference to identify potential collaborators.

4. Utilise Open Innovation Platforms:

- Engage with open innovation platforms like ***InnoCentive, NineSights,*** or ***Kaggle***. Utilise their databases to discover challenges and potential collaborators aligned with your organisation's needs.
- **Example:** Browse challenges on ***Wazoku InnoCentive*** related to specific technological problems or innovations sought by organisations.

5. Leverage Government and Industry Programs:

- Explore government agency websites for innovation programs and grants. Utilise industry-specific platforms like ***Innovate UK*** or ***SBIR/STTR*** databases for information on funding and collaboration opportunities.
- **Example:** Visit ***Innovate UK's*** website to find collaborative funding programs supporting technological innovation.

6. Analyse Venture Capital Investments:

- Utilise investment databases such as *Crunchbase, PitchBook,* or *CB Insights*. Analyse recent investment rounds in your industry to identify startups receiving significant funding.
- **Example:** Use *Crunchbase* to filter investment rounds in startups related to your technology domain.

7. Establish Corporate Innovation Programs:

- Create internal innovation programs with an accessible application process. Promote these programs through industry news, social media, and innovation-focused platforms.
- **Example:** Launch an innovation program with a dedicated webpage outlining collaboration opportunities for startups and external innovators.

8. Collaborate with Innovation Hubs:

- Explore innovation hubs' websites and directories.EG **https://www.techuk.org/** or **https://eda.gov/funding/programs/regional-technology-and-innovation-hubs/2023 or https://ukinnovationhub.ukri.org/**
- **Example:** Partner with a technology-focused co-working space and with the correct permissions utilise their member directory to identify potential collaborators.

9. Leverage Online Collaboration Platforms:

- Utilise platforms like *GitHub, GitLab,* or *Bitbucke*t for collaborative development. Engage with open-source communities, forums, and issue trackers to identify contributors aligned with your organisation's goals.
- **Example:** Contribute to an open-source project on *GitHub,* and explore collaboration opportunities with developers.

10. Establish Relationship Management Processes:

- Implement a Customer Relationship Management (CRM) system tailored for partnership management. Utilise tools like *Salesforce, HubSpot*, or *Zoho* CRM for tracking interactions, collaboration milestones, and ongoing relationship management.
- Utilise *Salesforce* to create a customised CRM module for managing and nurturing partnerships, tracking communication history, and collaboration progress.

By incorporating these detailed examples and leveraging specific data sources, CTOs can systematically navigate the partnership and collaboration landscape, ensuring a strategic and data-driven approach to fostering innovation and technological growth within their organisations.

2.8 Future-Proofing Strategies: Ensuring Technological Relevance in a Dynamic Landscape

As a CTO our decisions have far reaching consequences. Future proofing will ensure not just our current commercial relevance but sustained technological leadership of the organisation.

Embracing Technology Forecasting:

Future-proofing strategies begin with a deep understanding of technology forecasting. CTOs must actively engage with emerging technologies, industry reports, and technology roadmaps to anticipate trends and plan for integration into the organisational landscape.

Example: Monitoring advancements in quantum computing and understanding its potential applications can guide a CTO in preparing the organisation for future opportunities. At Airbus, the European aerospace giant, their CTO partnered with QuTech, a leading quantum research institute in the Netherlands, to explore how quantum computing could optimise aircraft routing. Traditional algorithms for flight planning can be computationally expensive and struggle with complex factors like weather patterns and air traffic control. This led to shorter flight paths, lower fuel usage and environmental benefits, improved operational efficiency and a competitive advantage.

Investing in Scalable Technologies:

Strategically investing in scalable technologies is a crucial element in future-proofing any business. This entails prioritising solutions that can adapt and evolve alongside the organisation's changing needs, thereby minimising the necessity for frequent overhauls. The benefits of selecting technologies that not only meet current requirements but also demonstrate flexibility in scaling up or down as needed are manifold. This forward-thinking approach empowers organisations to respond promptly to market trends, regulatory changes, and shifts in consumer behaviour. Scalable technologies offer the adaptability required to future-proof against uncertainties and challenges, providing a sturdy foundation for innovation and enabling organisations to navigate the intricacies of a rapidly evolving technological landscape with agility. Emphasising scalability positions businesses to thrive in an ever-changing digital environment.

The process of identifying and implementing scalable technologies involves an assessment of the organisation's current and future needs. A comprehensive technology audit, evaluating existing systems and infrastructure, helps pinpoint areas where scalability is crucial and ensures a clear understanding of the technological landscape. Engaging with key stakeholders, including end-users and department heads, provides a nuanced perspective on specific requirements and growth projections. Collaborative discussions yield valuable insights into potential pain points and areas where scalable technologies can enhance overall efficiency.

Prioritising Cybersecurity Resilience:

Cyber threats evolve, making cybersecurity a central aspect of future-proofing. CTOs must prioritise robust cybersecurity measures, including regular assessments, proactive threat intelligence, and the adoption of advanced security technologies.

Example: Implementing a cybersecurity framework like NIST or ISO 27001 and regularly conducting penetration testing enhances the organisation's resilience against evolving cyber threats.

Monitoring Societal and Cultural Shifts:

Technological relevance is intertwined with societal and cultural shifts. Future-proofing strategies should include monitoring changing demographics, consumer behaviours, and cultural preferences to align technology solutions with evolving societal trends.

Understanding social and cultural shifts is crucial as it allows organisations to stay ahead of the curve and anticipate changes in consumer behaviour. For instance, recognising a growing interest in sustainability or inclusivity may prompt businesses to adjust their marketing strategies, product offerings, or corporate initiatives to align with these evolving values. Failure to track social and cultural shifts can result in misaligned messaging, missed market opportunities, or even reputational risks. The ability to adapt swiftly to changing cultural landscapes enables organisations to connect with their audience authentically and build meaningful, long-lasting relationships.

To delve deeper into the significance of tracking social and cultural shifts, numerous online references provide valuable insights and guidance. Platforms like *Pew Research Centre (pewresearch.org)* offer extensive reports on societal trends and attitudes, providing a reliable source for understanding shifts in demographics, values, and behaviours. Additionally, publications like *Harvard Business Review (hbr.org)* frequently feature articles on cultural intelligence and market trends, offering practical advice on staying attuned to cultural shifts in the business context. By leveraging these references, organisations can augment their understanding of social and cultural dynamics, informing strategic decisions and fostering a more adaptive and culturally aware approach.

Example: Recognising the increasing emphasis on sustainability, a CTO may explore eco-friendly technologies and practices to align with changing consumer values. For example, Patagonia Outdoor Clothing went beyond acknowledging the trend in society's increasing concern for environmental issues, they integrated it into their core values and identity, right down to system choice and attracting like minded staff. They saw sustainability as not just a marketing trend but a cultural shift with growing momentum, driving significant competitive advantage.

Continual Evaluation and Adaptation:

Future-proofing is not a one-time effort but an ongoing process. CTOs should establish mechanisms for continual evaluation of technology strategies, regularly reassessing the relevance of existing technologies, and adapting to emerging trends.

Example: Implementing regular technology audits and trend assessments ensures that the organisation remains agile and responsive to the evolving technological landscape.

By embracing innovation, anticipating changes, prioritising cybersecurity, nurturing talent, and fostering a culture of adaptability, CTOs can position their organisations not just as current leaders but as resilient pioneers in the technological frontier.

Implementing Future-Proofing Strategies: A 10-Step Action Plan

Future-proofing strategies are crucial for maintaining technological relevance and adaptability in the face of rapid changes. This practical 10-step guide provides CTOs with actionable steps and detailed examples of data sources to identify and implement future-proofing strategies:

1. Conduct Technology Trend Analysis:

- Leverage industry reports from research firms such as *Gartner, Forrester,* and *IDC* to analyse current and emerging technology trends. Monitor technology-specific publications and blogs to gain insights into innovative solutions and disruptive technologies.
- **Example:** *Gartner's "Hype Cycle"* reports provide a comprehensive overview of emerging technologies and their expected maturity timelines.

2. Assess Regulatory and Compliance Landscapes:

- Regularly review updates from regulatory bodies and industry-specific compliance agencies. Utilise legal databases, government publications, and industry forums to stay informed about changes in data protection laws, cybersecurity regulations, and other relevant compliance requirements.
- **Example:** The official websites of regulatory bodies such as the *GDPR portal* or the *Federal Trade Commission (FTC)* provide up-to-date information on data protection and privacy regulations.

3. Engage in Technology Forecasting:

- Explore technology forecasting platforms and reports. Utilise resources like *MIT Technology Review, IEEE Spectrum,* and industry-specific forecasting publications to gain insights into upcoming technologies and their potential impact.
- **Example:** The "Top Emerging Technologies" report from the *World Economic Forum* outlines key technologies that are expected to shape the future.

4. Monitor Cybersecurity Threat Intelligence:

- Utilise cybersecurity threat intelligence platforms and services. Subscribe to cybersecurity threat feeds, participate in information-sharing platforms like *ISACs (Information Sharing and Analysis Centres),* and stay updated on cybersecurity news sources.
- **Example:** Threat intelligence platforms like *Recorded Future* or *ISACs* in sectors like finance or healthcare provide timely information on emerging cyber threats.

5. Collaborate with Industry Peers:

- Engage with industry consortia, forums, and collaborative platforms. Participate in technology alliances and collaborate with peers to share insights, discuss challenges, and collectively develop strategies for future-proofing.

- **Example:** Platforms like the ***Industrial Internet Consortium (IIC)*** bring together industry leaders to collaborate on the development and adoption of industrial internet technologies.

6. Leverage Data Analytics for Market Insights:

- Use data analytics tools to gain market insights. Analyse customer behaviour, market trends, and competitor activities using tools like ***Google Analytics***, social media analytics, and market research platforms.
- **Example:** ***Google Analytics*** provides detailed information about website traffic, user engagement, and conversion metrics, helping to identify changing customer preferences.

7. Stay Informed on Global Economic Trends:

- Monitor economic indicators and global economic reports. Utilise sources like the ***World Bank, International Monetary Fund (IMF),*** and financial news platforms to understand global economic trends that may impact the organisation.
- **Example:** ***The World Economic Outlook report*** from the IMF provides comprehensive insights into global economic trends.

8. Engage in Scenario Planning Workshops:

- Conduct scenario planning workshops within the organisation. Collaborate with cross-functional teams to envision different future states, identify potential challenges, and develop strategies to address uncertainties.
- **Example:** Scenario planning workshops often involve facilitated sessions with experts in strategic foresight and scenario development.

9. Establish Continuous Learning Programs:

- Implement continuous learning programs for the technology team. Encourage participation in online courses, workshops, and industry conferences to stay updated on new technologies, methodologies, and best practices.
- **Example:** Platforms like ***Coursera, Udacity,*** and industry-specific conferences offer opportunities for continuous learning and skill development.

10. Utilise Innovation Metrics and Key Performance Indicators (KPIs):

- Define innovation metrics and KPIs to measure the effectiveness of future-proofing efforts. Track indicators such as the adoption rate of new technologies, successful implementation of innovation projects, and the organisation's ability to adapt to change.
- **Example:** Key innovation metrics may include the percentage of revenue from new products, time-to-market for innovations, and employee engagement in innovation initiatives.

2.9 Competitive Analysis

A CTO needs to analyse not only direct competitors but also potential disruptors. Conducting regular competitive analyses helps in identifying gaps in technology adoption, potential areas for differentiation, and strategies employed by competitors. This intelligence informs the CTO's decision-making process, ensuring that technology initiatives align with broader business goals while maintaining a competitive edge.

Identifying and analysing competitors and potential disruptors in the market is a critical aspect of strategic planning for any business. One effective technique is conducting a thorough competitor analysis. This involves evaluating competitors' products, pricing strategies, market positioning, and customer reviews. Tools like **SEMrush** and **Ahrefs** aid in analysing competitors' online presence, including their website traffic, keywords, and backlink strategies. By benchmarking against industry peers, businesses can identify areas for improvement and potential areas of differentiation.

Monitoring industry trends and emerging technologies is another technique to identify potential disruptors. Staying abreast of innovations through industry reports, research publications, and technology forecasts provides insights into what could reshape the market. Platforms like CB Insights and Crunchbase track startup activities and funding, offering valuable information on emerging players and potential disruptors. Furthermore, participating in industry conferences, webinars, and forums facilitates networking and keeps businesses informed about new entrants and innovative ideas.

Utilising customer feedback and sentiment analysis is crucial for understanding how a brand is perceived relative to competitors. Social media listening tools like **Brandwatch** and **Mention** help in gauging customer sentiments and identifying areas where competitors may be falling short. Analysing customer reviews on platforms such as **Trustpilot** or **G2 Crowd** provides direct insights into the strengths and weaknesses of competitors from the customer's perspective. This technique allows businesses not only to refine their own offerings but also to identify opportunities to outperform competitors in areas that matter most to customers.

To further enhance competitive intelligence, businesses can leverage publicly available financial reports, earnings calls, and investor presentations. Platforms like Seeking Alpha and Yahoo Finance provide a wealth of financial information, helping businesses understand the financial health of competitors and potential disruptors. Analysing financial data provides insights into market share, profitability, and investment strategies, enabling businesses to anticipate competitive moves and disruptions. By combining these techniques, businesses can create a comprehensive understanding of their competitive landscape, empowering strategic decision-making and ensuring resilience in the face of potential disruptors.

In the above text I refer to a "disruptor". In this context it refers to a concept, product, or service that fundamentally alters the existing market landscape by challenging traditional approaches and norms. Disruptors often introduce groundbreaking ideas or technologies that redefine how industries operate, compelling established players to adapt or risk becoming obsolete. These innovations typically offer unique value propositions, such as increased efficiency, cost-effectiveness, or convenience, which resonate with consumers and drive widespread adoption. Disruptors can emerge from both established companies and startups, and they have the potential to revolutionise entire industries, creating new opportunities and reshaping competitive dynamics.

SWOT Analysis: Dissecting Strengths, Weaknesses, Opportunities, and Threats

It will be unlikely that you have reached this stage in your career without having worked on SWOT Analysis at least a hundred times, and because we have all used it so many times before it's sometimes easy to dismiss as something old fashioned, but SWOT analysis is akin to wielding a powerful magnifying glass for dissecting the competitive landscape. This method involves a thorough examination of an organisation's Strengths, Weaknesses, Opportunities, and Threats in the context of its competitors. Through this comprehensive assessment, we can gain a holistic understanding of the dynamics within the competitive environment. The analysis of strengths aids in identifying areas where the organisation excels and can potentially outperform competitors, while an exploration of weaknesses highlights vulnerabilities that may warrant strategic attention for improvement.

The SWOT analysis extends its reach into the realm of opportunities and threats. Pinpointing opportunities allows strategists to identify potential avenues for growth, innovation, or market expansion. Crafting strategies to harness these opportunities becomes integral to steering the organisation towards success. Simultaneously, recognising potential threats ensures a proactive approach to challenges that may arise from various sources, such as technological disruptions, regulatory changes, or emerging competition. Armed with this understanding we can formulate effective approaches that lever organisational strengths, address weaknesses, capitalise on opportunities, and mitigate potential threats. This approach provides a solid foundation for strategic decision-making and long-term success in a competitive landscape.

Technological Benchmarking

One approach to technological benchmarking involves evaluating the organisation's technological capabilities against industry best practices and standards. This can include analysing the efficiency and performance of IT infrastructure, evaluating the adoption of emerging technologies, and assessing the effectiveness of software applications. A thorough examination of competitors' technological strategies, product offerings, and digital initiatives provides valuable insights into industry norms and benchmarks, helping organisations understand where they stand in comparison.

To conduct technological benchmarking effectively, organisations can utilise various tools and techniques. Technology assessment frameworks such as the *Technology Capability Maturity Model (TCMM)* or the *Information Systems Research Framework (ISRF)* provide structured methodologies for evaluating technological capabilities. Additionally, industry reports, market research, and technology-focused publications offer external perspectives on prevailing technological trends and benchmarks.

Market Positioning and Brand Differentiation

Competitive analysis extends beyond features and capabilities—it encompasses market positioning and brand differentiation. By differentiating the organisation's brand from competitors, a CTO creates a unique selling proposition that resonates with target audiences.

One effective approach is to focus on unique value propositions that resonate with the target audience. This could involve highlighting specific features or attributes that set the brand apart, whether it's superior product quality, exceptional customer service, or a commitment to sustainability. For instance, in the UK automotive industry, Jaguar Land Rover differentiates itself through its focus on luxury, innovative design, and a rich British heritage, positioning the brand apart from other car manufacturers.

Crafting a compelling brand narrative is another key aspect of differentiation. This involves telling a story that resonates with consumers, evoking emotions and creating a memorable connection. A distinctive brand narrative communicates the brand's history, values, and mission in a way that engages and captivates the audience. An example from the UK retail sector is John Lewis, known for its heartwarming and emotionally resonant Christmas adverts. By focusing on storytelling and creating an emotional connection with consumers, John Lewis differentiates itself in the competitive retail landscape, reinforcing a positive and unique brand image.

Continuous Monitoring and Adaptation

Maintaining success in a competitive landscape necessitates organisations adopting vigilant monitoring strategies. To achieve this, businesses can implement effective mechanisms for continuous surveillance of competitors, market trends, and technological advancements. Leveraging advanced competitor tracking software, such as **SEMrush**, or **Crayon**, provides real-time updates on rivals' activities, product releases, and strategic manoeuvres. This proactive approach empowers the organisation to promptly adapt to changing market conditions, identifying potential threats or opportunities before they become critical. Routine analyses of market trends, facilitated by tools like **Brandwatch** or **SimilarWeb**, offer insights into shifting consumer preferences, emerging technologies, and industry developments, enabling the company to align its strategies accordingly.

In addition to technology-driven monitoring, organisations can explore cultivating networks of industry contacts and actively participate in relevant conferences and forums. Engaging with industry peers ensures a continuous influx of qualitative insights, keeping the organisation well-informed about emerging trends, best practices, and potential collaborations. This real-time, two-way flow of information, coupled with sophisticated competitor tracking tools, enhances the company's agility in responding to market dynamics. By fostering a culture of continuous learning and adaptation, the organisation ensures it remains at the forefront of innovation, maintaining a competitive edge in the long run.

Interrogating Competitive Analysis: A 10-Step Action Plan

Strategic competitive analysis is paramount for a Chief Technology Officer (CTO) to steer their organisation toward innovation and market leadership. This practical 10-step guide provides actionable steps, along with detailed examples of potential data sources, to help CTOs acquire and effectively interrogate competitive analysis:

1. Define Strategic Objectives:

- Clearly articulate the organisation's strategic objectives. Identify key focus areas, such as product innovation, market expansion, or technology differentiation, to tailor competitive analysis accordingly.
- **Example:** If the strategic objective is to outpace competitors in cloud computing services, focus the competitive analysis on their offerings, pricing models, and market positioning.

2. Identify Competitors:

- Compile a comprehensive list of direct and indirect competitors. Utilise industry reports, market research firms, and competitor analysis tools to identify organisations operating in the same or adjacent spaces.
- **Example:** Tools like SimilarWeb or SEMrush provide insights into competitors' online presence, traffic, and digital marketing strategies.

3. Gather Market Intelligence:

- Leverage industry reports, market studies, and subscription-based market intelligence platforms. Extract data on market trends, growth projections, and potential disruptors.
- **Example:** Gartner's Magic Quadrant reports offer in-depth analyses of market trends and the positioning of key players in various technology sectors.

4. Analyse Competitor Offerings:

- Scrutinise competitors' product and service offerings. Explore their features, functionalities, pricing models, and unique selling propositions (USPs).
- **Example:** Comparative analysis tools like Capterra or G2.Com provide insights into user reviews, feature comparisons, and pricing details for various products.

5. Evaluate Technological Maturity:

- Assess the technological maturity of competitors. Analyse their technology stacks, development methodologies, and the adoption of emerging technologies.
- **Example:** Platforms like StackShare or GitHub can reveal the technologies and tools competitors are using in their development processes.

6. Monitor Online Presence:

- Monitor competitors' online activities, including their website, social media channels, and content publications. Track their engagement, audience sentiment, and online strategies.
- **Example:** Social media listening tools like Brandwatch or Hootsuite can track competitors' mentions, audience interactions, and content performance.

7. Conduct SWOT Analysis:

- Perform a comprehensive SWOT analysis (Strengths, Weaknesses, Opportunities, Threats) for each key competitor. Identify their core competencies, vulnerabilities, potential growth areas, and external threats.
- **Example:** Tools like Miro or Lucidchart can facilitate collaborative SWOT analysis sessions with cross-functional teams.

8. Engage in Industry Forums and Conferences:

- Participate in industry-specific forums, conferences, and webinars. Extract insights from presentations, panel discussions, and networking opportunities to understand competitors' perspectives and strategies.
- **Example:** Industry-specific forums like Spiceworks or conferences such as CES provide platforms for networking and staying abreast of industry discussions.

9. Leverage Customer Feedback and Reviews:

- Analyse customer reviews, feedback forums, and support interactions for competitors. Understand customer pain points, satisfaction levels, and areas where competitors excel or fall short.
- **Example:** Customer review platforms like TrustRadius or Gartner Peer Insights offer a wealth of user-generated content for competitor analysis.

10. Utilise Competitive Intelligence Platforms:

- Explore dedicated competitive intelligence platforms. Tools like Crayon or Kompyte automate the collection of competitor data, providing real-time insights and alerts.
- **Example:** Crayon aggregates competitor information from diverse sources, offering a centralised platform for ongoing competitive analysis.

By following this guide and leveraging the specified examples of potential data sources, CTOs will have a better opportunity to interrogate competitive analysis effectively, informing strategic decision-making and positioning their organisations for technological excellence and market success.

2.10 Market Research and Customer Insights

Technology exists to serve a purpose, and that purpose is often defined by the needs and expectations of the end-users. It's not always easy to do with a host of competing financial commitments but a CTO should try to put aside a piece of his budget and invest time and resources in market research to understand customer behaviour, preferences, and pain points. Yes, you might expect a Marketing Department to do that on behalf of the organisation, but, from my experience, having a little budget of your own to confirm customer insights at various stages of product development is money well invested. By gathering customer insights, we can align technological developments with the demands of the market, ensuring that innovations are not just cutting-edge but also resonate with the intended audience. We can also understand if what we heard during the initial requirements collection is actually what our customers want. Customers sometimes say what they want, but when they start to see it, they can often change their minds, or markets shift opinion during development cycles.

The Landscape of Market Dynamics

Strategic market research commences with perceiving the market as a dynamic landscape, continually influenced by economic, cultural, and technological factors. A CTO immerses themselves in comprehensive market analysis, examining economic trends, regulatory landscapes, and cultural shifts that impact the technology demand.

Here are online resources that can assist in this endeavour:

1. **World Economic Forum (WEF):**

Website: ***World Economic Forum***: The WEF provides insights into global economic trends, technological developments, and discussions on the impact of technology on various industries.

2. **Statista:**

Website: ***Statista***: Statista offers a vast collection of statistics and studies on economic indicators, technology trends, and cultural shifts, providing valuable data for market analysis.

3. **Pew Research Centre:**

Website: ***Pew Research Centre***: Pew Research conducts surveys and studies on various topics, including technology adoption, social trends, and cultural shifts.

4. **Harvard Business Review:**

Website: ***Harvard Business Review***: HBR publishes articles, case studies, and insights on economic trends, regulatory changes, and the intersection of business and technology.

5. **TechCrunch:**

Website: ***TechCrunch***: TechCrunch provides news and analysis on the latest technology trends, startup ecosystems, and regulatory developments in the tech industry.

6. **The Economist:**

Website: ***The Economist***: The Economist covers global economic trends, regulatory changes, and in-depth analyses that can be valuable for understanding the broader business landscape.

7. **Deloitte Insights:**

Website: ***Deloitte Insights***: Deloitte publishes reports and insights on economic trends, industry-specific analyses, and the impact of technology on businesses.

8. **McKinsey & Company:**

Website: ***McKinsey & Company***: McKinsey offers research and insights on technology trends, digital transformation, and economic shifts.

9. **World Bank Data:**

Website: ***World Bank Data***: The World Bank provides economic data and research reports that can be valuable for understanding global economic trends.

10. **Regulations.gov:**

Website: ***Regulations.gov***: This platform allows access to U.S. federal regulatory information, making it a valuable resource for understanding regulatory landscapes.

11. **European Union Open Data Portal:**

Website: ***EU Open Data Portal***: For insights into European economic trends and regulatory information, the EU Open Data Portal provides a wealth of data.

12. **CIA World Factbook:**

Website: ***CIA World Factbook***: The CIA World Factbook offers comprehensive information on the economic, political, and cultural aspects of countries around the world.

Demographic and Psychographic Insights

Understanding the market requires looking at who might be interested in our organisation or products, and why. A CTO does this by studying two things: demographics (like age and location) and psychographics* (like interests and values) of the people we want to reach. By figuring out what people like and how they think, a CTO can create technology plans that connect with lots of different perspectives.

Checking out the Competition: No market study is complete without checking out what other companies are doing. A CTO compares their organisation to others, looking at what products they offer, where they stand in the market, and what their customers think. By understanding how they measure up against the competition, a CTO can find out what they're good at and where they can do even better.

Following Tech Trends: The world of technology is always changing, and a good CTO keeps an eye on the latest trends. They look at what new technologies people are using and how quickly they catch on. This includes looking at different groups and figuring out why some technologies become popular faster than others. By keeping up with these trends, a CTO makes sure their organisation is using the latest and greatest tech ideas.

For example, in marketing, knowing that a certain group of consumers values sustainability, enjoys outdoor activities, and has a preference for eco-friendly products provides valuable psychographic insights. This information allows businesses to create targeted campaigns and offerings that resonate with the values and preferences of that particular audience.

* Psychographics refer to the study and classification of people based on their attitudes, interests, values, lifestyles, and personality traits. Unlike demographics, which focus on observable characteristics like age, gender, and income, psychographics delve into the psychological and behavioural aspects of individuals. By understanding psychographics, marketers, researchers, and businesses gain insights into the motivations, preferences, and decision-making processes of their target audience.

Psychographic segmentation involves dividing a population into groups based on shared psychological characteristics. This segmentation can provide a more nuanced understanding of consumer behaviour and help tailor marketing strategies to specific audience segments. Psychographics can include factors such as hobbies, social values, lifestyle choices, opinions, and attitudes toward specific products or issues.

Tools: AI tools like *IBM Watson* or *Salesforce Einstein Analytics* can analyse large datasets to identify patterns and correlations, helping to uncover hidden psychographic insights. Platforms like **Reddit** or specialised forums related to specific interests or industries can be valuable sources for observing and understanding the discussions, preferences, and opinions of target audiences.

Customer Journey Mapping

Understanding customer insights extends beyond demographic data; it involves empathetic exploration of the customer journey. The CTO through a lens of commitment to customer satisfaction, actively engages in customer journey mapping. This entails tracing the steps customers take from initial awareness to post-purchase experiences. By pinpointing pain points, moments of delight, and areas for improvement in the customer journey, the organisation enhances its ability to deliver a seamless and satisfying user experience.

Customer journey mapping is a comprehensive process that involves tracing the steps customers take from the initial awareness of a product or service to their post-purchase experiences. By visualising and understanding each touchpoint in the customer's interaction with the brand, organisations can gain valuable insights into the customer experience. Identifying pain points, moments of delight, and areas for improvement throughout the customer journey enables businesses to strategically enhance the overall customer experience. This mapping typically covers awareness, consideration, purchase, retention, and advocacy stages, allowing businesses to comprehend the holistic journey their customers undertake.

The importance of customer journey mapping lies in its ability to provide a deep understanding of customer behaviours, needs, and preferences at each stage of interaction. Techniques for successful customer journey mapping involve a combination of quantitative and qualitative methods. Surveys, interviews, and direct feedback from customers help uncover their emotions, motivations, and pain points. Additionally, data analytics tools can provide quantitative insights into customer behaviour across various touchpoints. Once the journey map is developed, it becomes a powerful tool for improving sales, relationships, and engagement. By addressing pain points and enhancing moments of delight, businesses can streamline the customer journey, leading to increased satisfaction and loyalty.

To leverage customer journey mapping for improvement, businesses can implement targeted strategies at specific touchpoints. For instance, if a pain point is identified during the purchase stage, simplifying the checkout process or offering personalised discounts may enhance the experience. Similarly, if moments of delight occur during customer support interactions, replicating that positive experience across all touchpoints can contribute to improved engagement. Regularly revisiting and updating the customer journey map ensures that it remains reflective of evolving customer expectations and market dynamics. Through a continuous cycle of mapping, analysis, and improvement, businesses can create a customer-centric approach that not only drives sales but also strengthens relationships and fosters long-term engagement.

Several visual tools can be utilised to explain customer journey mapping and highlight the various touchpoints in a customer's interaction with a brand. One popular visual tool is *Lucidchart*. These are visual representations that illustrate the entire customer experience from initial awareness to post-purchase interactions. Customer journey maps often use a timeline or flowchart format to depict each stage, touchpoints, and customer emotions.

Another visual tool is Experience Maps, which are similar to customer journey maps. Experience maps focus on the emotional and physical aspects of a customer's interaction with a brand. They provide a visual overview of the customer's experience, emphasising emotional highs and lows. Experience maps can be created using tools like *Adobe Illustrator*, *Sketch*, *experiencemap.com*, or any graphic design software.

Touchpoint Diagrams are another effective tool, specifically highlighting individual touchpoints across the customer journey. Using symbols or icons, touchpoint diagrams visually represent each interaction point, making it easy to identify critical moments. Tools like *Microsoft PowerPoint*, *Canva*, or *draw.io* can be used to create touchpoint diagrams.

Flowcharts, being versatile tools, can effectively represent the sequential flow of a customer's journey. Each touchpoint, decision point, and potential pathway can be visually articulated in a flowchart. *Microsoft Visio*, *Google Drawings*, or even online platforms like *Gliffy* are suitable for creating flowcharts.

Lastly, Heatmaps can be utilised to emphasise the significance of certain touchpoints or stages in the customer journey. Heatmaps use colour gradients to highlight areas of higher or lower importance. Tools like *Hotjar* or *Crazy Egg* can generate heatmaps based on user interactions on digital platforms.

Also see Design Thinking Chapter 8

Data-Driven Decision-Making

Data analytics tools play a pivotal role in extracting actionable insights from large datasets in market research and data analysis. These tools enable businesses to mine vast datasets for patterns, trends, and correlations that can inform strategic decision-making. One powerful example is Google Analytics, a widely-used web analytics service. It allows businesses to track and analyse website traffic, user behaviour, and engagement metrics. By identifying patterns in user interactions, businesses can optimise their online presence, enhance user experience, and tailor marketing strategies based on real-time insights.

In addition to **Google Analytics**, tools like **Tableau** are instrumental in transforming raw data into visually appealing and understandable formats. Tableau enables users to create interactive and shareable dashboards, making it easier to identify trends and correlations within complex datasets. For instance, in market research, Tableau can be employed to visualise customer demographics, purchasing behaviour, and regional preferences. By leveraging such tools, businesses can move beyond mere data collection to gaining actionable insights that shape product development, marketing campaigns, and overall business strategies, ultimately enhancing their competitive edge in the market.

Strategic market research and understanding customers are crucial skills for a successful Chief Technology Officer. Imagine the market as a constantly changing picture, and a CTO needs to figure out things like who is buying what and why, what the competition is doing, and what new technologies people are getting into. They also look at how customers go through the process of buying and using products, listen to what customers have to say, and explore new opportunities. By using data to make smart decisions, a CTO helps the company not just survive in the market but also do well and come up with innovative ideas that customers will love. In the world of tech leadership, strategic market research makes sure everything works together smoothly and pushes the company towards success and staying relevant in the market.

Market Research and Customer Insights: 10 Step Action Plan

Acquiring comprehensive market research and understanding customer insights will help a CTO to shape successful technology strategies.

1. Define Research Objectives:

- Clearly outline the goals of your market research and customer insights initiatives. Whether it's understanding market trends, evaluating customer satisfaction, or identifying unmet needs, defining objectives guides the entire process.
- **Example:** If the objective is to enhance user experience, focus on gathering insights related to usability, feature preferences, and pain points.

2. Explore Market Research Platforms:

- Utilise market research platforms and databases. Platforms like **Statista, IBISWorld,** or **Nielsen** provide industry reports, market trends, and consumer behaviour data.
- **Example:** **Statista** offers a wide range of statistics and market studies across various industries.

3. Engage in Social Media Listening:

- Leverage social media listening tools to understand customer sentiments, preferences, and emerging trends. Tools like **Brandwatch, Hootsuite,** or **Sprout Social** can monitor mentions and discussions across platforms.
- **Example:** **Brandwatch** enables real-time tracking of brand mentions and sentiment analysis on social media.

4. Deploy Customer Surveys:

- Design and deploy customer surveys to gather direct feedback. Platforms like **SurveyMonkey, Typeform,** or **Google Forms** allow for easy creation and distribution of surveys.
- **Example:** **SurveyMonkey** provides customisable survey templates and robust analytics for customer feedback.

5. Utilise Customer Relationship Management (CRM) Data:

- Leverage CRM data to analyse customer interactions, preferences, and purchase behaviour. Platforms like **Salesforce, HubSpot,** or **Zoho CRM** offer insights into customer relationships.
- **Example:** **Salesforce** provides dashboards and reports to analyse customer data and interactions.

6. Monitor Online Reviews and Feedback:

- Regularly monitor online reviews on platforms such as **Trustpilot, Yelp,** or industry-specific review sites. Analyse feedback to understand customer satisfaction and areas for improvement.
- **Example:** *Trustpilot* aggregates customer reviews across various industries, offering insights into overall customer sentiment.

7. Conduct Competitor Benchmarking:

- Benchmark against competitors to understand your market position. Tools like **Crayon** or **Kompyte** automate competitor analysis, providing real-time data on product features and market strategies.
- **Example:** *Crayon* provides insights into competitor product positioning, messaging, and market share.

8. Engage in Customer Interviews and Focus Groups:

- Conduct one-on-one customer interviews or organise focus groups to gather qualitative insights. Tools like **Zoom** or **Microsoft Teams** facilitate virtual interviews and discussions.
- **Example:** *Zoom* offers features for conducting virtual focus groups and interviews with ease.

9. Analyse Web Analytics and User Behaviour:

- Use web analytics tools like **Google Analytics, Mixpanel,** or **Hotjar** to analyse user behaviour on digital platforms. Understand how users interact with your website or application.
- **Example:** *Google Analytics* provides detailed insights into website traffic, user engagement, and conversion metrics.

10. Stay Informed with Industry Reports and Conferences:

- Attend industry conferences and leverage industry reports. Conferences like **CES** or reports from industry associations provide insights into emerging technologies and market trends.
- **Example:** *CES (Consumer Electronics Show)* is a major conference that showcases cutting-edge technologies and trends across various industries.

By following these 10 steps and utilising the recommended products and data sources, CTOs can acquire and interrogate market research and customer insights effectively, empowering them to make informed technology decisions aligned with market demands and customer expectations.

Chapter 3.

Defining Technological Objectives

3.1 Crafting the Technological Vision: Defining Clear Objectives

Defining clear objectives is crucial when crafting the technological vision. This chapter is devoted to the strategic process of establishing technological goals that not only align with broader business objectives but also stimulate innovation, encourage collaboration, and fuel sustained growth. From the initiation of visionary goals to their methodical execution.

Summary of sections in this chapter:

3.1 Visionary Technological Leadership

At the core of defining technological objectives is visionary leadership. Crafting a technological vision that extends beyond immediate challenges, embracing long-term goals that align with the organisation's strategic vision is a primary function that CTO needs to nail down early in taking up position. This visionary outlook provides a North Star for the technology team, instilling a sense of purpose and direction that transcends day-to-day operations. This vision is also often the legacy that is remembered for a CTO's tenure.

3.2 Strategic Alignment with Business Goals

The synergy between technology and business goals is the cornerstone of effective leadership. Technological objectives need to be intricately aligned with overarching business objectives. This requires a deep understanding of the organisation's mission, market positioning, and growth trajectory. By forging a direct link between technological aspirations and business success, a CTO positions technology as a catalyst for achieving corporate objectives.

3.3 SMART Goal Setting

The art of defining technological objectives involves precision in goal-setting. The SMART framework—Specific, Measurable, Achievable, Relevant, and Time-bound is a good guide. Each objective should be crystal clear, quantifiable, realistically attainable, directly relevant to the organisation's strategic vision, and bound by a defined timeline.

This clarity not only guides the technology team but also facilitates effective evaluation and communication of progress.

3.4 Stakeholder Engagement and Alignment

Effective technological leadership goes beyond the technology team; it involves engaging and aligning diverse stakeholders, and cultivating relationships with key stakeholders, from executive leadership to end-users. By soliciting input, understanding concerns, and aligning technological objectives with the broader organisational ecosystem, a CTO ensures that the technology roadmap resonates with the entire spectrum of stakeholders.

3.5 Innovation as an Objective

Innovation is not just a byproduct; it's an objective in itself. Innovation must be at the forefront of technological goals, creating an environment where experimentation and creative problem-solving thrive. This involves fostering a culture that values novel ideas, encourages risk-taking, and views failure as a stepping stone toward innovation. This is a hard concept to sell to your CEO, so it needs tackling right up-front. I don't like the term "Fail-Fast", so try replacing with "Learn Fast". By making innovation a tangible objective, a CTO positions the organisation on the very cutting edge of technological advancement.

3.6 Metrics for Success

Defining technological objectives requires a robust measurement framework. Whether it's improved system performance, increased user engagement, or enhanced cybersecurity measures, these metrics become the yardstick by which the success of technological objectives is evaluated. This data-driven approach informs decision-making and ensures that objectives remain aligned with overarching goals.

3.7 Cultivating a Culture of Ownership

Ownership is the fuel that propels technological objectives from vision to reality. Success is dependent on fostering a culture where each member of the technology team takes ownership of their role in achieving the defined objectives. This involves empowering team members, providing a sense of purpose, and recognising contributions. A culture of ownership not only enhances motivation but also ensures that every individual is invested in the success of technological goals. I saw this post on X (Twitter) "When I talk to managers I get the feeling they're important. When I talk to leaders I get the feeling I'm important". It's a good distinction of the culture that needs to be cultivated and the CTO's position within it.

Defining technological objectives is a strategic art that requires a blend of visionary leadership, strategic alignment, precision in goal-setting, stakeholder engagement, a commitment to innovation, a robust measurement framework, and a culture of ownership. By crafting a clear and compelling technological vision, the CTO not only sets the course for the technology team but also transforms the organisation into a dynamic force in the ever-evolving landscape of technology.

3.1 Visionary Technological Leadership

It may be an old adage but, visionary leadership is the catalyst that propels organisations toward innovation and sustained success. The Chief Technology Officer plays a pivotal role in shaping a technological vision that not only addresses immediate challenges but also charts a course for long-term excellence, harmonising with the organisation's strategic goals. This section explores the key components and tangible examples of visionary technological leadership, supported by relevant data sources.

One thing that does need clarifying up front for this section is that you can be on the journey to employ visionary leadership qualities and techniques without calling yourself a "Visionary Technology Leader". It's a grandiose title that would just make your colleagues chuckle.

Crafting a Compelling Technological Vision:

Crafting a compelling technological vision that transcends the confines of day-to-day operations isn't necessarily easy or a natural skill for all CTO's, particularly early in their career. This involves envisioning the role of technology in the organisation's future and aligning it with broader business objectives. For example, if the organisational goal is to be a leader in sustainable practices, the CTO's vision might involve implementing cutting-edge green technologies and fostering a culture of innovation cantered around sustainability. Selling the vision is a step in its own right.

Aligning Technological Goals with Strategic Vision: Save or Earn.

Visionary leadership requires aligning technological goals with the overarching strategic vision of the organisation. Ensuring that technological initiatives directly contribute to achieving the organisation's long-term objectives is a top priority. This alignment ensures that the technology roadmap is not only in sync with current business objectives but is also future-proofed to accommodate anticipated growth and changes in the market landscape.

The conversation with the CEO about whether the organisation is in an "invest to save" or "invest to earn" phase plays a pivotal role in shaping the CTO's strategy. "Invest to save" typically involves technology initiatives aimed at optimising processes, reducing costs, and improving efficiency, which is usually the easiest to sell to the CFO. In contrast, "invest to earn" focuses on leveraging technology to create new revenue streams, develop innovative products, or enter untapped markets. While both approaches are valid, I would argue that the CTO's default position often leans towards "invest to earn" as technological advancements and innovations frequently drive competitive advantage and revenue generation. Establishing this foundational understanding is crucial as it sets the direction for the overall technology strategy and helps prevent misaligned expectations down the line.

Moreover, once the organisation's stance on the investment approach is clear, the CTO can strategically prioritise initiatives accordingly. Whether it's investing in cutting-edge technologies to create new revenue opportunities or implementing efficient systems to streamline operations and reduce costs, the chosen path significantly influences the allocation of resources and the overall success of technology-driven initiatives. By clearly defining and agreeing on the "invest to save" or "invest to earn" strategy, the CTO ensures that the technology roadmap aligns seamlessly with the broader organisational goals and fosters a cohesive and impactful technological vision.

"Invest to Save" and "Invest to Earn" represent two distinct approaches to constructing an IT strategy, each with its set of priorities, goals, and implications for an organisation.

Invest to Save: This approach centres on making strategic investments in technology with the primary goal of achieving cost savings and operational efficiencies. Organisations adopting an "Invest to Save" strategy often identify areas where technology can streamline processes, automate repetitive tasks, and reduce overall operational expenses. For example, investing in advanced data analytics tools may help optimise resource allocation, minimise waste, and enhance decision-making efficiency.

The emphasis is on long-term benefits that lead to a reduction in ongoing operational costs, creating a return on investment through increased efficiency rather than immediate revenue generation.

Invest to Earn: In contrast, the "Invest to Earn" strategy is focused on making IT investments that directly contribute to revenue generation and business growth. Organisations adopting this approach prioritise technology initiatives that enhance product offerings, improve customer experiences, or open new revenue streams. For instance, investing in e-commerce platforms, digital marketing technologies, or innovative product development can be part of an "Invest to Earn" strategy. The focus here is on leveraging technology as a proactive driver of revenue, seeking direct returns on the initial investment through increased sales, market expansion, or the creation of new business opportunities.

Comparison:

1. **Focus on Objectives:**

    1. *Invest to Save:* Focuses on optimising costs and improving operational efficiency.

    2. *Invest to Earn:* Concentrates on driving revenue growth and expanding business opportunities.

2. **Timeframe for Returns:**

    1. *Invest to Save:* Expects returns through cost savings over the long term.

    2. *Invest to Earn:* Anticipates more immediate returns through increased revenue and business growth.

3. **Risk Tolerance:**

    1. *Invest to Save:* Typically involves lower risk as it is often aimed at enhancing existing processes.

    2. *Invest to Earn:* May involve higher risk as it seeks to explore new markets, products, or technologies.

4. **Examples:**

    1. *Invest to Save:* Upgrading legacy systems, implementing process automation, or adopting cost-effective cloud solutions.

2. *Invest to Earn:* Developing new product features, expanding digital marketing efforts, or entering new markets through technology-driven strategies.

Ultimately, the choice between "Invest to Save" and "Invest to Earn" depends on the organisation's overall business objectives, risk appetite, and current market conditions. In some cases, a balanced strategy that incorporates elements of both approaches may be the most appropriate, allowing the organisation to achieve both efficiency improvements and revenue growth.

Inspiring a Sense of Purpose:

The CTO needs to inspire a sense of purpose within the technology team by clearly communicating the broader impact of their work. This involves connecting individual tasks to the overarching technological vision. For instance, if the vision emphasises advancing artificial intelligence for societal good, the CTO can highlight how each team member contributes to this larger goal through responsible development practices.

Data Source Example: Share case studies and success stories from organisations that have successfully aligned technological endeavours with a broader sense of purpose. Platforms like Harvard Business Review or McKinsey Insights often publish such case studies, providing valuable insights into the impact of visionary leadership.

Anticipating Future Technological Trends:

Visionary leadership involves staying ahead of the curve by anticipating future technological trends. A smart CTO actively engages with industry reports, emerging technologies, and innovation forums to shape a vision that is not only current but also future-proof. For example, if the industry is witnessing a shift towards edge computing, the CTO's vision might involve strategic investments in edge technologies.

See Section: **2.5 Embracing Emerging Technologies**

Data Source Example: Regularly consult technology forecasts and reports from reputable sources like Gartner, Forrester, or Deloitte. These reports offer insights into emerging technologies, adoption timelines, and potential implications for various industries.

Fostering a Culture of Innovation:

Visionary technological leadership extends beyond individual visions; it involves fostering a culture of innovation within the technology team. This includes creating avenues for continuous learning, experimentation, and ideation. For instance, a CTO might introduce innovation challenges or hackathons to encourage creative problem-solving among team members.

See Book 3

Data Source Example: Explore innovation indices and reports that highlight the most innovative companies globally. The Global Innovation Index (GII) or the Bloomberg Innovation Index can provide benchmarks and insights into fostering a culture of innovation within the organisation.

Measuring Progress and Adjusting Course:

Establishing metrics and key performance indicators (KPIs) to measure progress toward the technological vision is a key priority. Regular assessments and adjustments ensure that the vision remains relevant and aligned with organisational goals. For example, if the vision involves enhancing cybersecurity resilience, KPIs might include the reduction of security incidents and the successful implementation of proactive security measures.

See section 3.6 Metrics for Success and 4.7 Monitoring KPI's in a Technology Roadmap

Data Source Example: Leverage cybersecurity reports and threat intelligence platforms to assess the current state of cybersecurity landscapes. Reports from organisations like Verizon's Data Breach Investigations Report (DBIR) provide valuable insights into prevalent threats and industry best practices.

Visionary technological leadership is a dynamic and forward-thinking approach that requires a synthesis of strategic foresight, industry insights, and a commitment to innovation. By crafting a compelling technological vision, aligning goals strategically, inspiring a sense of purpose, anticipating future trends, fostering innovation, and implementing effective metrics, a smart CTO becomes a driving force in steering the organisation toward technological excellence and long-term success.

Shaping a Technological Vision: A 10 Step Action Plan

Creating a compelling technological vision involves a multifaceted approach that integrates strategic planning, communication strategies, and adaptability. This 10 step guide provides detailed insights into each step, combining key elements for crafting an impactful technological vision:

1. Define a Clear Vision Statement:

Develop a succinct and inspiring vision statement that encapsulates the essence of the technological vision. Ensure that it aligns seamlessly with organisational goals and resonates with a diverse range of stakeholders.

- **Example:** "Empower our organisation through innovative technology, driving us to redefine industry standards, foster growth, and pioneer transformative solutions."

2. Identify Industry Trends and Innovations:

- Stay abreast of emerging technologies and industry trends by leveraging reports from renowned research firms like Gartner and Forrester. Explore specialised publications catering to your industry for deeper insights.
- **Example:** Gartner's annual reports offer invaluable insights into transformative technologies shaping various industries, aiding in the identification of potential areas for innovation.

3. Collaborate with Cross-Functional Teams:

- Foster collaboration with cross-functional teams to gain a comprehensive understanding of business needs. Conduct ideation workshops that bring together diverse perspectives, ensuring the technological vision is holistic and inclusive.
- **Example:** Use collaborative platforms such as *Miro* or *Trello* to facilitate virtual brainstorming sessions where teams can contribute ideas and insights.

4. Develop a Technology Roadmap:

- Construct a detailed roadmap outlining specific initiatives, projects, and milestones. Break down the vision into actionable steps, assigning responsibilities and establishing realistic timelines for each phase.
- **Example:** A roadmap could include phrases like "Infrastructure Modernisation," "AI Integration," and "Cybersecurity Enhancement," each with granular tasks and corresponding timelines.

5. Foster Employee Engagement:

- Actively engage employees through workshops, town hall meetings, or virtual events to articulate the technological vision. Encourage questions, provide clarifications, and gather feedback to foster a sense of involvement.
- **Example:** Host webinars with interactive Q&A sessions to engage employees, allowing them to voice concerns, offer suggestions, and actively participate in the shaping of the technological vision.

6. Utilise Visual Aids and Infographics:

- Enhance communication by incorporating visually appealing aids and infographics. Leverage tools like Piktochart or Venngage to create graphics that convey complex technological concepts in an accessible manner.
- **Example:** Develop an infographic that visually illustrates key technological pillars, the roadmap, and expected outcomes, making it easier for diverse audiences to grasp the vision.

7. Establish Key Performance Indicators (KPIs):

- Define measurable KPIs that align with the technological vision. Utilise data analytics tools such as Google Analytics or Tableau to track and visualise progress against these KPIs.
- **Example:** KPIs could include metrics like "Time-to-Market for New Features," "Reduction in System Downtime," or "User Satisfaction Scores," providing tangible indicators of success.

8. Incorporate Flexibility and Adaptability:

- Recognise the dynamic nature of technology and business landscapes. Build flexibility into the vision, allowing for adjustments based on emerging opportunities or unforeseen challenges.
- **Example:** Include a dedicated review and adaptation phase in the roadmap, enabling the organisation to realign the technological vision with changing organisational or market dynamics.

9. Leverage Storytelling Techniques:

- Tell a compelling story around the technological vision. Utilise narrative techniques to create a journey that employees and stakeholders can connect with emotionally. Showcase the background, challenges, and aspirations associated with the vision.
- **Example:** Develop a storytelling video or written narrative that communicates the vision's origin, the hurdles it aims to overcome, and the transformative impact it seeks to achieve. Check out *Vyond*.

10. Maintain Open Communication Channels:

- Establish transparent communication channels for ongoing updates and feedback. Leverage collaboration tools like Slack or Microsoft Teams to facilitate real-time communication, ensuring that information flows seamlessly.
- **Example:** Create dedicated channels or forums where employees can ask questions, share insights, and stay informed about the latest developments related to the technological vision, fostering an environment of open dialogue.

3.2 Strategic Alignment with Business Goals - A Foundation for Technological Leadership

The symbiotic relationship between technology and business goals lies at the heart of successful digital leadership. This section explores the key facets and practical examples of strategic alignment with business goals.

In order to understand the "strategic alignment" and how the "Technological Vision" supports the "Corporate Vision" we will take a look at three examples:

Company: FutureLearn: Ed-Tech

Technological Vision: FutureLearn envisions leveraging cutting-edge technologies to redefine the landscape of global education. Through innovative Ed-Tech solutions, the company aims to create an ecosystem where learners worldwide can access high-quality online courses and degrees with unprecedented flexibility. The technological vision involves implementing advanced learning analytics, adaptive learning systems, and immersive educational experiences, ensuring a transformative and personalised journey for every user. By embracing technology, FutureLearn seeks to overcome geographical barriers and make education more inclusive and tailored to individual needs.

Comparison with Company Vision: FutureLearn's technological vision aligns seamlessly with its overarching company vision of "Transforming lives globally through accessible education". The focus on leveraging technology to enhance the educational experience directly supports the broader goal of breaking down barriers to learning. The technological vision serves as the instrumental means through which FutureLearn strives to achieve its company vision, demonstrating a strategic alignment between the company's overall mission and its specific technological aspirations.

Company: Revolut: Fin-Tech

Technological Vision: Revolut's technological vision revolves around pioneering advancements in financial technology to create a borderless and inclusive financial ecosystem. Through the integration of innovative FinTech solutions, Revolut aims to provide users with seamless and technologically advanced financial services. This includes features such as real-time payment processing, cryptocurrency accessibility, and personalised financial insights driven by artificial intelligence. The technological vision underscores Revolut's commitment to redefining traditional banking norms, offering users a modern and user-centric financial experience.

Comparison with Company Vision: Revolut's technological vision directly contributes to its overarching company vision of making financial services frictionless and accessible to all. The focus on cutting-edge FinTech aligns with the broader goal of revolutionising the way individuals and businesses manage their finances. The technological vision is not just a means to an end; it is a central element in realising Revolut's company vision, showcasing a strategic integration of technology to achieve transformative financial solutions.

Company: ASOS: Retail

Technological Vision: ASOS's technological vision centres on redefining the retail landscape through digital innovation. By leveraging state-of-the-art retail technology, ASOS aims to create a dynamic and personalised shopping experience for its customers. This includes implementing augmented reality for virtual try-ons, utilising data analytics for tailored recommendations, and optimising the overall e-commerce platform for seamless transactions. The technological vision reflects ASOS's commitment to setting industry standards through continuous technological advancements.

Comparison with Company Vision: ASOS's technological vision is intricately linked to its broader company vision of providing customers with a personalised and seamless shopping journey. The focus on employing cutting-edge technology aligns with the overarching goal of creating an engaging and customer-centric retail experience. ASOS's commitment to digital innovation is not just a component of its strategy; it is a critical element in achieving the company's vision of redefining online fashion retail.

Recommended Resources: "Global Sustainable Technology Market Outlook" from market research firm like *Technavio*

Understanding Organisational Mission:

Begin by delving deep into the organisation's mission, understanding its core values, and envisioning how technology can amplify and accelerate the fulfilment of that mission. For instance, if the mission revolves around providing accessible healthcare, the CTO may align technological goals with innovations like telemedicine platforms or healthcare analytics.

Data Source Example: Annual reports, mission statements, and official publications provide crucial insights into the organisation's mission and strategic priorities. Organisations often publish these documents on their official websites.

Mapping Market Positioning:

Strategic alignment requires a nuanced understanding of the organisation's market positioning. A CTO should analyse market trends, competitive landscapes, and consumer behaviours to align technological initiatives with the organisation's position in the market. If the organisation aims to be an industry leader in user experience, the CTO may focus on implementing cutting-edge UX/UI technologies.

Data Source Example: Market research reports from firms like Nielsen, Gartner, or Statista offer valuable insights into market dynamics, consumer preferences, and emerging trends.

Aligning with Growth Trajectory:

The growth trajectory of the organisation is a crucial element in strategic alignment. A visionary CTO anticipates the future needs of the business and aligns technological objectives accordingly. For instance, if the organisation is poised for international expansion, the CTO may prioritise scalable and globally adaptable technology solutions.

Data Source Example: Economic forecasts and industry-specific growth projections from sources like the International Monetary Fund (IMF) or industry associations provide valuable data to inform alignment strategies.

Integrating Technological and Business Roadmaps:

A seamless alignment requires integration between technological roadmaps and broader business strategies. The CTO ensures that every technological initiative is a strategic milestone on the overarching business roadmap. If the business roadmap involves diversification into new product lines, the CTO may focus on developing the necessary technological infrastructure to support such diversification.

Data Source Example: Project management tools, such as Jira or Trello, can be utilised to visually integrate and align technological and business roadmaps.

Leveraging Data-Driven Decision-Making:

Strategic alignment is enhanced through data-driven decision-making. The CTO utilises data analytics to inform technology decisions that directly contribute to achieving business goals. For example, leveraging customer data analytics may inform decisions on product development or marketing strategies.

Data Source Example: Business intelligence tools like Tableau or Power BI provide dynamic dashboards and reports, allowing CTOs to make informed decisions based on real-time data.

Demonstrating Value through Key Performance Indicators (KPIs):

Aligning technological objectives with business goals involves demonstrating tangible value. A smart CTO establishes Key Performance Indicators (KPIs) that directly reflect progress toward overarching business objectives. If the business goal is to enhance customer satisfaction, KPIs may include improvements in response time or user satisfaction scores. *See section 3.6 Metrics for Success and 4.7 Monitoring KPI's in a Technology Roadmap*

Data Source Example: Customer feedback platforms, surveys, and Net Promoter Scores (NPS) serve as valuable data sources for assessing customer satisfaction and aligning technological efforts accordingly.

Innovating for Competitive Advantage:

Strategic alignment is not just about meeting current business goals but also positioning the organisation for future competitiveness. The CTO should consider introducing innovative technologies that give the organisation a competitive edge. For instance, integrating artificial intelligence for predictive analytics might give the business a competitive advantage in decision-making.

Data Source Example: Innovation indices and reports from platforms like the Global Innovation Index (GII) provide insights into emerging technologies and best practices for maintaining a competitive edge.

Adapting to Regulatory and Compliance Requirements:

Ensuring that technological initiatives align with regulatory and compliance requirements, safeguarding the organisation's reputation and financial stability is key. If the business operates in a highly regulated industry, the CTO may prioritise cybersecurity measures to comply with data protection laws.

Data Source Example: Regular updates from regulatory bodies, industry-specific compliance agencies, and legal databases serve as critical data sources to ensure alignment with regulatory requirements.

Nurturing Cross-Functional Collaboration:

Effective strategic alignment extends beyond the technology team. A skilled CTO fosters collaboration with cross-functional teams to ensure that technological initiatives align seamlessly with broader business strategies. Collaboration tools like Slack or Microsoft Teams facilitate communication and coordination across different departments.

Data Source Example: Collaboration platforms often include features for document sharing, real-time communication, and project management, providing a comprehensive view of cross-functional efforts.

Measuring ROI and Continuous Optimisation:

Strategic alignment culminates in measurable return on investment (ROI). A smart CTO continuously monitors the performance of technological initiatives and optimises strategies to miximise ROI. If the business goal is cost reduction, the CTO may evaluate the efficiency of cloud infrastructure to optimise operational costs.

Data Source Example: Financial reports, cost-benefit analyses, and performance metrics tracked through enterprise resource planning (ERP) systems contribute to measuring ROI.

Strategic alignment with business goals is the linchpin of effective technological leadership. By understanding the organisation's mission, mapping market positioning, aligning with growth trajectories, integrating roadmaps, leveraging data-driven decision-making, demonstrating value through KPIs, innovating for competitive advantage, adapting to regulatory requirements, nurturing collaboration, and measuring ROI, a smart CTO not only positions technology as a catalyst for business success but also ensures sustained relevance and competitiveness in the digital landscape.

Technological Vision and Business Goals: A 10 Step Action Plan

Ensuring that the technological vision seamlessly aligns with the broader business goals is crucial for success. This comprehensive 10-step guide provides detailed insights into each step, focusing on strategic alignment between technological initiatives and overarching business objectives:

1. Establish Clear Business Objectives:

- Begin by thoroughly understanding and defining the organisation's overarching business objectives. Engage with key stakeholders, including executives and department heads, to ensure clarity and alignment.
- **Example:** If the business objective is to increase market share, the technological vision should support strategies that enhance product offerings, customer experience, or market reach.

2. Conduct a Business Impact Assessment:

- Evaluate how each component of the technological vision contributes to achieving specific business goals. Perform a comprehensive impact assessment to quantify the potential positive outcomes.
- **Example:** Assess how implementing a new customer relationship management (CRM) system aligns with the business goal of improving customer satisfaction and retention.

3. Facilitate Cross-Functional Collaboration:

- Foster collaboration between technology teams and other business units. Engage in open dialogues to understand the unique needs and priorities of each department, ensuring the technological vision is well-integrated.
- **Example:** Collaborate with the sales team to understand their customer relationship needs, aligning technology initiatives with their strategies.

4. Develop a Technology-Business Alignment Framework:

- Create a framework that clearly outlines how each technological initiative directly supports specific business objectives. This framework serves as a strategic guide for decision-making and resource allocation.
- **Example:** Map out how adopting cloud technology aligns with business objectives related to cost optimisation, scalability, and flexibility.

5. Establish Key Performance Indicators (KPIs) for Alignment:

- Define measurable KPIs that gauge the success of technological initiatives in alignment with business goals. Ensure these KPIs are regularly monitored and analysed.
- **Example:** If the business goal is to enhance operational efficiency, corresponding KPIs could include metrics like process cycle time reduction or resource utilisation improvement.

6. Ensure Scalability and Flexibility:

- Design the technological vision with scalability and flexibility in mind. Ensure that the chosen technologies and strategies can adapt to evolving business needs and growth.
- **Example:** Choose a cloud infrastructure that can seamlessly scale with increased demand or changes in business operations.

7. Conduct Regular Alignment Reviews:

- Schedule periodic reviews to assess the ongoing alignment between technological initiatives and business goals. Involve key stakeholders in these reviews to gather feedback and make necessary adjustments.
- **Example:** Conduct quarterly alignment reviews where representatives from various departments provide insights into the effectiveness of technology-business alignment.

8. Develop a Communication Strategy:

- Craft a communication strategy that effectively communicates how technological initiatives contribute to achieving business goals. Tailor messages for different stakeholders to ensure clarity and buy-in.
- **Example:** Use concise and visually appealing presentations to convey how adopting a new analytics tool directly supports sales growth targets.

9. Align IT Governance with Business Strategy:

- Ensure that IT governance structures are aligned with overall business strategy. This includes decision-making processes, risk management, and resource allocation.
- **Example:** Adjust IT governance policies to prioritise projects that have a direct impact on critical business objectives.

10. Foster a Culture of Continuous Alignment:

- Instil a culture where continuous alignment with business goals becomes ingrained in the organisational mindset. Encourage teams to regularly reassess and realign their technological efforts.
- **Example:** Introduce recognition programs that celebrate teams and individuals who successfully contribute to the strategic alignment of technological initiatives with business goals.

This 10 step guide emphasises the importance of strategic alignment, ensuring that technological initiatives consistently contribute to and enhance the achievement of broader business objectives.

3.3 SMART Goal Setting in Technological Leadership

The art of defining objectives is a crucial skill that distinguishes SMART Chief Technology Officers (CTOs). The CTO understands that the clarity and precision of goals play a pivotal role in guiding the technology team toward success. This chapter delves into the principles of SMART goal setting—Specific, Measurable, Achievable, Relevant, and Time-bound—and explores practical examples and relevant data sources that enhance the effectiveness of this framework.

Specific: Defining Clear Objectives

Specificity is the cornerstone of effective goal setting. It is important to ensure that technological objectives are explicitly defined, leaving no room for ambiguity. For instance, instead of setting a vague goal like "improve system performance," a specific goal might be "reduce application response time by 20% within the next quarter."

Utilising Project Management Tools
CTOs can use project management tools like Asana or Jira to create specific tasks and milestones, breaking down broader objectives into actionable items with clear deliverables and outcomes.

Measurable: Quantifying Progress and Success

Measurability enables the tracking of progress and success. A SMART CTO establishes quantifiable metrics that indicate the degree of goal achievement. Continuing with the previous example, if the goal is to reduce application response time, the CTO might set a measurable metric like "achieve an average response time of 2 seconds."

Analytics Platforms for Performance Metrics
Leverage analytics platforms such as New Relic or Google Analytics to measure and monitor specific performance metrics. These tools provide real-time data on application response times and other relevant performance indicators.

Achievable: Realistic and Attainable Goals

Setting achievable goals is essential for maintaining motivation and fostering a sense of accomplishment. A SMART CTO ensures that technological objectives are realistic within the given constraints. For instance, expecting a 50% reduction in response time might be unrealistic, but achieving a 20% improvement is feasible.

Historical Performance Data Analysis
Referencing historical performance data and benchmarks aids in setting achievable goals. Analyse past performance trends to establish realistic targets that align with the organisation's capabilities and resources.

Relevant: Aligning with Strategic Vision

Relevance ensures that technological goals directly contribute to the organisation's strategic vision. If the strategic vision involves enhancing customer experience, a relevant technological goal might focus on optimising user interfaces for key customer-facing applications.

Alignment with Strategic Initiatives
Regularly review the organisation's strategic initiatives and ensure that technological goals align with these initiatives. Strategic planning documents, such as annual reports or official strategic vision statements, serve as key references.

Time-bound: Setting a Defined Timeline

Establishing a clear timeline is essential for goal orientation and progress tracking while setting specific deadlines for achieving objectives. Following the earlier example, a time-bound goal might be to "reduce application response time by 20% within the next quarter."

Gantt Charts for Project Timelines
Gantt charts, available in project management tools like *Product Plan, Trello* or *Microsoft Project,* visually represent project timelines. They help in setting realistic deadlines and tracking progress over time.

Leveraging Data Sources for SMART Goal Setting:

To enhance the SMART goal-setting process, CTOs can leverage relevant data sources and platforms:

- **Market Intelligence Reports:** Industry reports and market intelligence platforms, such as *Statista* or *Gartner*, provide data on technological trends, benchmarks, and industry best practices. This information informs the specificity and relevance of technological goals.

- **User Feedback and Surveys:** Direct insights from users and stakeholders, gathered through surveys or feedback platforms like *UserTesting* or

SurveyMonkey, contribute to the measurable and achievable aspects of goal setting. User feedback guides the definition of specific improvements and realistic expectations.

- **Competitor Analysis Tools:** Platforms like Crayon or SimilarWeb provide data on competitors' technological strategies and performance metrics. This data aids in setting measurable goals that align with or surpass industry benchmarks.

- **Performance Monitoring Tools:** Utilise tools like *Dynatrace* or *Splunk* for real-time monitoring of technological performance. These tools generate data on specific metrics, supporting the measurable and time-bound aspects of SMART goals.

Monitoring Progress and Iterating:

SMART goal setting is an iterative process. A smart CTO continuously monitors progress using project management tools, performance analytics, and regular status updates. Data-driven insights from these sources facilitate adjustments to goals based on evolving circumstances and priorities.

The implementation of SMART goal setting in technological leadership empowers CTOs to drive focused and achievable outcomes. By utilising relevant examples and tapping into diverse data sources, CTOs can ensure that their technological objectives are not only SMART but also strategic contributors to the organisation's success.

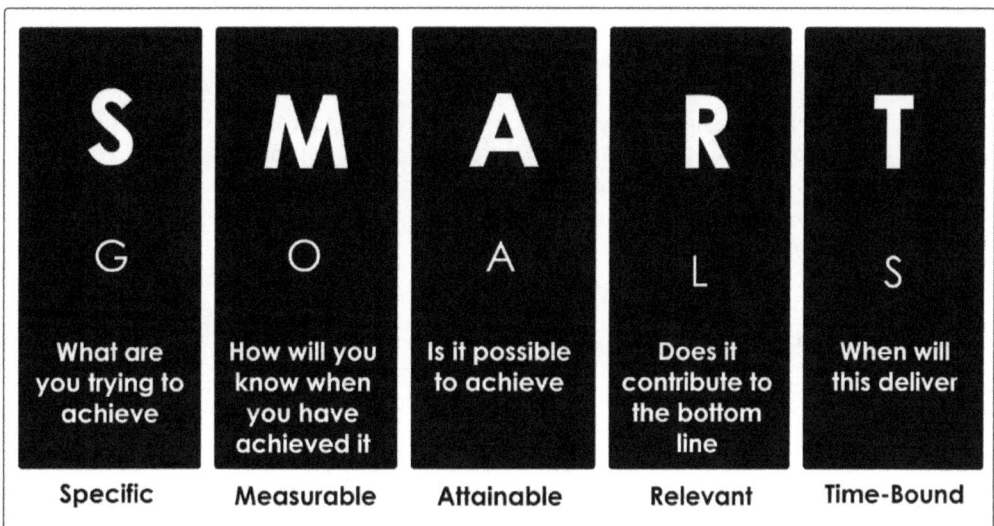

S	M	A	R	T
G	O	A	L	S
What are you trying to achieve	How will you know when you have achieved it	Is it possible to achieve	Does it contribute to the bottom line	When will this deliver
Specific	Measurable	Attainable	Relevant	Time-Bound

SMART Goals in Technological Leadership: A 10-Step Action Plan

Setting SMART goals (Specific, Measurable, Achievable, Relevant, Time-bound) is crucial for effective technological leadership. This 10-step guide provides practical insights into creating SMART goals in the context of technological leadership:

1. Define Specific Technological Objectives:

- Clearly articulate specific technological objectives that align with broader organisational goals. Ensure that each goal is well-defined, leaving no room for ambiguity.
- **Example:** Instead of a vague goal like "Improve cybersecurity," make it specific: "Implement multi-factor authentication across all user accounts within the next six months."

2. Establish Measurable Metrics:

- Identify quantifiable metrics that allow for objective measurement of progress. Define how success will be measured and set benchmarks for each metric.
- **Example:** If the goal is to enhance system performance, set a measurable metric such as "Reduce average system response time by 20% within the next quarter."

3. Ensure Achievability of Goals:

- Evaluate the feasibility of each goal by considering available resources, expertise, and potential challenges. Goals should be challenging but realistically attainable.
- **Example:** Assess whether the organisation has the necessary resources and expertise to implement a new technology and set a goal that stretches capabilities without being unattainable.

4. Align Goals with Organisational Relevance:

- Ensure that each goal directly contributes to the overall success and objectives of the organisation. Goals should align with the organisational mission and strategic priorities.
- **Example:** If the organisation prioritises customer satisfaction, a relevant goal might be "Implement a customer feedback system to increase satisfaction scores by 15% in the next fiscal year."

5. Set Time-Bound Milestones:

- Establish specific deadlines or milestones for each goal. This adds urgency and clarity to the timeline, facilitating effective monitoring and timely adjustments.
- **Example:** Instead of a vague timeframe, set a clear deadline: "Complete the migration to the cloud for all critical systems by the end of the second quarter."

6. Foster Collaboration in Goal Setting:

- Involve relevant stakeholders in the goal-setting process to gather diverse perspectives and ensure alignment with organisational strategies. Collaboration enhances commitment and understanding.
- **Example:** Convene cross-functional teams to collaboratively set goals, ensuring that each department's technological needs and perspectives are considered.

7. Prioritise High-Impact Goals:

- Prioritise goals based on their potential impact on organisational success. Focus on high-priority areas that align with strategic objectives.
- **Example:** Identify goals that have a direct impact on revenue growth, cost reduction, or customer satisfaction, aligning with the organisation's key priorities.

8. Regularly Review and Adjust Goals:

- Schedule regular reviews to assess goal progress and make adjustments as needed. Flexibility is crucial to adapt to evolving technological landscapes and unforeseen challenges.
- **Example:** Conduct quarterly reviews to evaluate goal progress, make adjustments based on emerging opportunities or obstacles, and ensure continued alignment with organisational objectives.

9. Leverage Technology for Monitoring:

- Utilise technology tools and platforms for monitoring and tracking goal progress. Implement dashboards or project management tools to provide real-time visibility.
- **Example:** Use project management tools like Jira or Trello to track progress, identify bottlenecks, and ensure timely completion of technological goals.

10. Communicate Clear Expectations:

- Effectively communicate the SMART goals and associated expectations to the entire team. Foster a shared understanding of each goal, its significance, and the collective effort required.
- **Example:** Conduct team meetings to discuss SMART goals, provide context on their importance, and encourage open dialogue for any clarifications or suggestions.

By following these 10 steps, CTOs can create SMART goals in technological leadership that are not only aligned with organisational objectives but also measurable, achievable, relevant, and time-bound, driving the organisation toward technological excellence.

3.4 Stakeholder Engagement and Alignment

The role of a Chief Technology Officer (CTO) extends far beyond the confines of the technology team. This chapter delves into the critical domain of stakeholder engagement and alignment, exploring how a skilled CTO cultivates relationships with diverse stakeholders and ensures that technological objectives resonate across the entire organisational spectrum.

Understanding the Stakeholder Landscape:

A CTO initiates the journey of stakeholder engagement by comprehensively understanding the stakeholder landscape. This involves identifying and categorising stakeholders, from executive leadership and department heads to end-users and external partners. For example, stakeholders may include C-suite executives, department heads, regulatory bodies, and customers.

Data Source Example: Utilise stakeholder mapping tools and frameworks to identify and categorise stakeholders based on their influence and interest. ***The Power/Interest Grid*** is one such tool that helps prioritise stakeholders. EG **https://www.improvementservice.org.uk/business-analysis-framework/consider-perspectives/powerinterest-grid**

Engaging Executive Leadership:

Effective stakeholder engagement begins at the top. A skilled CTO actively engages with executive leadership, presenting technological initiatives in alignment with broader business strategies.

Data Source Example: Executive leadership reports and strategic planning documents provide insights into the organisation's overarching goals and priorities. These documents are often available through internal communications or corporate intranets.

Collaborating with Department Heads:

It is important to recognise that just as it is important to align the technological vision and goals to the corporate vision and strategy, it is equally important to align technological objectives with other departmental goals. By collaborating closely with department heads, the CTO ensures that technological initiatives contribute directly to enhancing departmental efficiency and effectiveness. For example, collaboration with the marketing department may focus on implementing data analytics tools for targeted campaigns.

Data Source Example: Conduct regular meetings with department heads and leverage project management platforms like *Asana* or *Monday.com* to track and align technology initiatives with departmental goals.

Incorporating End-User Perspectives:

Stakeholder engagement is incomplete without considering the perspectives of end-users. A CTO should actively solicit input from end-users, understanding their needs, challenges, and expectations. This could involve conducting user surveys, focus groups, or usability testing.

Data Source Example: User feedback platforms, such as *UserTesting* or *Usabilla*, provide real-time insights into end-user experiences and preferences, guiding the refinement of technological solutions.

Aligning with Regulatory Bodies:

In certain industries, regulatory bodies play a crucial role as stakeholders. A CTO stays attuned to regulatory requirements, ensuring that technological initiatives align with industry regulations. For instance, if the organisation operates in healthcare, adherence to data protection laws and healthcare regulations is paramount.

Data Source Example: Regularly monitor updates from relevant regulatory bodies, subscribe to industry newsletters, and leverage legal databases to stay informed about evolving regulations.

Engaging External Partners and Suppliers:

For organisations with external partners and suppliers, stakeholder engagement extends beyond internal teams. A skilled CTO actively collaborates with external partners, fostering relationships that contribute to joint innovation and mutual success.

Data Source Example: Industry forums, collaborative platforms, and joint project management tools provide spaces for effective communication and collaboration with external partners.

Facilitating Cross-Functional Workshops:

A CTO orchestrates cross-functional workshops that bring together diverse stakeholders. These workshops serve as platforms for ideation, collaboration, and the alignment of technological objectives with broader organisational goals.

Data Source Example: Utilise collaboration tools like *Miro* or *MURAL* to facilitate virtual workshops, allowing stakeholders to contribute ideas, provide feedback, and align on technological strategies.

Measuring Stakeholder Satisfaction:

Quantifying stakeholder satisfaction is integral to assessing the success of engagement efforts. A skilled CTO establishes metrics, such as stakeholder satisfaction scores or feedback surveys, to gauge how well technological initiatives resonate with different stakeholder groups.

Data Source Example: Stakeholder satisfaction surveys conducted through platforms like *SurveyMonkey* or *Google Forms* provide quantitative and qualitative data on stakeholder perceptions.

Building a Communication Framework:

Effective stakeholder engagement requires a robust communication framework. A smart CTO establishes regular communication channels, such as newsletters, town hall meetings, or project update sessions, to keep stakeholders informed and engaged.

Data Source Example: Communication platforms like *Slack* or *Microsoft Teams* provide centralised channels for real-time communication, ensuring stakeholders are consistently informed and engaged.

Leveraging Data for Stakeholder Insights:

Data-driven insights into stakeholder behaviours and preferences enhance engagement strategies. A skilled CTO utilises data analytics tools to understand stakeholder interactions, tailor communication strategies, and refine technological initiatives based on stakeholder insights.

Data Source Example: Customer Relationship Management (CRM) platforms, such as *Salesforce* or *HubSpot*, offer data analytics features to track stakeholder interactions and preferences.

Stakeholder engagement and alignment are pivotal elements of effective technological leadership. By understanding the stakeholder landscape, engaging executive leadership, collaborating with department heads, incorporating end-user perspectives, aligning with regulatory bodies, engaging external partners, facilitating cross-functional workshops, measuring stakeholder satisfaction, building a communication framework, and leveraging data for insights, a smart CTO orchestrates a harmonious integration of technology within the entire organisational ecosystem.

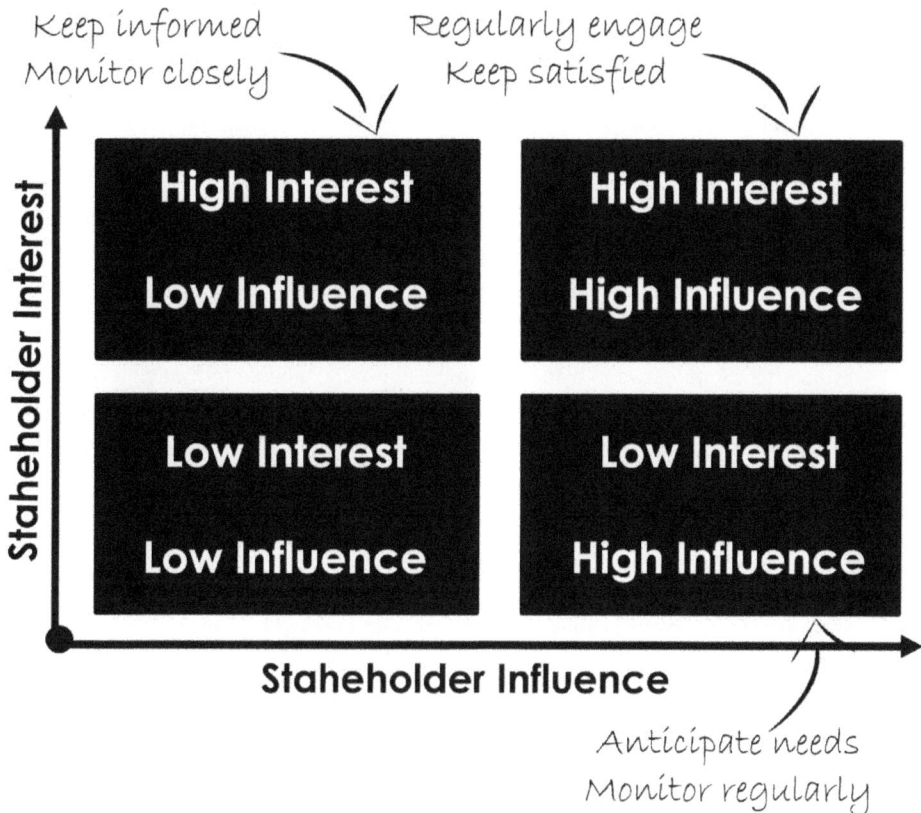

Keep informed
Monitor closely

Regularly engage
Keep satisfied

Stakeholder Interest

High Interest	High Interest
Low Influence	High Influence
Low Interest	Low Interest
Low Influence	High Influence

Staheholder Influence

Anticipate needs
Monitor regularly

Stakeholder Engagement and Alignment: A 10-Step Action Plan

Creating stakeholder engagement and alignment is pivotal for the success of technological initiatives. This practical 10-step guide provides detailed insights into each step, offering a comprehensive approach to foster engagement and ensure alignment with stakeholders:

1. Identify Key Stakeholders:

- Begin by identifying all relevant stakeholders, including internal teams, executives, customers, and external partners. Develop a comprehensive stakeholder map to ensure no crucial parties are overlooked.
- **Example:** Internal stakeholders may include development teams, marketing, and operations, while external stakeholders could be customers, regulatory bodies, or industry associations.

2. Understand Stakeholder Needs and Expectations:

- Conduct thorough assessments to understand the unique needs and expectations of each stakeholder group. Gather feedback through surveys, interviews, and workshops to gain insights into their perspectives.
- **Example:** Use customer feedback surveys, one-on-one meetings with internal teams, and collaborative workshops to capture diverse stakeholder expectations.

3. Develop a Clear Communication Plan:

- Establish a communication plan that outlines how and when information will be shared with stakeholders. Ensure transparency and consistency in communication channels to build trust.
- **Example:** Utilise a combination of emails, newsletters, and regular update meetings to keep stakeholders informed about technological developments and progress.

4. Align Technological Goals with Organisational Objectives:

- Ensure that technological goals directly align with overarching organisational objectives. Clearly articulate how technological initiatives contribute to the achievement of broader business goals.
- **Example:** If the organisation aims to increase market share, align technological goals to support initiatives that enhance product innovation or customer experience.

5. Facilitate Cross-Functional Collaboration:

- Encourage collaboration among cross-functional teams to break down silos and ensure a holistic understanding of technological initiatives. Leverage collaborative platforms to facilitate communication.
- **Example:** Use platforms like *Microsoft Teams* or *Slack* to create dedicated channels where teams can share updates, ask questions, and collaborate on projects.

6. Tailor Communication to Different Stakeholder Groups:

- Customise communication strategies for different stakeholder groups based on their level of technical expertise and specific interests. Avoid jargon when communicating with non-technical stakeholders.
- **Example:** While providing detailed technical updates to development teams, present high-level summaries to executives using non-technical language.

7. Establish Feedback Mechanisms:

- Set up structured feedback mechanisms to allow stakeholders to express their opinions and concerns. Use surveys, suggestion boxes, or regular feedback sessions to gather insights.
- **Example:** Implement a quarterly feedback survey for internal teams and conduct customer feedback sessions after new technological features or updates.

8. Incorporate Stakeholder Feedback:

- Actively incorporate valuable feedback into technological decision-making processes. Demonstrate a commitment to listening and adapting based on stakeholder input.
- **Example:** If customers express concerns about a user interface, prioritise updates and improvements in the next development sprint.

9. Align Technology Roadmap with Stakeholder Priorities:

- Align the technology roadmap with the priorities of key stakeholders. Ensure that milestones and deliverables resonate with their needs, fostering a sense of shared purpose.
- **Example:** If customer satisfaction is a priority, prioritise features or improvements that directly enhance the user experience.

10. Celebrate Successes and Acknowledge Challenges:

- Acknowledge and celebrate achievements and milestones, keeping stakeholders informed of successes. Similarly, be transparent about challenges, outlining plans for resolution and lessons learned.
- **Example:** Host a virtual celebration after the successful launch of a new product feature and communicate openly about any technical challenges faced during development.

This comprehensive 10-step guide provides a strategic approach to stakeholder engagement and alignment, ensuring that technological initiatives are well-understood, supported, and integrated into the broader goals of the organisation.

3.5 Elevating Innovation as a Core Objective

Innovation is a strategic objective that propels organisations toward sustained growth and competitive advantage. It is not a passive outcome but an intentional goal. This section explores the dimensions of elevating innovation as a core objective, delving into practical examples referencing relevant data sources that illuminate the path to technological advancement.

A more detailed dive into Innovation can be found in Book 3.

Making Innovation Tangible:

Innovation is more than a buzzword; it's a tangible goal that a CTO sets for the technology team. By framing innovation as an <u>explicit objective</u>, the CTO communicates the strategic importance of creativity, continuous improvement and more importantly - step-change technology leaps. For example, if the business goal is to enhance customer engagement, the CTO might set an innovation objective to explore and implement emerging technologies like augmented reality for interactive customer experiences.

Here might be a good time to say that I'm **NOT** the biggest fan of the term - continuous improvement. Don't get me wrong refining or removing steps within a customer journey can have a huge impact to the success of an engagement or purchasing decision, but at a higher level the principle of aiming for continuous improvement I think makes us lazy. We continue with a process in the way we have always done it and just make it a bit more efficient here and tweak it a bit there. I think it prevents us looking at the business or challenge completely afresh and using imagination to see how we might re-imagine that whole solution or outcome. Investigate the chapter on Imagination within Book 3.

Data Source Example: Utilise innovation indices such as the *Global Innovation Index (GII)* to benchmark the organisation against global innovators. These indices often provide insights into the innovation ecosystem, enabling the CTO to align objectives with industry benchmarks.

Fostering an Experimental Environment:

Advancing innovation requires cultivating an environment that encourages experimentation. A CTO plays a crucial role in promoting a culture where team members feel empowered to explore new ideas, technologies, and methodologies. This involves more than just endorsing innovation as a concept; it necessitates the establishment of practical frameworks for experimentation. For instance, implementing regular hackathons or innovation sprints can be instrumental in fostering an experimental mindset within the technology team. These events provide dedicated time and space for team members to collaborate, brainstorm, and prototype new solutions, encouraging a hands-on approach to innovation. By creating a structured yet flexible space for experimentation, the CTO ensures that innovative ideas can flourish and be integrated into the organisation's broader technological landscape.

Moreover, embracing experimentation is not just about tolerating failure; it's about recognising it as a valuable learning opportunity. A culture that encourages experimentation acknowledges that not every idea will lead to immediate success. Instead, failures are seen as stepping stones towards refinement and improvement. A CTO who actively promotes a positive perspective on failure ensures that the team is not discouraged by setbacks but rather motivated to iterate and innovate persistently. This mindset shift contributes significantly to the continuous improvement and evolution of the technology team's capabilities, fostering resilience and adaptability in the face of technological challenges.

In addition to internal experimentation, collaboration with external partners, startups, or industry experts can inject fresh perspectives and diverse approaches into the innovation process. Establishing open channels for dialogue and collaboration beyond the organisation's boundaries allows the technology team to tap into external expertise and stay attuned to industry trends. This external engagement contributes to a dynamic innovation ecosystem, ensuring that the technology team remains well-positioned to integrate the latest advancements and maintain a competitive edge in the ever-evolving technology landscape.

Data Source Example: Explore case studies of successful innovation programs in organisations similar to yours. Platforms like **Harvard Business Review** or **McKinsey** Insights often publish insights into innovative practices and their impact on organisational culture.

Embracing Creative Problem-Solving:

Innovation flourishes when creative problem-solving is championed. A smart CTO encourages the technology team to approach challenges with fresh perspectives and inventive solutions. If a business challenge involves optimising supply chain logistics, the CTO may encourage the team to re-think what processes or solutions might be explored. Yes, we may investigate emerging technologies like blockchain for transparent and efficient supply chain management, but before that let's explore what the ideal outcome would be before we investigate the technology to deliver it.

Data Source Example: Industry-specific reports from research firms such as **Gartner** or

Forrester can provide insights into creative solutions adopted by peer organisations to address common industry challenges.

Valuing Novel Ideas:

A culture of innovation is rooted in the appreciation of novel ideas. A CTO actively seeks and values input from team members at all levels. This can involve implementing idea-sharing platforms, where employees can contribute innovative concepts related to technology or process improvements.

Data Source Example: Explore platforms like *IdeaScale* or *Spigit/ Idea Place* that facilitate idea crowdsourcing and innovation management. These tools often provide analytics on the volume and quality of submitted ideas.

Encouraging Risk-Taking:

Innovation often involves stepping into the unknown, and a CTO should encourage calculated risk-taking. This can manifest in the exploration of emerging technologies, unconventional partnerships, or bold product development initiatives. For instance, if the organisation aims to lead in smart city solutions, the CTO may advocate for experimental projects in IoT and urban technology or investigate and then partner with inner city pressure groups.

Data Source Example: Understand how risk can be mitigated and creatively sold to senior management. Investigate GARP (Global Assoc of Risk Professionals). Their website has some interesting ideas and approaches. Also consider academic sites that may cover risk for specific verticals like IEEE Xplore for Environmental and PubMed for Biomedical.

Learn Fast: Viewing Failure as a Stepping Stone

A critical aspect of innovation is reframing failure as a stepping stone rather than an endpoint. A CTO fosters a culture where failures are viewed as valuable learning experiences. This can involve post-mortem analyses of projects, emphasising lessons learned, and encouraging resilience. When you are sitting with your CEO, perhaps think about the language that will best be received. Citing failure might not land well, whereas "Research and Learning" as step towards success will be more appropriate.

Data Source Example: Industry-specific blogs, podcasts, or webinars often feature discussions on embracing failure as a part of the innovation process. For example, platforms like *TED Talks* often host speakers who share their insights on overcoming setbacks in the pursuit of innovation. Search for "8 ways to fuel innovation".

Leveraging Design Thinking:

Design thinking is a powerful methodology for driving innovation. A visionary CTO integrates design thinking principles into the technology development process. This involves empathising with end-users, defining problem statements, ideating, prototyping, and testing solutions iteratively. Read - **Design Thinking & User-Centric Design in Chapter 8**

Data Source Example: Books and articles by design thinking experts like *Tim Brown* or *IDEO's* publications offer comprehensive insights into the principles and applications of design thinking in innovation.

Measuring Innovation Impact:

To elevate innovation as a core objective, we need to establish metrics to measure its impact. This can include indicators like the number of successful product launches, improvements in operational efficiency resulting from innovative processes, or increased employee engagement in innovation initiatives. More-so you need to show financial impact that is directly attributable to the innovation being presented.

Establishing clear key performance indicators (KPIs) tied to financial outcomes is essential. These KPIs could include metrics such as increased revenue, cost savings, or improved profitability directly associated with the innovation. For instance, if introducing a new product, track the uptick in sales and revenue generated over a specific period, highlighting the innovation's contribution to the financial growth.

Secondly, conducting a robust cost-benefit analysis can provide a comprehensive overview of the financial implications of the innovation. This involves quantifying both the upfront investment in the innovation and the subsequent financial returns. By meticulously accounting for expenses, including research and development costs, implementation expenditures, and ongoing operational expenses, alongside the measurable financial gains, stakeholders can clearly see the net financial impact of the innovation.

Furthermore, presenting a comparative analysis between the pre-innovation and post-innovation financial scenarios adds depth to the demonstration. By showcasing the financial landscape before and after the innovation's implementation, the tangible effects become more evident. This comparative approach can illuminate changes in revenue streams, cost structures, and overall financial performance, providing a compelling narrative that underscores the innovation's direct influence on the organisation's financial health. Effective storytelling and visual representations, such as graphs or charts, can enhance the clarity and impact of this comparative analysis, making it more accessible to a diverse audience of stakeholders.

Data Source Example: Utilise innovation management platforms that offer analytics and reporting features. Platforms like *Brightidea* or *Qmarkets* provide tools to measure the effectiveness of innovation initiatives.

Participating in Innovation Ecosystems:

Innovation thrives in collaboration. Actively engaging with external innovation ecosystems, participating in industry forums, collaborations with startups, and partnerships with research institutions can deliver benefits across the network. This broadens the scope of innovative ideas and fosters a culture of continuous learning.

Data Source Example: Participate in innovation conferences, such as the *Consumer Electronics Show (CES)* or industry-specific innovation summits. These events provide exposure to cutting-edge technologies and networking opportunities with innovative thought leaders.

Aligning Innovation with Market Trends:

We covered Understanding the Business Landscape in a previous chapter, but this equally applies to Innovation. Innovation moves from R&D to strategic when aligned with market trends. As a CTO we constantly need to monitor industry reports, technology forecasts, and market dynamics to ensure that innovation objectives are in sync with the evolving needs of the market. This does not have to be a time-consuming activity. Spend an hour looking at your social media feeds, I can guarantee it isn't something you have done for a long while as we all forget about it. You are probably getting very relevant information for that job you had 5-10 years ago. Select innovation preferences or join innovation groups. For example, on LinkedIn search and join groups where the field of innovation aligns with your market sector.

Data Source Example: Leverage market research reports from reputable firms to stay informed about emerging technologies and market trends. Platforms like *CB Insights* or *Frost & Sullivan* offer valuable insights into technological innovations shaping various industries.

Elevating innovation as a core objective requires intentional leadership and a commitment to creating an environment where creativity thrives. By making innovation tangible, fostering experimentation, embracing creative problem-solving, valuing novel ideas, encouraging risk-taking, viewing failure as a stepping stone, leveraging design thinking, measuring impact, participating in innovation ecosystems, and aligning with market trends, a CTO positions the organisation on the cutting edge of technological advancement, with the aim to drive sustained growth and success.

Measuring the Impact of Digital Innovation

Metrics	Examples
Net Promoter Score (NPS)	Increase in customer satisfaction and loyalty
Customer Acquisition Cost (CAC)	Reduction in cost to acquire new customers
Customer Lifetime Value (CLTV)	Increase in average revenue generated per customer
Customer Effort Score (CES)	Reduction in effort required for customers to complete tasks
Time to Resolution (TTR)	Faster resolution of customer issues
Process Cycle Time	Reduction in time it takes to complete key processes
Error Rates	Decrease in errors within processes
Employee Productivity	Improvement in employee output and efficiency
Resource Utilisation	Increased utilisation of resources like equipment or software
Cost Savings	Reduction in operational costs through automation or digital solutions
Revenue Growth	Increase in overall revenue generated
Profit Margin	Improvement in profitability through cost savings or increased revenue
Return on Investment (ROI)	Measurable financial return on investments in digital initiatives
Market Share	Growth in market share compared to competitors
Brand Value	Increased brand recognition and reputation
Number of New Ideas Generated	Increased rate of innovation and new ideas within the organisation
Employee Engagement	Improved employee engagement in innovation and problem-solving
Time to Market	Reduced time to launch new products or services
Number of Skills Acquired	Increased employee skillset through digital learning and training
Failure Rate of Experiments	Controlled experimentation to learn from successes and failures

Elevating Innovation as a Core Objective: A 10 Step Action Plan

Incorporating innovation as a core objective is vital for organisations striving to stay ahead in the dynamic technological landscape. This practical 10-step guide provides detailed insights into each step, offering a comprehensive approach for Chief Technology Officers (CTOs) to foster and elevate innovation within their organisations:

1. Establish a Clear Innovation Vision:
- Define a clear and inspiring innovation vision that aligns with the organisation's overall goals. Ensure that it encapsulates the spirit of creativity, experimentation, and continuous improvement.
- **Example:** "Pioneer transformative solutions through a culture of relentless innovation, driving industry advancements and ensuring long-term sustainability."

2. Cultivate a Culture of Curiosity:
- Foster a culture that encourages curiosity and inquisitiveness. Develop programs, such as innovation challenges or dedicated time for exploration, to ignite and channel the curiosity of your teams.
- **Example:** Implement a "Curiosity Hour" where employees can dedicate a set time each week to explore new technologies, ideas, or industry trends.

3. Establish Cross-Functional Innovation Teams:
- Form cross-functional teams that bring together diverse skills and perspectives. Encourage collaboration between departments to generate innovative solutions that span various facets of the organisation.
- **Example:** Create innovation teams with members from engineering, marketing, and customer support to address challenges holistically.

4. Invest in Continuous Learning Initiatives:
- Provide opportunities for ongoing learning and skill development. Invest in training programs, workshops, and certifications to keep your teams updated on the latest technologies and industry best practices.
- **Example:** Sponsor online courses, workshops, or industry conferences to ensure that your teams stay informed about cutting-edge developments. Investigate *The Marlow Business School course* schedule.

5. Encourage Experimentation and Risk-Taking:
- Create a safe space for experimentation and risk-taking. Acknowledge that not all endeavours will succeed but emphasise the value of learning from failures and iterating on ideas.
- **Example:** Establish an "Innovation Lab" where teams can experiment with new technologies or approaches without the fear of immediate consequences.

6. Implement a Structured Innovation Process:
- Develop a structured innovation process that guides teams from idea generation to implementation. Include stages for ideation, feasibility analysis, prototyping, and testing.
- **Example:** Adopt methodologies like Design Thinking or Agile Innovation to provide a framework for structured and iterative innovation.

7. Leverage Technology Scouting:
- Engage in technology scouting to identify external innovations, startups, or emerging technologies that could complement or disrupt your industry. Collaborate with external partners to bring fresh perspectives.
- **Example:** Partner with innovation hubs, startup accelerators, or industry conferences to scout for potential technologies or ideas.

8. Encourage Employee Intrapreneurship:
- Empower employees to act as intrapreneurs by allowing them to pursue innovative projects within the organisation. Create avenues for employees to pitch and lead initiatives that align with the innovation vision.
- **Example:** Establish an "Innovation Fund" that allocates resources for employees to develop and lead their innovative projects.

9. Foster Open Communication Channels:
- Establish open communication channels that facilitate the sharing of ideas, insights, and feedback. Create platforms where employees can contribute to the innovation discourse and collaborate seamlessly.
- **Example:** Implement digital collaboration tools such as Slack or Microsoft Teams to enable real-time communication and idea-sharing.

10. Measure and Recognise Innovation:
- Define key performance indicators (KPIs) to measure innovation effectiveness. Acknowledge and reward teams and individuals who contribute significantly to innovation, fostering a culture where innovation is celebrated.
- **Example:** KPIs could include metrics such as the number of implemented innovative ideas, time-to-market for new innovations, or the impact of innovations on business outcomes.

This comprehensive 10-step guide provides actionable strategies for CTOs to elevate innovation as a core objective within their organisations, fostering a culture of creativity, experimentation, and continuous improvement.

3.6 Metrics for Success in Technological Leadership

Success does not have to be a subjective notion, but a quantifiable achievement measured through well-defined metrics. As a Chief Technology Officer we already understand the importance of robust measurement frameworks to gauge the impact of technological operations, but we need to additionally apply that degree of rigor to our technological leadership. This section explores the significance of metrics for success, providing examples and suggesting relevant data sources for a comprehensive approach.

Before diving into metrics, the CTO begins by establishing clear technological objectives. These objectives should align with the overarching business goals and contribute directly to the organisation's success. For instance, if the business goal is to enhance customer experience, a technological objective might involve reducing application response time. Reducing by how much, and by when, by what cost and with what anticipated impact, including financial.

Data Source Example: Internal stakeholder discussions, market research, and customer feedback are initial data sources to inform the establishment of clear technological objectives.

Identifying Key Performance Indicators (KPIs):

Key Performance Indicators (KPIs) are the heartbeat of a metrics-driven approach. A CTO identifies KPIs that directly reflect the success of technological objectives. If the objective is to enhance system performance, KPIs may include response time, system uptime, and error rates.

Explore section 4.7 Monitoring KPI's in a Technology Roadmap

System Performance Metrics:

For objectives related to system performance, metrics such as response time, throughput, and error rates become critical. These metrics provide a tangible assessment of how well the technology infrastructure is meeting performance goals.

Firstly, it is essential to establish clear performance objectives. Define acceptable response time thresholds based on user expectations and the nature of the application. For instance, in a web-based system, users may expect pages to load within a specific timeframe. Concurrently, determine the desired throughput, indicating the system's capacity to handle a certain volume of transactions or requests per unit of time. Additionally, establish acceptable error rates, specifying the maximum tolerable percentage of errors or failures in system interactions.

Once performance objectives are set, employing effective monitoring tools is crucial for measurement. Implement performance monitoring tools that can track system response times, throughput rates, and error occurrences in real-time. Regularly collect and analyse performance data under various conditions to identify trends and potential bottlenecks. Conduct systematic testing, including load testing and stress testing, to simulate heavy usage scenarios and observe how the system responds. Utilise key performance indicators (KPIs) to measure and report on response times, throughput levels, and error rates. By continuously monitoring and analysing these metrics, organisations can proactively identify performance issues, optimise system configurations, and ensure that the IT system meets or exceeds the defined performance objectives. Regularly revisiting and adjusting these parameters in response to evolving user requirements and technological advancements is essential for maintaining optimal system performance over time.

Data Source Example: Monitoring tools like *New Relic, Datadog,* or *Azure Monitor* offer real-time insights into system performance metrics.

User Engagement Metrics:

If the objective is to increase user engagement, setting and measuring key metrics such as active users, session duration, and feature adoption rate is crucial for evaluating the success and impact of a digital product or service. Firstly, defining clear benchmarks and goals for each metric is essential. Identify what constitutes an active user, set realistic targets for session durations, and establish expectations for feature adoption rates. These benchmarks should align with the overarching objectives of the product or service, ensuring that the metrics measured directly contribute to the desired outcomes. For example, if the goal is to increase user engagement, active users and session duration targets should reflect that objective.

To measure these metrics effectively, leverage analytics tools that provide insights into user behaviour and interactions with the product. Platforms like Google Analytics, Mixpanel, or Hotjar offer robust features to track active users, session durations, and feature adoption. Regularly monitor these metrics and conduct A/B testing or user surveys to gain deeper insights into user preferences and behaviours. Additionally, setting up custom dashboards or reports within these analytics tools can facilitate quick and accessible assessments, allowing teams to track progress against established benchmarks and make data-driven decisions to enhance user engagement and feature adoption. Regularly reviewing and adjusting these benchmarks based on evolving business goals and user expectations ensures that the metrics remain relevant and aligned with the product's ongoing success.

Data Source Example: Analytics platforms like *Google Analytics, Mixpanel,* or *Amplitude* provide detailed user engagement metrics and behaviour analysis.

Cybersecurity Measures:

Enhancing cybersecurity measures requires specific metrics to assess the effectiveness of security protocols. Metrics such as incident response time, detection rates, and vulnerability resolution time help gauge the organisation's cybersecurity posture.

Setting and measuring cybersecurity incident response metrics is vital for evaluating an organisation's ability to effectively manage and mitigate potential threats. Firstly, establishing clear benchmarks for incident response time is critical. This metric assesses how swiftly an organisation can detect and respond to a cybersecurity incident. Setting a target response time, such as within minutes or hours, helps define the urgency and efficiency required to address security incidents promptly. Regularly reviewing and adjusting these benchmarks based on the evolving threat landscape and the organisation's capabilities ensures that response times remain effective and aligned with current cybersecurity needs.

Detection rates, another essential metric, measure the organisation's ability to identify and recognise security incidents accurately. To enhance detection rates, organisations can implement advanced threat detection tools, conduct regular security awareness training for staff, and employ proactive monitoring techniques. Regularly assessing and benchmarking detection rates against industry standards or historical data provides insights into the effectiveness of the organisation's detection capabilities. Implementing key performance indicators (KPIs) for detection rates ensures ongoing improvement and a proactive approach to identifying and addressing potential threats before they escalate.

Vulnerability resolution time is a crucial metric that evaluates how quickly an organisation can address and remediate identified vulnerabilities. Establishing a target resolution time, typically measured in days or weeks, helps prioritise and streamline the vulnerability management process. Regularly measuring and reporting on the time taken to resolve vulnerabilities aids in identifying bottlenecks, refining processes, and improving the overall cybersecurity posture. Integrating automation tools and workflows can expedite vulnerability resolution, reducing the window of opportunity for potential exploitation. Continuous monitoring and reporting on these metrics contribute to a more robust and responsive cybersecurity incident management framework, aligning the organisation with best practices and industry standards.

Data Source Example: Cybersecurity platforms like *CrowdStrike, Splunk,* or *Qualys* offer metrics and reports on security incidents and vulnerabilities.

Cost-Efficiency Metrics:

For objectives related to cost efficiency, metrics like total cost of ownership, return on investment (ROI), and resource utilisation rates are crucial. These metrics ensure that technological initiatives align with financial goals.

Setting and measuring total cost of ownership (TCO), return on investment (ROI), and resource utilisation rates are integral components of effective financial management for any innovation or business initiative. To establish a comprehensive TCO, start by identifying all relevant costs associated with the innovation, including initial investments, ongoing operational expenses, and any potential hidden costs. This encompasses not only direct monetary expenditures but also factors such as training, maintenance, and potential downtime. Once the TCO components are identified, implementing a robust tracking and reporting system is crucial to continuously monitor and measure these costs over time, ensuring a comprehensive understanding of the total financial commitment.

Measuring return on investment involves quantifying the financial gains generated by the innovation. ROI is calculated by comparing the net gain from the investment (revenue, cost savings, or other financial benefits) against the initial investment cost. Regularly evaluating the ROI at predefined intervals provides insights into the effectiveness of the innovation over time. This iterative assessment allows stakeholders to make informed decisions about the continued viability of the innovation and whether adjustments or optimisations are necessary to maximise its financial impact. Resource utilisation rates gauge how efficiently resources, such as manpower, technology, or equipment, are deployed in the context of the innovation. Establishing baseline metrics for resource utilisation and regularly monitoring these rates helps identify potential bottlenecks, inefficiencies, or underutilised assets. This process enables proactive adjustments to optimise resource allocation, ensuring that the innovation operates at peak efficiency. Utilising project management tools, time tracking software, and periodic assessments allows organisations to maintain a real-time understanding of resource utilisation rates and make informed decisions to enhance overall efficiency and financial performance.

Data Source Example: Financial reports, accounting software, and project management tools contribute to the collection of cost-efficiency metrics.

Customer Satisfaction Metrics:

If the technological objective is to enhance customer satisfaction, metrics like *Net Promoter Score (NPS)*, customer feedback scores, and support ticket resolution time are indicative of success.

To establish an effective NPS, begin by crafting a straightforward survey that asks customers how likely they are to recommend your product or service to others. Responses are usually measured on a scale from 0 to 10, categorising customers as promoters (9-10), passives (7-8), or detractors (0-6). Calculate the NPS by subtracting the percentage of detractors from the percentage of promoters. Regularly surveying a representative sample of customers allows for ongoing measurement and trend analysis, helping to gauge changes in customer sentiment over time and identify areas for improvement.

Customer feedback scores provide valuable insights into specific aspects of your products or services. Develop targeted surveys that capture customer opinions on various touchpoints, such as purchasing processes, product quality, or customer service interactions. Assign numerical values to responses to quantify feedback, and then aggregate the scores to derive an overall customer feedback score. This approach allows businesses to pinpoint areas requiring attention and prioritise enhancements based on customer priorities. Regularly monitoring and analysing customer feedback scores helps in maintaining a customer-centric focus and continuously refining offerings to meet evolving customer expectations.

Support ticket resolution is a critical metric for assessing the efficiency of customer support services. To set and measure this metric, define clear benchmarks for acceptable resolution times based on the complexity of issues. Track the time taken from ticket submission to resolution and categorise resolutions based on their effectiveness. Regularly analyse support ticket resolution metrics to identify patterns, address recurring issues, and streamline support processes. Utilising customer feedback within this context can also provide additional context on the quality of the resolution process. Continuous improvement in support ticket resolution contributes to enhanced customer satisfaction and loyalty.

Data Source Example: Customer relationship management (CRM) systems, customer survey tools, and social media monitoring platforms provide valuable customer satisfaction metrics.

Innovation Adoption Metrics:

For objectives related to innovation adoption, metrics like adoption rates, feature usage patterns, and time-to-adoption become instrumental. These metrics track how quickly users embrace new technologies or features.

Setting and measuring adoption rates, feature usage patterns, and time-to-adoption are critical aspects of evaluating the success and impact of an innovation. Firstly, establishing clear benchmarks and goals for adoption rates is essential. Define what constitutes successful adoption within a specified timeframe and identify key milestones. Utilise analytics tools to track user engagement and monitor the number of users actively using the innovation. Adoption rates can be measured by assessing the percentage of the target audience or user base that has embraced the innovation. Regularly analyse these rates over time, identifying trends and patterns that can inform adjustments to strategies or features to enhance adoption.

Secondly, understanding feature usage patterns provides valuable insights into user behaviour and preferences. Implement analytics tools that allow for the tracking of specific feature usage, enabling a granular understanding of which components of the innovation are most impactful. Monitor which features are gaining traction, how frequently they are accessed, and the duration of user engagement. By segmenting feature usage data, it becomes possible to identify popular functionalities, potential areas for improvement, and even unexplored features that may require additional user education or promotion. This detailed feature-level analysis helps refine the innovation, ensuring that it aligns with user needs and preferences.

In addition to adoption rates and feature usage patterns, assessing time-to-adoption provides a temporal dimension to the evaluation process. Measure how quickly users embrace the innovation from its initial introduction. A shorter time-to-adoption indicates a more seamless integration into the user base. Analysing this metric allows for the identification of potential bottlenecks or barriers to adoption that may need addressing. By continually refining strategies based on adoption rates, feature usage patterns, and time-to-adoption, organisations can ensure that their innovations resonate with users and deliver sustained value over time.

Scalability Metrics:

When scalability is a key objective, metrics like system load handling, resource scalability, and response time under load are crucial. These metrics ensure that the technology can grow seamlessly with the organisation.

Firstly, setting clear benchmarks for system load handling involves determining the maximum volume of concurrent users, transactions, or data inputs the system should be capable of managing without compromising performance. By defining these parameters, organisations can create realistic scenarios that simulate real-world usage conditions. Load testing tools, such as Apache JMeter or LoadRunner, can then be employed to systematically increase the load on the system, allowing for the identification of bottlenecks, potential failures, or degraded response times.

In tandem with load handling, resource scalability should be a primary focus. This involves assessing how well the system can adapt to increased demand by adding resources such as servers, storage, or network capacity. Cloud-based solutions like AWS Auto Scaling or Kubernetes can aid in dynamically adjusting resources based on system load. The effectiveness of resource scalability can be measured by observing how the system maintains performance levels and response times as demand fluctuates. Evaluating the system's ability to seamlessly scale resources up or down in response to varying loads ensures optimal performance during both peak and normal operating conditions.

Measuring response time under load is crucial for evaluating the system's responsiveness and user experience. It involves assessing the time it takes for the system to process a request and provide a corresponding output under varying levels of load. Monitoring tools, such as New Relic or AppDynamics, can capture response times and help identify any degradation or delays in system performance. Setting acceptable response time thresholds and continuously measuring against them during load testing enables organisations to proactively address performance issues, optimise system architecture, and enhance overall user satisfaction.

Data Source Example: Load testing tools and performance monitoring platforms provide data on scalability metrics.

Continuous Improvement Metrics:

I know in previous chapters I have been critical of Continuous Improvement as a concept due to it potentially distracting us from re-imagining the solution, only tweaking the status quo, however we can't dismiss it in the right context, especially when related to iterative development cycles as part of Customer Journey Mapping, feedback loop effectiveness, and time-to-market for new features.

Data Source Example: Agile project management tools, version control systems, and continuous integration/delivery (CI/CD) platforms contribute to continuous improvement metrics.

Leveraging Data-Driven Decision-Making:

The data collected from these metrics should not exist in isolation but instead be seen as interconnected components within a comprehensive analytical framework. Adopting a holistic perspective on data involves understanding the intricate relationships and dependencies between different metrics. For example, when assessing Net Promoter Score (NPS) alongside customer feedback scores, a nuanced understanding of how these metrics interact provides a more complete picture of customer satisfaction. By recognising patterns and correlations, organisations can uncover valuable insights that go beyond the surface-level interpretation of individual metrics.

A truly data-driven approach necessitates not only the collection and analysis of metrics but also the derivation of actionable insights. This involves delving deeper into the data to understand the root causes behind specific trends or anomalies. For instance, if there's a sudden dip in NPS, a thorough analysis might reveal that it correlates with a recent change in customer support processes. Identifying such correlations allows organisations to take targeted actions, whether it involves refining internal processes, addressing specific pain points in the customer journey, or enhancing product features. The goal is to transform raw data into strategic intelligence that guides informed decision-making.

Moreover, to harness the full potential of a data-driven approach, organisations should implement a robust feedback loop. This loop involves not only analysing metrics but also incorporating the insights gained into operational adjustments and improvements. For instance, if customer feedback highlights issues with product usability, a responsive feedback loop ensures that these insights lead to tangible changes in design or user experience. By closing the loop between data analysis and proactive decision-making, organisations create a continuous cycle of improvement, ensuring that their strategies remain adaptive and responsive to evolving customer needs and market dynamics.

Data Source Example: Business intelligence platforms like *Tableau* or *Power BI* facilitate the visualisation and analysis of data collected from various metrics.

Adjusting Objectives Based on Metrics:

Metrics are not static; they evolve with the organisation's goals and external factors. We need to adjust technological objectives based on the insights derived from ongoing metric analysis.

While some metrics remain foundational, their interpretation and significance can change as the business landscape transforms. For instance, the key performance indicators (KPIs) that were once solely focused on operational efficiency might pivot towards customer-centric metrics as the organisation places greater emphasis on enhancing the overall customer experience. This adaptability of metrics ensures they align with the strategic direction of the business, reflecting its current priorities and objectives.

The fluid nature of metrics requires us to lever ongoing analysis to inform technological objectives. Through scrutiny of metric trends, the CTO gains valuable insights into the performance and impact of existing technology initiatives. This data-driven approach enables us to make informed decisions on refining or redirecting technological strategies, ensuring that they remain in harmony with the evolving goals of the organisation. For instance, if the analysis indicates that a particular technology investment is not yielding the expected results, the CTO should reallocate resources or introduce new solutions based on emerging trends and industry best practices.

Moreover, the iterative nature of metric analysis allows the CTO to consider how best to use continuous improvement within the technology department or deep dive into the desired outcome. Regular evaluations of metrics create a feedback loop, prompting the identification of areas for enhancement and optimisation. This proactive stance ensures that technological initiatives not only align with the current organisational objectives but also position the company to navigate future challenges and capitalise on emerging opportunities. Metric analysis is therefore not merely a performance evaluation tool but a compass guiding our technology strategies towards success.

Data Source Example: Threat intelligence feeds, security incident reports, and collaboration with cybersecurity experts inform adjustments to cybersecurity objectives.

Success Metrics in Technological Leadership: A 10 Step Action Plan

Crafting meaningful metrics for success is vital for assessing and enhancing technological leadership. This practical 10-step guide provides detailed insights into each step, offering a comprehensive approach to creating metrics that align with organisational objectives:

1. Define Organisational Objectives:

- Begin by understanding and defining the broader organisational objectives. Collaborate with executive leadership to identify key strategic goals that technology initiatives should support.
- **Example:** If the organisational objective is to enhance customer satisfaction, the technology metric could focus on reducing application response time to improve user experience.

2. Align Technological Goals:

- Align technological goals with organisational objectives. Identify specific areas where technology can contribute to achieving these objectives and create corresponding performance indicators.
- **Example:** If scalability is crucial for business growth, a metric could involve tracking the system's ability to handle increased user loads without performance degradation.

3. Develop SMART Metrics: See Chapter 3.3

- Formulate Specific, Measurable, Achievable, Relevant, and Time-bound (SMART) metrics. Ensure that each metric provides clear and quantifiable insights into the progress and impact of technological initiatives.
- **Example:** A SMART metric for security could be reducing the number of security incidents by 20% within the next six months.

4. Identify Key Performance Indicators (KPIs):

- Identify KPIs that directly reflect the success of technological leadership. These could include metrics related to system performance, user satisfaction, cybersecurity, and innovation.
- **Example:** KPIs might encompass metrics like system uptime, response times, customer satisfaction scores, and the number of successful technology innovations.

5. Quantify Innovation Impact:

- Develop metrics that quantify the impact of technological innovations on business processes and outcomes. This could involve measuring the adoption rates of new technologies or the efficiency gains achieved.
- **Example:** Track the percentage increase in operational efficiency resulting from the implementation of innovative technologies.

6. Implement Continuous Monitoring:

- Establish a system for continuous monitoring of metrics. Leverage monitoring tools and dashboards to keep real-time track of key indicators, facilitating prompt response to deviations.
- **Example:** Utilise monitoring tools like Prometheus or Grafana to create dashboards that display real-time performance metrics.

7. Foster Data-Driven Decision Making:

- Foster a culture of data-driven decision-making within the technology team. Ensure that insights from metrics inform strategic decisions and guide ongoing improvements.
- **Example:** Regularly conduct data review sessions where team members analyse metrics to identify areas for optimisation and innovation.

8. Conduct Benchmarking Analysis:

- Compare technological metrics against industry benchmarks and competitors. This provides context and helps set ambitious yet realistic goals for technological leadership.
- **Example:** Benchmark system uptime against industry standards to determine if the organisation is meeting or exceeding performance expectations.

9. Involve Stakeholders in Metric Definition:

- Involve key stakeholders, including cross-functional teams and end-users, in the definition of metrics. Their input ensures that metrics align with the overall needs and expectations of the organisation.
- **Example:** Hold regular workshops or surveys to gather feedback on the relevance and effectiveness of chosen metrics.

10. Periodic Review and Adaptation:

- Establish a periodic review process for metrics. Adapt metrics based on changing organisational goals, technology landscape, and emerging opportunities or challenges.
- **Example:** Conduct quarterly reviews to ensure that metrics remain aligned with organisational objectives and adjust them as needed.

This 10-step guide empowers CTOs to create metrics for success in technological leadership that are not only aligned with organisational goals but also flexible, actionable, and reflective of the dynamic nature of technology and business environments.

3.7 Cultivating a Culture of Ownership for Technological Excellence

The transformative power of a culture of ownership cannot be overstated. Ownership by everyone within the organisation but especially the C-Suite is crucial to turn technological visions into tangible realities.

This section delves into the intricacies of cultivating a culture of ownership within a technology team, supported by practical examples and insights drawn from relevant data sources.

Empowering Team Members:

Cultivating ownership begins with empowering individual team members to take initiative and responsibility for their work. A skilled CTO encourages autonomy and provides the necessary tools and resources for team members to excel in their roles. For example, providing access to advanced training programs or certifications empowers team members to enhance their skills and take ownership of their professional development.

Data Source Example: Employee feedback platforms, such as *Glassdoor* or internal surveys, offer insights into the perceived level of empowerment within the team.

Providing a Clear Sense of Purpose:

Ownership flourishes in an environment where team members have a clear understanding of how their contributions contribute to the larger mission and goals of the organisation. The CTO communicates the broader purpose behind technological initiatives, fostering a sense of meaning in the day-to-day tasks. For instance, if the organisation aims to revolutionise accessibility through technology, team members can take pride in their role in creating inclusive solutions.

Data Source Example: Utilise employee engagement surveys to gauge the team's alignment with organisational goals and their perception of the purpose behind their work.

Recognising Contributions:

Acknowledging and celebrating individual and team achievements is a cornerstone of a culture of ownership. A skilled CTO actively recognises and rewards contributions, creating a positive feedback loop that reinforces a sense of ownership. Recognition can take various forms, from public acknowledgments in team meetings to more formalised awards or incentives.

I was always taught praise in public, criticise in private. I don't believe anyone comes to work with the aim to do a poor job, they may just be in the wrong job, or they need better direction to that which they have been given. Demotivation in an individual or losing valuable staff is more often down to the style of leadership that sits above them.

Data Source Example: Performance management tools and platforms often provide data on individual and team accomplishments, contributing to a comprehensive view of contributions within the organisation.

Here are some examples of widely used and effective tools:

1. *BambooHR:*

    1. BambooHR offers a comprehensive HR software solution, including performance management features. It allows organisations to set goals, track employee progress, and conduct performance reviews. The platform also facilitates continuous feedback and supports customisable performance metrics.

2. *15Five:*

    1. 15Five focuses on employee feedback and engagement. It enables managers to gather weekly insights from team members in just 15 minutes. The tool emphasises continuous communication, goal setting, and tracking employee performance against objectives.

3. *Glint:*

    1. Glint, now a part of LinkedIn, is an employee engagement and performance management platform. It provides real-time analytics, employee surveys, and insights to help organisations understand and improve employee engagement and performance.

4. *Lattice:*

    1. Lattice is a performance management platform that supports goal setting, continuous feedback, and performance reviews. It allows

organisations to align individual goals with company objectives, fostering transparency and accountability.

5. *Workday Performance Management:*

    1. Workday's Performance Management module is part of its broader suite of HR and finance solutions. It offers features for goal setting, performance reviews, and feedback, providing a unified platform for workforce management.

6. *SuccessFactors by SAP:*

    1. SuccessFactors is an integrated human capital management (HCM) solution that includes performance management features. It enables organisations to set goals, conduct reviews, and provide continuous feedback, contributing to overall talent development.

7. *Kudos:*

    1. Kudos is a platform that focuses on employee recognition and engagement. While not a traditional performance management tool, it plays a crucial role in acknowledging and rewarding employee achievements, contributing to a positive work culture.

8. *Reflektive:*

    1. Reflektive is a performance management and employee engagement platform that emphasises real-time feedback, goal tracking, and analytics. It aims to foster a culture of continuous improvement and development.

9. *Trakstar:*

    1. Trakstar is a performance management tool that supports goal setting, performance reviews, and 360-degree feedback. It is designed to streamline the performance management process and enhance communication between managers and employees.

10. *Reviewsnap:*

    1. Reviewsnap is a cloud-based performance management software that facilitates goal setting, employee appraisals, and development planning. It is known for its user-friendly interface and customisable performance review templates.

Ownership and the C-Suite

Establishing a culture of ownership within the C-Suite requires deliberate efforts to align leaders with the organisation's technological objectives. First and foremost, fostering open communication channels is crucial. We need to engage in transparent discussions about the organisation's technological vision, its challenges, and the role each member of the C-Suite plays in achieving these goals. This transparency cultivates a shared understanding, making it easier for C-Suite members to take ownership of specific aspects of the technological roadmap.

Leading by example is another effective technique in instilling a culture of ownership among the C-Suite. When top executives visibly demonstrate their commitment to technological initiatives, it sets a powerful precedent for the rest of the organisation. This can involve C-Suite members actively participating in technology-related projects, openly sharing their insights and experiences, and taking responsibility for both successes and setbacks. By showcasing a strong sense of ownership, leaders encourage their peers to embrace a similar mindset, creating a domino effect throughout the C-Suite.

Implementing clear accountability structures and performance metrics tied to technological objectives is instrumental in embedding a culture of ownership. When C-Suite members have well-defined roles, responsibilities, and metrics that align with the organisation's technological vision, they are more likely to take ownership of their contributions. Regularly reviewing and acknowledging achievements, even small wins, reinforces the culture of ownership by recognising and celebrating the efforts of each C-Suite member. Additionally, providing opportunities for ongoing learning and development ensures that leaders stay informed about emerging technologies, fostering a proactive and informed approach to ownership within the C-Suite.

Even though a "Machiavellian" solution may feel very tempting when someone appears to de-rail or undermine a project, (I have read the book The CIO in Wolves Clothing!) the best approach is clearly defining and documenting the authority and accountability of each C-Suite member involved in the project and attempting a framework for collaboration. This structure ensures that decision-making is transparent, well-informed, and aligned with the project's success. There is no reason why a workshop cannot be run mid-way during a project or development to re-set responsibilities and ownership.

Encouraging Innovation and Initiative:

Ownership thrives in an environment that encourages innovation and the pursuit of new ideas. The CTO fosters a culture where team members feel empowered to propose innovative solutions and take the initiative to address challenges. For example, establishing regular brainstorming sessions or innovation forums provides a platform for team members to contribute their ideas. See detailed section in Book 3

Data Source Example: Innovation indices and reports, such as those from the *Global Innovation Index (GII),* can offer benchmarks and insights into fostering innovation within the organisation.

Aligning Personal Development with Organisational Goals:

Ownership extends beyond daily tasks to encompass personal and professional growth aligned with organisational goals. As CTO's, line managers and mentors we need to be proactive in supporting team members setting and achieving career development goals that align with both individual aspirations and the overall success of the technology team and the organisation. Engaging in regular one-on-one discussions with team members to comprehend their career aspirations, strengths, and areas for growth enables us to gain insights into each team member's professional goals, enabling them to tailor career development plans that align with both personal aspirations and the strategic needs of the organisation.

Furthermore, the CTO plays a pivotal role in providing mentorship and guidance to team members as they navigate their career paths. This includes offering advice on skill development, recommending relevant training or educational opportunities, and creating a supportive environment for team members to undertake challenging projects that contribute to their professional growth. By actively investing in the career development of team members, the CTO not only enhances individual skill sets but also contributes to the overall strength and capabilities of the technology team.

Experience, mentoring, and education are deemed crucial components in this journey, as they collectively build a foundation for continuous learning and advancement. It is recommended that team members explore professional development programmes, like those offered by Cambridge or the Marlow Business School, to further augment their skills and contribute significantly to the organisation's technological prowess. It is also worth reminding team members that it is they who own and manage their career development (not their manager or the organisation) and that they should plan (or request) the steps where they can gain the experience they need, ask for mentoring, and put forward appropriate training with a rationale of the benefit that will result.

- **Data Source Example:** Talent management platforms often provide data on employee skill development, training completion, and career progression, offering insights into the alignment of personal development with organisational goals.

- *Example:* **Bonusly, Achievers**: These platforms allow employees and managers to acknowledge and reward outstanding performance. They often include features like peer recognition, gamification, and rewards programs to boost morale.

Building a Collaborative Environment:

Ownership is amplified in a collaborative environment where team members feel interconnected and are collectively invested in the success of technological goals. A CTO fosters collaboration through open communication channels, cross-functional initiatives, and shared accountability. For instance, establishing collaborative project spaces using tools like *Microsoft Teams* or *Slack* encourages real-time communication and mutual support.

Balancing Autonomy and Accountability:

Ownership is most effective when team members strike a balance between autonomy and accountability. It is good to set clear expectations, as that provides the autonomy to execute tasks, and holds individuals accountable for their outcomes. This balance ensures that team members feel a sense of ownership without sacrificing accountability.

Nurturing a Learning Culture:

Implementing a balanced approach to ownership within a team involves several key strategies. Firstly, it's essential to establish clear expectations for each team member, outlining their responsibilities, goals, and the broader objectives of the project or initiative. This clarity provides team members with the autonomy to execute tasks in alignment with their skill sets and areas of expertise. Clear expectations also act as a guiding framework, allowing individuals to understand their role within the team and the impact of their contributions on the overall project success.

Simultaneously, accountability measures must be in place to ensure that team members are held responsible for their outcomes. Regular check-ins, progress reviews, and performance assessments contribute to a culture of accountability, ensuring that each team member is aware of their role in achieving collective goals. The balance between autonomy and accountability is further reinforced by fostering an open communication environment where team members feel comfortable seeking guidance when needed. This approach ensures that the sense of ownership is cultivated without compromising individual responsibility, creating a team dynamic where autonomy and accountability coexist harmoniously.

Demonstrating Leadership by Example:

Establishing a culture of ownership is fundamentally driven by the actions and decisions of leaders, starting with the Chief Technology Officer (CTO). The CTO assumes a leadership role by exemplifying ownership in their actions and decisions, setting a standard for the entire team to follow. This involves the CTO actively participating in projects, showcasing personal commitment to the team's objectives, and openly communicating expectations while maintaining high standards. Demonstrating accountability in both successes and setbacks, the CTO establishes a foundation for a culture of ownership within the team.

To reinforce this culture, it is worth introducing a mentorship program within the technology team. Pairing experienced team members with those who are less seasoned fosters knowledge transfer and collaborative learning. Through mentorship we can encourage seasoned professionals to share insights on ownership, instilling a sense of responsibility in their mentees. Regular check-ins, discussions on project ownership, and collaborative problem-solving sessions become mechanisms for embedding a culture of ownership throughout the team.

Additionally, celebrating individual and team achievements is a powerful strategy for the CTO. Recognition of instances where team members exhibit exemplary ownership behaviours can be done through formal recognition programs, team-wide communications, or public acknowledgment during team meetings. By publicly acknowledging and celebrating ownership, the CTO reinforces the value placed on this quality within the team, inspiring others to embrace similar behaviours. Creating a positive feedback loop around ownership not only motivates individuals but also fortifies the collective culture of ownership within the technology team.

Data Source Example: Leadership effectiveness surveys and 360-degree feedback mechanisms offer insights into leadership behaviours and their impact on the perception of ownership within the team.

Cultivating a culture of ownership is a nuanced and deliberate process that requires ongoing commitment and thoughtful leadership. By empowering team members, providing a clear sense of purpose, recognising contributions, encouraging innovation, aligning personal development, fostering collaboration, balancing autonomy and accountability, nurturing a learning culture, and demonstrating leadership by example, a smart CTO establishes a foundation for technological excellence driven by a culture of ownership.

Creating a Culture of Ownership: A 10 Step Action Plan

Creating a culture of ownership within your technological teams is pivotal for fostering innovation, accountability, and continuous improvement. This practical 10-step guide provides detailed insights into each step, offering a comprehensive approach to cultivating a culture of ownership for technological excellence:

1. Establish Clear Expectations:

- Clearly communicate expectations regarding roles, responsibilities, and the significance of each team member's contribution. Emphasise the importance of taking ownership of tasks and projects.
- **Example:** Clearly outline in job descriptions and project charters the expected level of ownership for each team member, fostering a shared understanding from the outset.

2. Promote Transparent Communication:

- Foster an environment of open communication where team members feel comfortable expressing ideas, concerns, and feedback. Implement tools like Slack or Microsoft Teams for real-time collaboration.
- **Example:** Create dedicated channels for project discussions and encourage team members to share progress updates, challenges, and ideas openly.

3. Empower Decision-Making:

- Empower team members to make decisions within their areas of expertise. Encourage autonomy in problem-solving and decision-making processes.
- **Example:** Implement a decentralised decision-making structure, allowing team members to take ownership of decisions related to their specific domains.

4. Recognise and Celebrate Achievements:

- Acknowledge and celebrate individual and team achievements. Regularly recognise contributions and milestones, fostering a sense of pride and ownership.
- **Example:** Establish a recognition program that highlights exceptional performance, whether it's delivering a successful project, solving a challenging problem, or going above and beyond in their responsibilities.

5. Foster Continuous Learning:

- Promote a culture of continuous learning and improvement. Provide opportunities for skill development, training, and access to resources that empower team members to excel in their roles.
- **Example:** Implement regular training sessions, workshops, and subscriptions to online learning platforms to enhance the skills of team members.

6. Encourage Collaboration and Knowledge Sharing:

- Create a collaborative environment where team members share knowledge and insights freely. Establish platforms or forums for cross-team collaboration and idea exchange.
- **Example:** Implement regular knowledge-sharing sessions, where team members present their learnings, experiences, and best practices to the broader team.

7. Align Individual Goals with Organisational Objectives:

- Ensure that individual goals align with broader organisational objectives. Facilitate discussions to connect the dots between individual contributions and the overall success of the organisation.
- **Example:** Conduct regular goal-setting sessions that link individual performance objectives with the strategic goals of the department or organisation.

8. Provide Resources and Support:

- Ensure that teams have the necessary resources and support to excel in their roles. Remove obstacles and provide the tools and technologies needed for success.
- **Example:** Invest in state-of-the-art technologies, provide adequate training, and ensure teams have access to the resources required for optimal performance.

9. Emphasise Accountability for Results:

- Foster a sense of accountability for results. Encourage teams to take ownership not only of their tasks but also the overall outcomes of projects and initiatives.
- **Example:** Implement a results-driven performance management system that rewards accountability and achievement of key objectives.

10. Lead by Example:

- Demonstrate a strong sense of ownership and accountability as a leader. Model the behaviour and values you wish to instil in your team.
- **Example:** Actively participate in projects, take responsibility for decisions, and showcase a commitment to excellence, setting the standard for the entire team.

This 10-step guide offers actionable strategies for CTOs to cultivate a culture of ownership, emphasising the importance of clear communication, empowerment, recognition, continuous learning, and leadership by example.

Chapter 4.

Creating a Technology Roadmap

Charting the Course: Creating a Technology Roadmap

A technology roadmap is the strategic blueprint that transforms a Chief Technology Officer's (CTO) vision into tangible actions. In this chapter, we explore the art of crafting a comprehensive technology roadmap that not only aligns with business objectives but also guides the technology team toward successful implementation. From defining milestones to fostering agility, the creation of a technology roadmap is a critical phase in the journey of technological leadership.

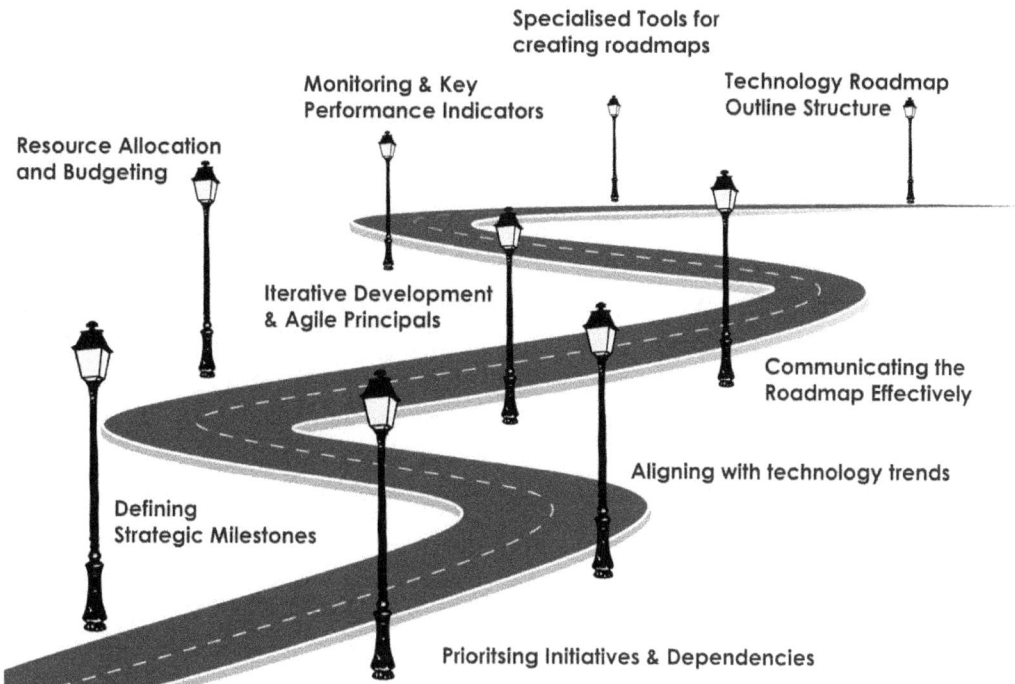

Specialised Tools for
creating roadmaps

Monitoring & Key
Performance Indicators

Technology Roadmap
Outline Structure

Resource Allocation
and Budgeting

Iterative Development
& Agile Principals

Communicating the
Roadmap Effectively

Aligning with technology trends

Defining
Strategic Milestones

Prioritsing Initiatives & Dependencies

Here is a summary of the items in this chapter:

4.1 Defining Strategic Milestones

The creation of a technology roadmap begins with the identification of strategic milestones and breaking down overarching technological objectives into actionable steps, creating a roadmap that serves as a timeline for achieving each milestone. These milestones should be aligned with the organisation's broader strategic goals, providing a clear path toward the realisation of the CTO's vision.

4.2 Prioritising Initiatives and Dependencies

Not all initiatives are created equal, and dependencies can shape the trajectory of progress.

Identifying dependencies between different projects is a crucial for ensuring seamless coordination and successful execution across interconnected initiatives. One effective technique is the use of Dependency Mapping. This involves creating visual representations or charts that illustrate the relationships and dependencies among various projects. By mapping out these dependencies, project managers gain a comprehensive understanding of how different tasks and deliverables rely on one another. This visual aid helps in anticipating potential bottlenecks, resource conflicts, or schedule misalignments, enabling proactive management to mitigate risks and enhance overall project efficiency.

Another valuable technique is the Dependency Structure Matrix (DSM). DSM is a method for visually representing and analysing the dependencies between different elements within a system or, in this context, projects. The matrix allows us to identify and prioritise dependencies, offering insights into the critical relationships that must be managed closely. DSM simplifies the visualisation of complex interdependencies, making it easier to identify potential points of contention or areas where adjustments may be needed. By employing these techniques, project managers can proactively manage dependencies, streamline workflows, and enhance the overall success of multiple projects operating within the same organisational landscape.
https://dsmweb.org/

4.3 Aligning with Technological Trends

The technology landscape is dynamic, with trends shaping and reshaping the terrain. Keeping abreast of technological trends must be built into the roadmap. This involves evaluating emerging technologies, understanding their potential impact, and ensuring that the roadmap remains adaptable to the evolving landscape. By aligning with technological trends, the roadmap becomes a flexible guide that navigates the organisation toward the cutting edge. *See section 2.5 Embracing Emerging Technologies*

4.4 Resource Allocation and Budgeting

Execution of a technology roadmap is contingent upon effective resource allocation and budgeting.

Aligning a budget with the technology roadmap is integral to effective technology leadership. As a Chief Technology Officer, meticulous cost estimations for each project phase, incorporating personnel, materials, software, and external services, are key. Drawing insights from historical data ensures precise projections, with a contingency fund accounting for unforeseen contingencies. Effective communication and collaboration with technology and finance teams are crucial for coherence in budgetary understanding. Regular reviews and updates maintain adaptability to evolving requirements, technological challenges, or emerging opportunities. This dynamic budgeting strategy, marked by transparency and flexibility, allows the budget to dynamically respond to the changing technological landscape while staying aligned with the overarching technology roadmap.

4.5 Iterative Development and Agile Principles

A successful technology roadmap embraces the principles of iterative development and agility. Recognising that the technological landscape is subject to change, a CTO structures the roadmap to accommodate feedback loops, allowing for adjustments and refinements. By infusing agile principles into the development process, the technology roadmap becomes a dynamic tool that adapts to challenges and exploits opportunities as they arise.

4.6 Communicating the Roadmap Effectively

Communication is the linchpin of successful roadmap implementation. A key skill for any CTO to master is not only creating the roadmap but in communicating it effectively to both the technology team and key stakeholders. This involves translating technical jargon into a language that resonates with diverse audiences, fostering understanding and buy-in. A well-communicated roadmap aligns the entire organisation toward a common vision and purpose.

4.7 Monitoring and Key Performance Indicators (KPIs)

A technology roadmap is not a static document; it is a living, breathing guide that demands constant monitoring. Establishing the key performance indicators (KPIs) to assess the progress of initiatives and ensure alignment with strategic objectives, is good to do right at the very beginning of the roadmap design. By instituting a robust monitoring mechanism, the CTO can identify deviations, mitigate risks, and celebrate successes as the organisation traverses the roadmap. *See Section 3.6 Metrics for success*

4.8 Specialised Tools for creating Roadmaps

There are several specialised roadmap software tools designed to create visually engaging presentations for communicating project plans, strategies, and technology roadmaps.

4.9 Technology Roadmap Outline Structure

This sample structure has been proven to be a successful outline for the roadmap.

The creation of a technology roadmap is a strategic undertaking and key component of your strategy. It requires meticulous planning, adaptability, and effective communication. By defining strategic milestones, prioritising initiatives, aligning with technological trends, managing resources judiciously, embracing agile principles, communicating effectively, and monitoring progress through KPIs, a smart CTO transforms vision into action. In the symphony of technological leadership, the roadmap is the score that guides the orchestra toward a harmonious and successful performance.

4.1 Defining Strategic Milestones: A Cornerstone of Technology Roadmap Planning

The ability to chart a course and define clear strategic milestones is foundational to successful navigation and progress. This section explores the profound importance of defining strategic milestones within the broader context of creating a technology roadmap. By understanding and leveraging these milestones, CTOs can guide their teams toward achieving overarching business objectives with precision and purpose.

Guiding Progress with Clear Direction:

Strategic milestones act as navigational markers, offering a clear path for the technology team to follow. They crystallise the broader business strategy into actionable steps, providing a roadmap with a defined direction.

These milestones should not merely be technical achievements but should align with specific business outcomes. For example, if the overarching goal is to enhance customer satisfaction, a strategic milestone could be the successful implementation of a customer relationship management (CRM) system.

Enhancing Resource Allocation, Efficiency, and Stakeholder Confidence:

Clear milestones play a pivotal role in promoting judicious resource allocation and ensuring optimal efficiency. CTOs can strategically allocate resources based on the criticality and impact of each milestone. For instance, if a milestone involves the integration of a new CRM system, resources can be directed towards comprehensive training programmes to maximise staff utilisation, prevent inefficiencies, and ensure a smooth transition.

Simultaneously, these strategic milestones serve as visible markers of a well-thought-out plan, building confidence and trust among stakeholders. Regular reports or presentations tied to milestones provide tangible evidence of progress and success. Demonstrating the correlation between achieved milestones and overall business objectives helps build confidence in the CTO's strategic planning.

Facilitating Informed Decision-Making and Ensuring Long-Term Vision Alignment:

Strategic milestones also function as decision-making tools, offering a framework for evaluating progress and assessing risks. Informed decisions aligned with broader organisational priorities can be made based on milestone evaluations. Milestones act as checkpoints for evaluating the success of ongoing initiatives, providing valuable insights for subsequent data-driven strategies.

Moreover, these milestones prevent the roadmap from deviating from the long-term vision. Each milestone represents a deliberate step toward overarching goals, ensuring alignment and preventing drift. Regular strategic reviews integrated into the milestone schedule assess alignment with long-term goals. For instance, if the long-term vision involves sustainability leadership, milestones should reflect eco-friendly technology integrations or initiatives.

Encouraging Cross-Functional Collaboration:

Strategic milestones become touchpoints for fostering collaboration between technology teams and other business functions. They provide common ground for aligning efforts and objectives, with cross-functional workshops organised around specific milestones. For example, if a milestone involves the implementation of an enterprise resource planning (ERP) system, workshops could include representatives from finance, operations, and IT to ensure holistic collaboration.

Furthermore, strategic milestones create a culture of continuous improvement by presenting opportunities for reflection, learning, and optimisation. Post-implementation reviews following major milestones contribute to refining processes and enhancing overall efficiency. Whether conducted internally or externally, these reviews support the evolution of the technology roadmap alongside organisational needs.

10-Step Action Plan for Defining Measurable Strategic Milestones.

Defining strategic milestones is a crucial aspect of creating a technology roadmap that aligns with business goals. This enhanced 10-step plan, provides actionable guidance for CTOs to articulate and implement milestones effectively, incorporating measurable indicators for success.

1. Understand Business Objectives:

- **Action:** Engage in in-depth discussions with key stakeholders to understand and quantify overarching business goals.
- **Example:** Business Goal - Increase Market Share; Measurable Milestone - Achieve a 15% YoY increase in market share within the next 18 months.

2. Conduct a Technology Audit:

- **Action:** Evaluate the current technology landscape to identify areas for improvement and quantify the impact on business outcomes.
- **Example:** Business Goal - Enhance Customer Engagement; Measurable Milestone - Increase customer satisfaction scores by 20% within six months through technology improvements.

3. Align Milestones with Phases of Development:

- **Action:** Break down the technology roadmap into distinct phases, aligning each with a specific business objective and measurable outcome.
- **Example:** Business Goal - Improve Operational Efficiency; Measurable Milestone - Decrease project turnaround time by 25% after the implementation of an integrated project management system.

4. Prioritise Milestones Based on Impact:

- **Action:** Prioritise milestones according to their potential impact on business outcomes, quantifying the expected benefits.
- **Example:** Business Goal - Enhance Customer Experience; Measurable Milestone - Increase Net Promoter Score (NPS) by 10 points within 12 months through the implementation of personalised customer interactions.

5. Involve Cross-Functional Teams:

- **Action:** Collaborate with cross-functional teams to ensure that milestones are representative of various departmental needs and have quantifiable benefits.
- **Example:** Business Goal - Boost Marketing Effectiveness; Measurable Milestone - Achieve a 30% increase in conversion rates through collaboration between IT and Marketing on data-driven marketing strategies.

6. Set Measurable Success Criteria:

- **Action:** Define specific, measurable criteria to determine the success of each milestone, establishing quantifiable targets.
- **Example:** Business Goal - Strengthen Data Security; Measurable Milestone - Reduce security incidents by 25% within six months through the implementation of an advanced security protocol.

7. Establish Milestone Timelines:

- **Action:** Assign realistic timelines to each milestone, considering dependencies and resource availability, and measure progress against deadlines.
- **Example:** Business Goal - Migrate to Cloud Infrastructure for Scalability; Measurable Milestone - Complete migration and achieve full system functionality within the planned timeline of eight months.

8. Ensure Milestones Are Communicated Effectively:

- **Action:** Develop a communication plan to keep all stakeholders informed about milestone progress, using quantifiable metrics where possible.
- **Example:** Business Goal - Enhance Stakeholder Communication; Measurable Milestone - Increase stakeholder satisfaction scores by 15% through consistent communication and transparency.

9. Monitor and Adapt:

- **Action:** Implement a robust monitoring system to track progress regularly, adjusting milestones in response to changes in business priorities or external factors.
- **Example:** Business Goal - Respond to Market Shifts; Measurable Milestone - Adapt product development milestones based on real-time market feedback and achieve a 10% increase in product relevance.

10. Conduct Post-Milestone Reviews:

- **Action:** Conduct thorough reviews after each milestone to assess the effectiveness of implemented changes and gather insights for continuous improvement, using measurable feedback.
- **Example:** Business Goal - Improve Customer Feedback System; Measurable Milestone - Increase customer feedback response rates by 20% within three months and analyse feedback to implement improvements in subsequent milestones.

By incorporating measurable milestones, CTOs can ensure that progress is not only defined but also quantifiable, allowing for continuous improvement and demonstrating the impact of technology initiatives on overarching business objectives.

4.2 Prioritising Initiatives and Dependencies: The Cornerstone of a Robust Technology Roadmap

The effectiveness of a roadmap hinges on the meticulous prioritisation of initiatives and dependencies. This section delves into the critical importance of this process, exploring the nuances of effective prioritisation.

Aligning with Organisational Objectives:

Prioritisation commences with a profound understanding of organisational objectives. Aligning technology initiatives with these objectives ensures that the roadmap is not merely a technical plan but a strategic tool driving the overall success of the business.

To clearly illustrate how a Technology Roadmap aligns with organisational objectives, it is essential to integrate a specific section within the roadmap explicitly outlining this alignment. Begin by articulating the overarching organisational goals and objectives, encompassing business growth targets, efficiency improvements, customer experience enhancements, or any other strategic priorities. Providing a concise and focused overview of these objectives lays the groundwork for showcasing the roadmap's alignment.

In the subsequent section, establish a direct link between each phase or milestone in the Technology Roadmap and specific organisational objectives. For each significant technology initiative or development, elucidate how it contributes to the achievement of overarching goals. Employ clear and measurable metrics to establish the connection between the technology roadmap and organisational success. Whether it involves streamlining processes, enhancing product innovation, or improving customer satisfaction, the roadmap should act as a visual guide, illustrating how each technological advancement aligns with and propels the organisation toward its predefined objectives.

To enhance clarity, incorporate visual elements such as charts, graphs, or timelines that highlight the correlation between roadmap milestones and organisational objectives. A visual representation can effectively communicate the strategic alignment, making it easier for stakeholders to comprehend the roadmap's significance in achieving broader goals. Consider including key performance indicators (KPIs) that will be tracked throughout the implementation, providing a tangible way to measure progress and success against the overarching organisational objectives. This dedicated section should serve as a strategic bridge, connecting the technological advancements outlined in the roadmap with the broader aspirations and success metrics of the organisation.

Example: If the organisation's primary goal is market expansion, prioritising initiatives that enhance scalability and global accessibility takes precedence. This alignment ensures that technological efforts directly contribute to overarching business objectives.

Assessing Business Impact:

Every technology initiative should undergo a thorough assessment based on its potential impact on the business. Prioritise projects that promise substantial returns, whether in terms of revenue growth, cost reduction, or improved customer satisfaction.

In the Technology Roadmap, it is crucial to allocate a dedicated section for evaluating and prioritising technology initiatives based on their potential impact on the business. This entails conducting a comprehensive assessment to gauge the expected outcomes and benefits of each initiative. Establishing a structured evaluation framework involves considering factors such as revenue growth, cost reduction, and enhanced customer satisfaction. By systematically weighing these criteria, businesses can make informed decisions about prioritising projects, ensuring that resources are directed towards initiatives with the greatest potential for positive impact.

To effectively communicate this prioritisation strategy in the Technology Roadmap, it is essential to articulate the specific metrics and key performance indicators (KPIs) used in the assessment process. For instance, projects with the potential to significantly contribute to revenue growth might be evaluated based on projected sales figures, market expansion opportunities, or potential new customer acquisitions. Similarly, initiatives aimed at cost reduction could be assessed through an analysis of expected operational efficiencies, resource optimisation, or streamlined processes. Clear and quantifiable metrics provide a transparent basis for prioritisation, allowing stakeholders to understand the rationale behind the chosen order of technology initiatives.

Furthermore, the Technology Roadmap should emphasise the dynamic nature of this assessment process. Business landscapes evolve, and technological advancements may impact the feasibility and potential impact of different initiatives. Therefore, it is crucial to underscore the importance of periodic reviews and adjustments to the prioritisation framework. This adaptability ensures that the Technology Roadmap remains responsive to changing business needs and technological advancements, allowing organisations to stay agile and make strategic adjustments as necessary.

Example: A customer-facing mobile app might have a higher priority if the business is aiming to enhance its digital presence and customer engagement. This prioritisation is rooted in the tangible impact on key business metrics.

Criteria	Metrics	Weighting (%)
Revenue Growth	Projected Sales Increase	20
	Market Expansion Opportunities	10
	New Customer Acquisition Potential	5
Cost Reduction	Expected Operational Efficiency Gains	5
	Resource Optimization	5
	Streamlined Processes	3
Customer Satisfaction	Improved User Experience	17
	Enhanced Customer Support	5
	Positive Customer Feedback	10
Strategic Alignment	Alignment with Business Objectives	5
	Contribution to Long-Term Strategy	15

In this example above, each criterion is accompanied by specific metrics that provide measurable insights into the impact of a technology initiative. The weighting percentages indicate the relative importance of each criterion in the overall assessment. Project teams can use this framework to systematically evaluate and prioritise initiatives based on their potential business impact. The metrics help in quantifying and qualifying the expected outcomes, facilitating a more informed decision-making process.

The Dependency Structure Matrix (DSM)

DSM is a robust management tool intended to visually represent and analyse the dependencies between different elements within a system or a set of projects. The principle behind DSM is based on organising and displaying complex interdependencies in a matrix format. The matrix captures relationships between tasks, components, or projects, providing a clear and concise overview of how they depend on each other. The DSM principle is rooted in simplifying the representation of intricate dependencies, making it easier for project managers to identify critical relationships and manage interdependencies effectively.

One significant advantage of using DSM is its ability to streamline project planning and execution. By visualising dependencies in a matrix, we can quickly identify the sequencing of tasks, potential bottlenecks, and critical paths. This leads to enhanced project efficiency, as teams can focus on resolving dependencies that have the most significant impact on the overall project timeline. Additionally, DSM helps in avoiding delays and miscommunications by providing a transparent view of the relationships between different elements, facilitating more informed decision-making.

Various techniques are involved in implementing DSM effectively. Initially, creating a comprehensive list of tasks, components, or projects is crucial. Once identified, relationships and dependencies between these elements are documented within the matrix. Software tools designed for DSM can automate this process, allowing for easy manipulation and adjustment as project requirements evolve. Regularly updating and reviewing the DSM throughout the project lifecycle ensures that it remains a dynamic and accurate representation of dependencies, aiding in proactive management and successful project delivery.

Dependencies can significantly influence project timelines and outcomes. Thoroughly map out dependencies between different initiatives to ensure a realistic and achievable roadmap. Recognising and addressing dependencies in the prioritisation process is critical for maintaining project coherence.

Example: If a database migration is a prerequisite for implementing a new analytics tool, recognising and prioritising this dependency becomes crucial. This strategic approach prevents bottlenecks and ensures a smoother execution of the overall roadmap.

Balancing Short-term Wins and Long-term Strategy:

Striking a balance between short-term wins and long-term strategic goals is pivotal. Prioritise initiatives that offer quick wins to maintain momentum while also advancing projects critical for the organisation's future.

Achieving a delicate equilibrium between short-term wins and long-term strategic goals is pivotal for the sustained success and growth of an organisation. Short-term wins encompass immediate, tangible achievements that deliver quick results and positive outcomes. In contrast, long-term strategic goals involve overarching objectives contributing to the organisation's vision and future sustainability. Prioritising initiatives that offer quick wins is crucial to maintaining momentum and keeping stakeholders engaged. These quick wins can include efficiency improvements, cost savings, or the successful implementation of smaller projects that deliver visible and immediate benefits.

However, focusing solely on short-term wins without considering long-term strategic goals can be shortsighted. It's imperative to advance projects critical for the organisation's future, even if they may not yield immediate results. These long-term initiatives often involve significant investments, research and development, and changes to the organisational structure. Examples of long-term strategic goals could include adopting transformative technologies, entering new markets, or investing in employee training and development to build a future-ready workforce.

For instance, a manufacturing company might prioritise a short-term win by implementing lean processes to increase operational efficiency and reduce costs. Simultaneously, it could invest in long-term strategic goals, such as researching and developing innovative products that align with emerging market trends. Striking this balance ensures that the organisation not only addresses immediate challenges and opportunities but also positions itself for sustained growth and competitiveness in the ever-evolving business landscape.

In the technology sector, a company might focus on quick wins by optimising its existing software systems to enhance user experience and boost customer satisfaction. Simultaneously, it could be investing in long-term strategic initiatives like developing cutting-edge artificial intelligence applications or exploring partnerships to stay at the forefront of industry innovation. Balancing short-term wins with long-term strategic goals allows organisations to navigate the present while preparing for a future where they remain adaptable and relevant.

Example: Implementing a customer feedback tool for immediate insights may be balanced with a long-term initiative to overhaul the entire CRM system. This balanced approach ensures a mix of immediate benefits and sustained, strategic growth.

Mitigating Risks:

Identifying and prioritising high-risk initiatives early in the roadmap helps in developing mitigation strategies. Prioritise riskier projects with proactive risk management plans to minimise potential negative impacts.

Innovative Risk Navigation Methodology for Technology Roadmap Development: A Strategic Framework

The "Proactive Risk Innovation Framework" is a methodology from the Marlow Business school designed to identify and mitigate risks early in the development of a Technology Roadmap, ensuring a balanced approach that fosters innovation while maintaining effective risk management.

1. **Early Identification and Evaluation:** The framework initiates with a thorough risk identification phase, involving cross-functional collaboration among key stakeholders. Utilising brainstorming sessions, risk workshops, and historical analysis, potential risks are identified and categorised based on their potential impact and probability. Each identified risk is then evaluated through a structured risk assessment process, considering factors such as project complexity, technology maturity, and external dependencies. This evaluation provides a nuanced understanding of the risks' potential implications, enabling the team to prioritise effectively.

2. **Dynamic Tracking and Continuous Monitoring:** To ensure proactive risk management, a robust tracking mechanism is implemented. Risks are assigned ownership, and a dynamic tracking system, often facilitated by dedicated software tools, is employed to monitor risk indicators, triggers, and potential warning signs. Regular check-ins and status updates are integrated into project meetings to facilitate continuous monitoring. This real-time visibility ensures that any emerging risks are swiftly identified, allowing for timely interventions and adjustments in the Technology Roadmap.

3. **Adaptive Risk Planning for High-Risk Initiatives:** Recognising that high-risk initiatives often carry the potential for substantial commercial benefits, the framework advocates an adaptive risk planning approach. Rather than adopting an overly risk-averse stance, teams are encouraged to embrace higher-risk projects with a strategic risk management plan in place. This plan includes comprehensive risk mitigation strategies, contingency plans, and a well-defined risk tolerance threshold. By addressing the uncertainties associated with high-risk initiatives proactively, organisations can navigate challenges, unlocking innovation and potential commercial success.

4. **Iterative Risk Updates:** The framework promotes an iterative approach to risk management. Regular updates and reviews are embedded into the Technology Roadmap development process, allowing for the continuous refinement of risk assessments and mitigation plans. Lessons learned from previous projects are systematically incorporated, contributing to a culture of continuous improvement in risk management practices.

By adopting the Proactive Risk Innovation Framework, organisations can strike a delicate balance between innovation and risk management. This approach empowers teams to identify, evaluate, and mitigate risks early in the Technology Roadmap development, fostering an environment where calculated risks contribute to meaningful innovation and commercial success.

The Risk Register

The development of a comprehensive Risk Register is crucial for organisations, projects, products, and departments, acting as a foundational tool in proactive risk management. At the organisational level, a Risk Register provides a holistic view of potential risks that could impact the entire enterprise, enabling leadership to make informed decisions about resource allocation, strategic planning, and overall risk appetite. By centralising risk information, organisations can cultivate a culture of risk awareness, promoting transparency and accountability across all levels.

In the context of projects, a well-maintained Risk Register is instrumental in identifying, assessing, and mitigating potential threats that may affect project timelines, budgets, and deliverables. By systematically documenting risks, we can anticipate challenges, allocate resources effectively, and implement mitigation strategies. This proactive approach not only enhances project success rates but also contributes to stakeholder confidence and satisfaction.

Similarly, at the product level, maintaining a Risk Register is essential for understanding and managing uncertainties that could impact the product's development, launch, and lifecycle. By systematically cataloguing potential risks, product managers can make informed decisions, prioritise features or enhancements, and ensure the delivery of a resilient and competitive product. A robust Risk Register supports the product team in navigating market uncertainties and evolving customer demands.

At the departmental level, a dedicated Risk Register enables managers to assess and address risks specific to their area of responsibility. This targeted approach allows departments to align risk management strategies with organisational objectives, fostering a culture of accountability and proactive problem-solving. By consistently updating and reviewing the Risk Register, departments can adapt to changing circumstances and contribute to the overall resilience of the organisation.

Example: A cybersecurity enhancement initiative might take precedence due to its inherent risk mitigation implications. Prioritising such initiatives showcases a commitment to safeguarding the organisation's technological assets and ensuring operational resilience.

Ensuring the accuracy, relevance, and visibility of the Risk Register within an organisation is a vital process that involves several key steps to uphold its effectiveness. Here's a comprehensive process for keeping the Risk Register up to date, regularly reviewed, and visible at senior levels.

This process can ensure the Risk register is kept current and the right people have visibility.

1. **Identification and Documentation:**
    1. Team members, project managers, and department heads actively identify and document potential risks, leveraging their expertise and experience.
    2. A centralised platform or software tool is used to record risks, ensuring consistency and ease of access. A suggested tool can be found in the appendix.

2. **Regular Reviews and Updates:**
    1. A designated Risk Management Team conducts regular reviews of the Risk Register, typically on a predetermined schedule (e.g., monthly or quarterly).
    2. Team members contribute updates based on evolving project dynamics, emerging risks, or changes in the external environment.
    3. Historical data and lessons learned from previous projects are incorporated to enhance the accuracy of risk assessments.

3. **Risk Assessment and Analysis:**
    1. Risks are assessed for likelihood, impact, and overall severity using predetermined criteria.
    2. The Risk Management Team collaborates to analyse new risks and reassess existing ones, considering the evolving nature of the organisation and its projects.

4. **Mitigation Strategy Development:**
    1. For each identified risk, mitigation strategies are developed or updated, outlining actionable steps to minimise the potential impact.
    2. Strategies may involve changes to project plans, resource allocation, or the implementation of contingency measures.

5. **Communication and Collaboration:**
    1. Regular communication channels are established to keep team members informed about updates to the Risk Register.
    2. Nest the project and product and department risk registers, bringing the priority 1 and 2 risks into the corporate Risk Register.

6. **Escalation to Senior Levels:**
    1. Risks with significant impact or those requiring senior management attention, priority 1, are escalated to higher levels of the organisation immediately rather than wait for monthly consolidation of risk.
    2. Executive summaries or reports are generated to present a consolidated view of the Risk Register to senior leaders.

7. **Integration with Decision-Making Processes:**
    1. The Risk Register is integrated into decision-making processes, ensuring that risk considerations are taken into account when making strategic or operational decisions.
    2. Senior leaders actively review and discuss the Risk Register during key decision meetings and Risk should be an agenda item on the board meetings.

8. **Continuous Training and Awareness:**
    1. Ongoing training programmes and awareness initiatives are conducted to ensure that team members understand the importance of the Risk Register and how to contribute effectively.
    2. Training extends to senior leaders to enhance their understanding of the risks associated with ongoing projects and operations.

See example Risk Register in Appendix

Soliciting Stakeholder Input:

An effective approach to soliciting stakeholder input in the prioritisation process is to foster inclusive engagement across various levels of the organisation. Recognising that prioritisation is not a solitary task, the first step involves identifying key stakeholders who hold valuable insights into different aspects of the projects or initiatives. This group may include department heads, end-users, executives, and individuals with domain expertise. Establishing clear communication channels and forums for collaboration is crucial, ensuring that stakeholders can openly share their perspectives on the importance and urgency of various tasks.

Once the stakeholders are identified, a systematic engagement process is initiated. This involves conducting structured meetings, surveys, or workshops where stakeholders can provide their input on prioritisation criteria, project goals, and potential challenges. Facilitators guide discussions to extract valuable insights, considering both the strategic objectives of the organisation and the specific needs and expectations of individual departments or user groups. This inclusive approach not only gathers diverse viewpoints but also fosters a sense of ownership and alignment among stakeholders, leading to more informed and collectively endorsed prioritisation decisions.

Example: In a manufacturing-focused organisation, involving the production team in prioritising initiatives can provide critical insights into operational needs. This collaborative approach enhances the relevance and effectiveness of the prioritisation process.

Continuous Monitoring and Adjusting:

Regularly monitor the progress of initiatives against the roadmap. Be prepared to adjust priorities based on changing circumstances, new market trends, or unexpected challenges.

An effective strategy involves establishing a systematic review process at predetermined intervals, ensuring that key stakeholders convene to assess the status of ongoing initiatives. This process should encompass a comprehensive evaluation of project milestones, resource allocation, and adherence to timelines outlined in the roadmap. By employing key performance indicators (KPIs) as discussed in other sections, and project management tools, organisations can obtain real-time insights into the progress of initiatives, facilitating data-driven decision-making.

Inherent in this approach is the flexibility to adjust priorities based on changing circumstances, emerging market trends, or unforeseen challenges. Regular review meetings serve as forums for discussing the impact of external factors on the roadmap and making informed decisions regarding the recalibration of priorities. An agile mindset is crucial, empowering teams to swiftly adapt to evolving circumstances and seize new opportunities. By remaining attuned to the broader business environment, organisations can ensure that their initiatives align with strategic objectives and are positioned for success in the face of dynamic challenges.

Example: If a competitor introduces a game-changing technology, the roadmap might need urgent adjustment to stay competitive. Continuous monitoring and adjustment ensure that the roadmap remains agile and aligned with the evolving business landscape.

Communicating Priorities Effectively:

Clear communication of prioritised initiatives and dependencies is essential for organisational alignment. Ensure that all stakeholders understand the reasoning behind the prioritisation decisions.

Sample Communications Plan: Prioritised Initiatives and Dependencies

Objective: The primary objective of this communications plan is to ensure clear, consistent, and transparent communication regarding prioritised initiatives and dependencies within the organisation. By facilitating understanding among stakeholders, this plan aims to foster organisational alignment, promote collaboration, and provide insights into the rationale behind prioritisation decisions.

Audience:
1. **Executive Leadership Team:** Regular briefings for the leadership team to communicate high-level priorities, dependencies, and strategic considerations.
2. **Project Managers and Teams:** Detailed communications outlining specific initiatives, their prioritisation, and dependencies, tailored to project managers and teams responsible for execution.
3. **Cross-Functional Teams:** Targeted communications for cross-functional teams to enhance collaboration and facilitate a shared understanding of the interconnectedness of initiatives.
4. **Individual Contributors:** Regular updates for individual contributors to keep them informed about the broader context of prioritised initiatives and dependencies affecting their work.

Channels:
1. **Executive Briefings:** Monthly or quarterly executive briefings delivered by the CEO in conjunction with the CTO or relevant executives to disseminate high-level information on prioritised initiatives and dependencies. This is important to show this is not a "technology" priority, but a organisation imperative.
2. **Project Updates:** Bi-weekly or monthly project update meetings, webinars, or written updates for project managers and teams, providing detailed insights into prioritisation decisions.
3. **Collaboration Platforms:** Utilise collaboration tools, such as Slack channels or Microsoft Teams, to facilitate ongoing discussions and address queries related to prioritised initiatives.
4. **Town Hall Meetings:** Conduct regular town hall meetings to address cross-functional teams, promoting a shared understanding of dependencies and

encouraging open communication.

5. **Email Newsletters:** Periodic email newsletters summarising key updates, priority shifts, and dependencies for individual contributors to stay informed.

Content:

1. **Rationale Behind Prioritisation:** Clearly articulate the strategic goals guiding prioritisation decisions and how they align with the organisation's vision.
2. **Initiative Details:** Provide comprehensive information about each prioritised initiative, including objectives, timelines, and potential impact on other projects.
3. **Dependencies Overview:** Clearly outline the dependencies between initiatives, illustrating how each contributes to the overall success of the organisation.
4. **Feedback Mechanism:** Establish a feedback mechanism, such as regular Q&A sessions or dedicated communication channels, to address stakeholder queries and concerns.

Responsibilities:

1. **Communication Team:** Draft and disseminate communications across various channels, ensuring consistency and accuracy of information.
2. **Executive Sponsors:** Executives or leaders responsible for specific initiatives should actively participate in communication efforts to provide context and answer questions.
3. **Project Managers:** Collaborate with the communication team to ensure project-specific information is accurately conveyed to relevant teams.

The idea behind this communications plan, is for the organisation to create a transparent and collaborative environment, enabling stakeholders at all levels to comprehend the prioritisation decisions and actively contribute to the successful execution of initiatives.

Prioritising initiatives and dependencies is not a one-time activity but an ongoing strategic process that requires a blend of business acumen, risk management, and adaptability. By carefully navigating these considerations technology leaders can create a Technology Roadmap that not only aligns with organisational goals but also propels the business towards technological excellence and sustained success

10-Step Action Plan for Prioritising Initiatives and Dependencies

Creating an effective Technology Roadmap requires a meticulous process of prioritising initiatives and dependencies. This practical 10-step guide, provides a detailed approach for technology leaders to navigate this crucial aspect:

1. Organisational Alignment:

- **Action:** Begin by aligning technology initiatives with organisational objectives.
- **Example:** If the business aims for market expansion, prioritise initiatives enhancing scalability and global accessibility.

2. Business Impact Assessment:

- **Action:** Assess each technology initiative based on its potential impact on the business.
- **Example:** Prioritise projects promising substantial returns, be it revenue growth, cost reduction, or enhanced customer satisfaction.

3. Dependency Mapping:

- **Action:** Thoroughly map out dependencies between different initiatives for a realistic and achievable roadmap.
- **Example:** Recognise and prioritise dependencies, ensuring a smoother execution of the overall roadmap.

4. Short-term Wins vs Long-term Strategy:

- **Action:** Strike a balance between short-term wins and long-term strategic goals.
- **Example:** Balance implementing a customer feedback tool for immediate insights with a long-term initiative to overhaul the entire CRM system.

5. Resource Capacity Evaluation:

- **Action:** Realistically assess available resources, including budget, skilled personnel, and time.
- **Example:** Prioritise initiatives aligning with resource capacity to avoid overcommitment and burnout.

6. Risk Prioritisation:

- **Action:** Identify and prioritise high-risk initiatives, developing proactive risk management plans.
- **Example:** Prioritise a cybersecurity enhancement initiative for its inherent risk mitigation implications.

7. Iterative and Adaptive Planning:

- **Action:** Adopt an iterative approach to planning, allowing for adaptive prioritisation.
- **Example:** Reassess prioritised initiatives in response to sudden shifts in market trends for continued relevance.

8. Stakeholder Engagement:

- **Action:** Engage key stakeholders to gather insights and perspectives for a more holistic view of priorities.
- **Example:** Involve the production team in manufacturing-focused organisations for critical operational insights.

9. Continuous Monitoring and Adjustment:

- **Action:** Regularly monitor the progress of initiatives against the roadmap and adjust priorities as needed.
- **Example:** Adjust the roadmap in response to competitors introducing game-changing technologies.

10. Effective Communication:

- **Action:** Communicate prioritised initiatives and dependencies clearly to ensure organisational alignment.
- **Example:** Host regular town hall meetings or send newsletters to enhance transparency and understanding of prioritisation choices.

This 10-step process provides a comprehensive and practical guide for CTOs and technology leaders, facilitating the creation of a Technology Roadmap that not only aligns with organisational goals but also propels the business towards technological excellence and sustained success.

4.3 Aligning with Technological Trends

Also review section 2.5 Emerging Technologies

The success of an organisation hinges on its ability to not only stay abreast of current technological trends but to proactively align its strategies with these advancements. This section explores the critical importance of aligning with technological trends within the broader context of creating a comprehensive Technology Roadmap.

Foresight for Strategic Planning

Embracing technological trends is not merely a response to the current state of affairs; it is an exercise in foresight. A Technology Roadmap that aligns with current and emerging trends enables an organisation to anticipate shifts in the technological landscape. By weaving these trends into strategic planning, a business can position itself as an innovator rather than a follower, gaining a competitive edge in the market.
Example: Utilise sources like *__Gartner's Technology Trends__*, which provides in-depth analyses of emerging technologies, enabling CTOs to foresee trends and integrate them into long-term strategic planning.

Enhancing Competitive Advantage

In a rapidly changing technological ecosystem, aligning with trends is synonymous with maintaining a competitive advantage. By adopting emerging technologies that resonate with market demands, organisations can differentiate themselves from competitors. This alignment fosters agility, enabling businesses to respond promptly to market shifts and deliver products or services that meet the evolving needs of their customers.
Example: Regularly monitor industry-specific online publications such as *__TechCrunch__* or *__Wired__* to identify how competitors are leveraging emerging technologies, allowing your organisation to stay ahead in the innovation curve.

Adaptability and Resilience

The pace of technological change necessitates a culture of adaptability. Aligning with trends ensures that an organisation's Technology Roadmap is dynamic and responsive. This adaptability becomes a cornerstone of resilience, allowing the business to weather disruptions, unforeseen challenges, or sudden shifts in customer preferences.
Example: Leverage real-time news sources such as *__Tech News websites__* or industry-specific blogs to stay informed about the latest technological developments, fostering an environment of adaptability within the organisation.

Customer-Centric Innovation

Also review section 2.6

Technological trends often originate from a deep understanding of customer needs. Aligning with these trends implies a commitment to customer-centric innovation. By integrating technologies that resonate with the end user, businesses can enhance the customer experience, build brand loyalty, and gain a deeper understanding of market dynamics.

Example: Engage with customer feedback platforms, social media channels, and online forums to gather insights into customer preferences and pain points, shaping the integration of technology with a customer-centric approach.

Enabling Scalability

A well-aligned Technology Roadmap incorporates scalable technologies that can grow with the business. Whether it's cloud computing, scalable architectures, or modular systems, trends that support scalability ensure that technological investments remain relevant as the organisation expands. This foresight prevents the need for frequent overhauls and minimises technical debt.

Example: Regularly review reports from cloud service providers like **AWS** or **Azure,** assessing new features and tools that facilitate scalability and aligning them with the organisation's growth trajectory.

Risk Mitigation through Compliance

Technological trends are often intertwined with regulatory and compliance considerations. Aligning with these trends ensures that a Technology Roadmap factors in the necessary compliance measures. This proactive approach mitigates risks associated with non-compliance, safeguarding the organisation against legal challenges and reputational damage.

Example: Stay informed about evolving data protection laws and cybersecurity regulations through regulatory websites such as the ***Information Commissioner's Office (ICO),*** ensuring that your Technology Roadmap aligns with the latest compliance requirements.

Facilitating Cross-Functional Collaboration

Aligning with technological trends is not solely the responsibility of the technology department. It necessitates collaboration across functions. A well-crafted Technology Roadmap that aligns with trends fosters cross-functional collaboration. By involving diverse perspectives, from marketing to operations, businesses can ensure that technology aligns with broader organisational goals.

Example: Leverage collaboration platforms like **_Microsoft Teams_** or **_Slack_** to facilitate cross-functional communication, fostering a collaborative environment where insights from different departments contribute to technological alignment.

Tapping into Innovation Ecosystems

Technological trends often emerge from vibrant innovation ecosystems. A Technology Roadmap aligned with trends involves tapping into these ecosystems. This could mean collaborating with startups, engaging in industry forums, or participating in open-source communities. Such engagement provides access to cutting-edge ideas, fostering a culture of continuous innovation.

Example: Participate in industry-specific forums like **_GitHub_** or collaborate with startup incubators, creating a symbiotic relationship that allows your organisation to tap into emerging technologies.

Cultivating a Future-Oriented Mindset

Aligning with technological trends is not a one-time exercise but a mindset that embraces the future. It requires ongoing monitoring, evaluation, and adjustment. A Technology Roadmap embedded in this mindset cultivates a culture of continuous learning, positioning the organisation as a proactive player in the ever-evolving technological landscape.

Example: Encourage teams to subscribe to relevant newsletters, follow thought leaders on social media platforms, and participate in webinars to stay abreast of evolving technologies, fostering a continuous learning culture.

Strategic Resource Allocation

Resources are finite, and aligning with technological trends assists in their strategic allocation. By understanding which trends are most relevant to the business, organisations can allocate resources—financial, human, and technological—effectively. This targeted investment ensures that the Technology Roadmap remains both impactful and sustainable.

Example: Leverage project management tools such as **_Jira_** or **_Asana_** to allocate resources efficiently, ensuring that teams are aligned with the identified technological trends and strategic goals of the organisation.

10 Step Action Plan: Aligning Technological Trends

To strategically align your organisation with technological trends, follow this detailed and measurable 10-point action plan:

1. Establish Trend Monitoring Mechanisms:
- **Action:**
  - Implement tools like ***Google Alerts, Feedly,*** or specialised trend monitoring platforms.
  - Subscribe to relevant industry publications, blogs, and newsletters.
- **Measurement:**
  - Track the number of new trends identified monthly.
  - Evaluate the accuracy of trend predictions over time.

2. Create a Cross-Functional Trend Analysis Team:
- **Action:**
  - Form a team comprising representatives from IT, marketing, operations, and customer service.
  - Conduct training sessions to equip team members with trend analysis skills.
- **Measurement:**
  - Assess the team's ability to provide diverse insights and actionable recommendations.

3. Implement Regular Trend Review Meetings:
- **Action:**
  - Schedule bi-monthly meetings to discuss identified trends.
  - Encourage team members to present findings and insights.
- **Measurement:**
  - Measure the number of actionable insights generated from these meetings.
  - Track the implementation rate of insights into ongoing projects.

4. Engage in Industry-Specific Forums and Networks:
- **Action:**
  - Identify and participate in at least two industry-specific forums or networks.
  - Encourage team members to actively contribute to discussions.
- **Measurement:**
  - Evaluate the quality and relevance of insights gained from these engagements.
  - Monitor the number of valuable connections made through participation.

5. Utilise Trend Analysis Tools:
- **Action:**
  - Invest in tools such as ***Trend Hunter, Statista***, or industry-specific analytics platforms.
  - Provide training for team members on effectively using these tools.
- **Measurement:**
  - Track the accuracy of predictions made using these tools.
  - Measure the time saved in trend analysis with the implementation of these tools.

6. Encourage Continuous Learning:

*   **Action:**
    *   Host regular workshops, webinars, and knowledge-sharing sessions.
    *   Create an internal knowledge repository for employees to access learning materials.
*   **Measurement:**
    *   Monitor the participation rate in learning initiatives.
    *   Assess the knowledge gain through periodic assessments or quizzes.

7. Incorporate Customer Feedback on Technology:

*   **Action:**
    *   Implement surveys and feedback mechanisms to understand customer expectations.
    *   Establish a dedicated channel for customer feedback related to technology.
*   **Measurement:**
    *   Track the number of customer-driven technological enhancements implemented.
    *   Evaluate customer satisfaction scores related to technology-driven improvements.

8. Regularly Assess Competitor Technological Initiatives:

*   **Action:**
    *   Conduct quarterly reviews of competitors' technological advancements.
    *   Establish a process for gathering competitive intelligence.
*   **Measurement:**
    *   Benchmark your organisation against competitors based on technology adoption.
    *   Identify areas where your organisation outperforms or lags behind competitors.

9. Allocate Resources for Technological Exploration:

*   **Action:**
    *   Dedicate a percentage of the budget specifically for experimenting with emerging technologies.
    *   Create an innovation fund to support small-scale technology projects.
*   **Measurement:**
    *   Evaluate the success rate of technology experiments and their impact.
    *   Track the return on investment (ROI) for technology-driven initiatives.

10. Establish a Technological Trend KPI Dashboard:

*   **Action:**
    *   Develop a dashboard integrating key performance indicators (KPIs) related to technological trends.
    *   Include metrics like trend identification accuracy, implementation rates, and customer satisfaction.
*   **Measurement:**
    *   Regularly assess the dashboard for insights and adjust strategies accordingly.
    *   Use the dashboard to communicate progress and trends to stakeholders.

Additional Resources for Technological Trend Alignment:

1. ***TechCrunch:*** A leading technology media property, dedicated to obsessively profiling startups, reviewing new Internet products, and breaking tech news.

2. ***Wired:*** Offers in-depth coverage of current and future trends in technology, business, and culture.

3. ***Gartner's Technology Trends:*** Gartner's official source for technology trends and insights.

4. ***Trend Hunter:*** A platform that showcases innovative ideas and trends across various industries.

5. ***Statista:*** Provides statistical data on a wide range of topics, including technology trends and market forecasts.

6. ***Google Alerts:*** A free service that delivers email updates of the latest relevant Google search results based on chosen queries.

7. ***Feedly:*** A content aggregation tool that allows you to subscribe to various sources, keeping you informed about industry trends.

8. ***GitHub:*** A platform for developers to collaborate on and discover new projects, often indicating emerging trends in software development.

9. ***Tech News websites:*** Provides up-to-date news on the latest technology trends and developments.

10. ***Information Commissioner's Office (ICO):*** The official UK regulatory authority for data protection and privacy, essential for staying informed about evolving data protection laws.

4.4 Enhancing Resource Allocation and Efficiency: A Strategic Imperative in Creating a Technology Roadmap

Crafting a robust technology roadmap is not only about envisioning the future but also about mastering the art of resource allocation and operational efficiency. This section delves into the pivotal role that enhancing resource allocation and efficiency plays in the process of creating a technology roadmap.

Considering Resource Availability:

A realistic assessment of available resources, including budget, skilled personnel, and time, is paramount.

A comprehensive evaluation of available resources, which includes budget, skilled personnel, and time, is crucial for project, product and organisational success. Without a thorough understanding of the resources at hand, there is an increased risk of overcommitment, leading to budgetary constraints, strained personnel, and missed deadlines. A realistic assessment ensures that the organisation operates within its means, promoting a sustainable approach to project execution. This evaluation also allows decision-makers to strategically allocate resources, identifying potential gaps and implementing proactive solutions.

Prioritising initiatives in line with the organisation's resource capacity is a strategic imperative to prevent overcommitment and burnout among the workforce. In the absence of realistic resource planning, there is a tendency to undertake too many initiatives simultaneously, stretching the available talent pool thin and compromising the quality of work. By aligning initiatives with resource capacity, organisations can optimise their workforce, leveraging specialised skills where needed and avoiding unnecessary strain on teams. This approach contributes to a healthier work environment, reduces the risk of employee burnout, and enhances overall team morale and productivity.

Moreover, prioritising initiatives based on resource capacity ensures that the organisation can respond effectively to unexpected challenges or opportunities. Flexibility in resource allocation becomes possible when initiatives are aligned with the organisation's actual capabilities. This adaptability is crucial in dynamic business environments where shifting priorities and unforeseen circumstances are commonplace. By avoiding overcommitment, organisations can navigate changes more smoothly, maintain project quality, and build a resilient foundation for sustained success.

Strategic Alignment with Business Goals:

A technology roadmap serves as a blueprint for achieving business objectives. Effective resource allocation ensures that the technological initiatives outlined in the roadmap align strategically with overarching business goals.

By comprehensively aligning technology initiatives with business goals, organisations can optimise resource allocation to focus on projects that directly contribute to the company's success. This alignment enhances the efficiency of resource utilisation by avoiding investments in non-strategic areas.

Effectively allocating resources involves identifying the specific needs of projects, understanding the skill sets required, and aligning those needs with the organisation's overarching objectives. This targeted allocation ensures that the right resources are deployed to the right tasks, optimising efficiency and fostering innovation.

Acquiring and hiring expert resources further enhance an organisation's strategic capabilities. Expertise in emerging technologies, industry trends, and specialised domains empowers the organisation to stay at the forefront of innovation. A strategic approach involves identifying the skills and knowledge gaps within the existing team and proactively hiring individuals who bring diverse expertise to the table. Whether through hiring permanent staff, engaging consultants, or fostering partnerships, integrating top-tier talent ensures that the organisation possesses the necessary skills to tackle challenges and pursue strategic initiatives.

Identifying skills and knowledge gaps within existing teams is a crucial step in ensuring the continual growth and adaptability of an organisation. Conducting comprehensive skills assessments involves evaluating the current competencies of team members against the skills required for present and future projects. This can be achieved through one-on-one interviews, self-assessment surveys, and performance evaluations. Additionally, regular feedback sessions and discussions about career aspirations can shed light on individual aspirations and areas for development.

Various tools and online resources can aid in this process. Skill assessment platforms, such as ***Skillsoft*** and ***LinkedIn Learning***, offer a range of courses and assessments to evaluate and enhance existing skills. Internal knowledge-sharing platforms and collaborative tools, such as Confluence or Microsoft Teams, facilitate team members in sharing their expertise and learning from one another. Additionally, regular engagement with industry-specific forums, webinars, and conferences can help teams stay abreast of emerging trends and technologies, ensuring that skill gaps are identified and addressed proactively.

Skill Assessment Platforms:

1. *LinkedIn Learning:* Provides assessments and courses covering a broad range of skills.

2. *Skillsoft:* Offers eLearning solutions with skill assessments across various domains.

3. *HackerRank:* Focuses on coding challenges for assessing programming skills.

4. *Codewars:* Provides coding challenges known as "kata" to practice and assess programming abilities.

5. *Marlow Business school:* Assesses acquired career skills as part of its CTO Programme.

6. *InterviewBit:* Prepares individuals for technical interviews with assessments and problem-solving exercises.

7. *Pramp:* Facilitates live interview practice sessions with peers to assess technical skills.

8. *Pluralsight:* Provides expert-led courses and assessments in software development and IT operations.

9. *TestDome:* Allows creating custom technical assessments to evaluate coding and problem-solving skills.

10. *Codility:* Focuses on assessing coding skills with customisable tests.

11. *Vervoe:* Assesses not only technical but also soft skills through simulations and real-world scenarios.

12. *Pymetrics:* Uses neuroscience-based games to assess cognitive and soft skills.

13. *Crucible (by Atlassian):* Facilitates code reviews, allowing teams to assess code quality and collaboration skills.

Ultimately, resource allocation and strategic hiring contribute to building a dynamic and adaptive workforce that aligns with the organisation's strategic goals. By investing in the right people and ensuring that resources are deployed effectively, organisations can enhance their capacity to innovate, respond to market changes, and achieve long-term success in a rapidly evolving technological landscape.

Optimising Budget Utilisation:

Financial resources are finite, and efficient resource allocation demands judicious budget management. A technology roadmap should reflect a clear understanding of budget constraints while ensuring the optimal use of available funds.

By strategically allocating resources according to budget constraints, organisations can avoid unnecessary expenditure on less impactful initiatives. This approach enhances efficiency by directing financial resources towards projects that deliver the most significant value within the allocated budget.

Formulating a pragmatic budget that aligns seamlessly with the technology roadmap is a critical facet of effective technology leadership. One approach involves conducting a thorough cost estimation for each project phase, encompassing personnel expenditures, materials, software, equipment, and external services. Drawing insights from historical data or industry benchmarks adds precision to cost projections. Moreover, it is imperative to factor in unforeseen contingencies by incorporating a buffer or contingency fund, acknowledging the dynamic nature of technological projects.

Effective communication and collaboration across key stakeholders, including technology teams and financial counterparts, ensure coherence in budgetary understanding, harmonising financial allocations with technological priorities. Regular reviews and updates of the budget throughout the technology project's lifecycle emerge as essential practices, facilitating adaptability to evolving requirements, technological challenges, or emerging opportunities. Upholding transparency and flexibility, this dynamic budgeting strategy serves as a pivotal tool for a Chief Technology Officer, enabling the budget to dynamically respond to the changing technological landscape while remaining aligned with the overarching technology roadmap. Refer to Chapter 13 for in-depth insights into Financial Management and Budgeting.

Agility and Adaptability:

The technology landscape is dynamic, and an agile approach is crucial for responding to changing circumstances. Efficient resource allocation allows for flexibility in adapting the roadmap in response to emerging technologies, market shifts, or unforeseen challenges.

Agile resource allocation involves the ability to reallocate resources swiftly based on changing priorities or unexpected developments. This adaptability ensures that the technology roadmap remains relevant and resilient in the face of evolving.

Preparing a team with versatile skills to seamlessly switch resources between different projects or products is essential for enhancing organisational agility and adaptability. Firstly, fostering a culture of cross-training within the team is crucial. This involves providing opportunities for team members to diversify their skill sets by learning about various aspects of the projects or products they work on. Cross-training enables team members to acquire a broader understanding of the organisation's technology landscape, making it easier to transition between projects when the need arises.

Secondly, implementing a knowledge-sharing platform can significantly contribute to skill preparation. This platform could include documentation, wikis, or internal forums where team members can share insights, best practices, and lessons learned from their experiences. Encouraging collaboration through regular knowledge-sharing sessions or workshops ensures that team members are well-versed in each other's expertise, creating a knowledge reservoir that facilitates smooth transitions between projects.

Lastly, adopting a mindset of continuous learning and professional development is key. Encouraging team members to stay updated on industry trends, attend relevant training sessions, and pursue certifications keeps their skills current and adaptable. This proactive approach not only equips individuals with the latest knowledge but also cultivates a team culture that embraces change and welcomes new challenges. In the fast-paced realm of technology, a team with well-rounded skills and a commitment to continuous learning is better prepared to navigate shifts in project requirements or explore new opportunities.

These products that can serve as knowledge-sharing platforms, facilitating skill preparation and collaboration within teams:

Confluence by Atlassian:

Confluence is a collaboration and documentation tool that allows teams to create, share, and collaborate on content. It includes features such as wikis, pages, and spaces, making it easy for team members to document insights, best practices, and lessons learned. The platform integrates seamlessly with other Atlassian products, fostering a comprehensive collaborative environment.

Microsoft SharePoint:

SharePoint is a web-based collaboration platform that integrates with Microsoft 365. It offers features like document libraries, wikis, and discussion forums, providing a centralised space for team members to share knowledge. SharePoint allows for the creation of intranet sites, enabling teams to organise and collaborate on information relevant to their projects.

Notion:

Notion is an all-in-one workspace that combines notetaking, collaboration, and knowledge-sharing features. Teams can create wikis, databases, and documentation within a shared space. Notion's flexibility allows for the creation of a customised knowledge base, making it a versatile platform for sharing insights and best practices.

Trello:

Trello is a visual project management tool that uses boards, lists, and cards to organise tasks. While primarily a project management tool, Trello's flexibility allows teams to create boards with documentation, share insights, and maintain collaborative spaces. It's particularly useful for teams that prefer a visual and intuitive approach to knowledge sharing.

Slack:

Slack is a popular team collaboration platform that facilitates real-time communication through channels. While known for its messaging features, Slack also supports integrations with various apps, including document-sharing tools and wikis. Teams can use Slack to share information, discuss best practices, and maintain an organised archive of conversations.

Technology Lifecycle Management:

Technologies have lifecycles and managing them strategically is essential for resource optimisation. A technology roadmap should incorporate plans for the assessment, adoption, and retirement of technologies to enhance overall efficiency.

Technology Lifecycle Management (TLM) is a strategic framework that acknowledges the dynamic nature of technologies and products, recognising that they undergo distinct lifecycles from introduction to eventual retirement. Just as living organisms progress through stages of birth, growth, maturity, and decline, technologies follow a similar cycle. TLM involves the systematic planning, assessment, adoption, and retirement of technologies to optimise resource utilisation and enhance overall efficiency within an organisation.

The process begins with the assessment of emerging technologies, evaluating their potential impact and relevance to the organisation's goals. This anticipatory phase allows decision-makers to align technology adoption with the overall technology roadmap. By introducing new technologies strategically, organisations can leverage cutting-edge tools and methodologies, ensuring that resources are invested when the technologies are at their peak relevance and effectiveness.

Equally crucial in the TLM process is the retirement of outdated or redundant technologies. All technologies eventually reach a point where their value diminishes due to factors like advancements in the industry, changes in business requirements, or the emergence of more efficient alternatives. Integrating retirement plans into the technology roadmap enables organisations to proactively phase out obsolete technologies, preventing the accumulation of technical debt and associated inefficiencies.

By aligning technology adoption and retirement with the overall roadmap, organisations establish a proactive lifecycle management approach. This strategic perspective ensures that resources are consistently directed towards technologies that align with the organisation's goals while mitigating the risks and inefficiencies associated with relying on outdated or redundant technologies. Technology Lifecycle Management, when integrated into the broader organisational strategy, becomes a powerful tool for resource optimisation and maintaining a competitive edge in the ever-evolving technology landscape.

These tools and online resources may be valuable for preparation for Technology Lifecycle Management (TLM):

Gartner Technology Adoption Curve:

Gartner provides a comprehensive Technology Adoption Curve that illustrates the lifecycle stages of emerging technologies. This resource can help in understanding where different technologies stand in their maturity and adoption, aiding in strategic planning.

ITIL (Information Technology Infrastructure Library):

ITIL offers a set of practices for IT service management, including a framework for lifecycle management. It provides guidance on how to align IT services with business needs and effectively manage the entire lifecycle of IT services and technologies.

NIST Cybersecurity Framework:

The National Institute of Standards and Technology (NIST) Cybersecurity Framework includes guidelines for managing the lifecycle of cybersecurity technologies. It provides a structured approach to identify, protect, detect, respond to, and recover from cyber threats.

TechCrunch and Wired:

Stay informed about the latest technology trends and advancements by following reputable technology news sources like TechCrunch and Wired. These platforms offer insights into emerging technologies and can aid in keeping abreast of industry developments.

Enterprise Architecture Tools (e.g., Sparx Systems Enterprise Architect):

Enterprise Architecture tools assist in visualising and managing the structure and operation of an organisation. These tools can aid in creating holistic views of technology landscapes, supporting TLM initiatives.

TOGAF (The Open Group Architecture Framework):

TOGAF is an enterprise architecture methodology and framework that can guide organisations in managing the lifecycle of their technology architecture. It offers a structured approach to align business goals with technology strategies.

Technology Evaluation Centres (TEC):

TEC provides research and evaluation tools for comparing various enterprise software solutions. This can be valuable when assessing the adoption or replacement of specific technologies in the TLM process.

Utilising a combination of these tools and resources can enhance the effectiveness of Technology Lifecycle Management by providing strategic insights, practical frameworks, and up-to-date information on emerging technologies and industry trends.

Enhancing resource allocation and efficiency is a cornerstone of successful technology roadmap creation. By strategically aligning initiatives with business goals, prioritising for maximum impact, optimising budget utilisation, fostering agility, skillfully utilising the workforce, managing technology lifecycles, organisations can cultivate a culture of operational excellence that propels them towards technological success.

10 Step Action Plan for Enhancing Resource Efficiency

Efficient resource allocation is pivotal in crafting a technology roadmap that aligns with business goals and maximises impact. This practical 10-step plan outlines actionable strategies for enhancing resource allocation and efficiency within the strategic imperative of creating a technology roadmap:

1. Establish Clear Business-Technology Alignment:

- **Action:** Conduct a thorough analysis of business goals and objectives.
- **Implementation:** Ensure that every technological initiative outlined in the roadmap is aligned with specific business outcomes.
- **Example:** If the business aims for cost reduction, prioritise initiatives that enhance operational efficiency or streamline processes.

2. Conduct Comprehensive Prioritisation:

- **Action:** Evaluate the potential impact and dependencies of each initiative.
- **Implementation:** Prioritise projects based on strategic importance, potential business impact, and dependencies on other initiatives.
- **Example:** Initiatives critical to revenue generation or customer satisfaction should be prioritised over secondary projects.

3. Implement Budget-Focused Resource Allocation:

- **Action:** Assess available budget constraints and financial resources.
- **Implementation:** Align resource allocation with the budget, ensuring optimal utilisation of funds.
- **Example:** Allocate resources judiciously, avoiding unnecessary expenditure on non-strategic or low-impact projects.

4. Foster an Agile and Adaptive Approach:

- **Action:** Acknowledge the dynamic nature of technology landscapes.
- **Implementation:** Build flexibility into resource allocation plans, allowing for quick adaptation to changing circumstances or emerging opportunities.
- **Example:** Allocate resources in short sprints, reassessing priorities and adjusting allocations as needed.

5. Skilful Workforce Alignment:

- **Action:** Assess the skill sets within the technology team.
- **Implementation:** Align skilled personnel with projects that match their expertise, ensuring optimal workforce utilisation.
- **Example:** Assign data analytics projects to team members with a strong background in data science.

6. Integrate Technology Lifecycle Management:

- **Action:** Review existing technologies and their lifecycles.
- **Implementation:** Develop a strategy for the assessment, adoption, and retirement of technologies within the roadmap.
- **Example:** Regularly assess the relevance of existing technologies, plan for seamless adoption of new tools, and phase out outdated systems.

7. Implement Robust Performance Monitoring:

- **Action:** Define key performance indicators (KPIs) for each initiative.
- **Implementation:** Regularly monitor and assess the performance of ongoing projects against established KPIs.
- **Example:** Track KPIs such as project timelines, budget adherence, and user satisfaction to gauge performance.

8. Develop Comprehensive Risk Mitigation Strategies:

- **Action:** Identify potential risks associated with each initiative.
- **Implementation:** Develop contingency plans and mitigation strategies to address identified risks.
- **Example:** If a key team member is at risk of leaving, have a succession plan in place to ensure project continuity.

9. Foster Stakeholder Engagement:

- **Action:** Identify key stakeholders involved in or affected by the technology roadmap.
- **Implementation:** Establish regular communication channels to keep stakeholders informed and engaged.
- **Example:** Hold quarterly briefings or town hall meetings to update stakeholders on progress, challenges, and upcoming initiatives.

10. Implement Continuous Evaluation and Feedback Loops:

- **Action:** Set up mechanisms for continuous evaluation.
- **Implementation:** Establish regular feedback loops, incorporating insights from team members, stakeholders, and external factors.
- **Example:** Conduct monthly retrospectives to gather feedback, assess project progress, and identify areas for improvement.

By aligning initiatives with business goals, prioritising strategically, managing budgets effectively, fostering agility, aligning the workforce skilfully, integrating technology lifecycle management, implementing performance monitoring, developing risk mitigation strategies, engaging stakeholders, and maintaining continuous evaluation, organisations can optimise resource utilisation and set the course for technological success.

4.5 Iterative Development & Agile Principles

Also review chapter 7.3

The adoption of iterative development and Agile principles serves as a transformative strategy for CTOs endeavouring to craft resilient and adaptive technology roadmaps. This section delves into the nuanced importance of embracing an iterative and Agile approach in the roadmap creation process, offering a deeper exploration into key aspects and practical insights.

Adapting to Change and Uncertainty:

Adapting to change and uncertainty is a fundamental aspect of navigating the technology landscape. The tech industry is inherently volatile, marked by continuous shifts in technology trends, evolving user expectations, and dynamic market dynamics. To effectively navigate this uncertainty, organisations need to embrace iterative development methodologies and Agile principles. These methodologies empower teams to respond swiftly and adeptly to changes in project requirements, ensuring that the technology roadmap remains flexible and responsive.

The iterative development process, integral to Agile methodologies, promotes incremental and continuous advancements. By breaking down larger projects into smaller, manageable components, organisations can respond to evolving circumstances with greater ease. Agile principles encourage collaboration, adaptability, and a customer-centric approach, fostering an environment where teams can swiftly pivot in response to emerging opportunities or challenges. This iterative and Agile mindset allows organisations to embrace change as a constant, adapting their technology strategies to align with evolving industry landscapes.

In the face of uncertainty, organisations that adopt an Agile approach benefit from enhanced resilience. The iterative nature of Agile development allows for ongoing feedback loops and adjustments, enabling teams to course-correct as needed. By fostering a culture that values adaptability and continuous improvement, organisations can proactively navigate uncertainties, ensuring that their technology roadmap not only endures change but thrives in the midst of it. This adaptability becomes a strategic advantage, positioning organisations to seize emerging opportunities and deliver technology solutions that remain aligned with evolving needs and expectations.
Example: Suppose a fintech company is developing a new payment feature. An Agile approach allows the team to pivot swiftly if market trends or regulatory requirements change during the development process.

Agile Ceremonies: Embrace Agile ceremonies such as Sprint Planning, where the team collaboratively plans the upcoming work, and Sprint Review, where stakeholders provide feedback on the delivered increment.

Online Resources: Explore websites like *__Agile Alliance__* and *__Scrum.org__* for in-depth information on Agile ceremonies and best practices.

Feedback Loops:

Iterative development fosters continuous feedback loops between stakeholders, end-users, and development teams. Regular feedback ensures that the evolving technology roadmap remains aligned with business goals and user needs, enhancing overall project success.

Example: Through Agile ceremonies like Sprint Retrospectives, teams reflect on their processes, identify improvements, and continuously refine their approach based on feedback.

Agile Ceremonies: Leverage Sprint Retrospectives to gather insights on team performance and identify areas for improvement.

Online Resources: Platforms like *__Atlassian Agile Coach__* and *__Scrum Alliance__* offer resources on fostering feedback loops within Agile teams.

Prioritisation of Value:

Agile principles emphasise delivering high-value features early and consistently throughout the development process. This ensures that the technology roadmap prioritises elements that bring the most significant value to the business and end-users.

By giving precedence to high-value features, Agile development ensures that the most critical functionalities are integrated early in the development process. This approach recognises that not all features hold equal significance for the business and end-users. Instead, it directs resources towards those functionalities that offer the greatest value, whether in terms of revenue generation, customer satisfaction, or strategic business objectives.

The iterative nature of Agile development facilitates a continuous feedback loop, allowing stakeholders to assess and refine high-value features throughout the development lifecycle. This adaptability ensures that the technology roadmap remains responsive to evolving business needs and user expectations. Furthermore, by consistently delivering valuable features, Agile methodologies foster a dynamic development environment, enabling organisations to swiftly respond to market changes and competitive dynamics. This iterative and customer-focused approach is pivotal in building products that align closely with user requirements and business goals.

Prioritising high-value features in an Agile framework not only accelerates time-to-market but also enhances overall project success. By addressing critical functionalities early in the development cycle, organisations can secure faster returns on investment and garner quicker user adoption. This strategic alignment with business priorities establishes a foundation for sustained value delivery, allowing the technology roadmap to evolve iteratively, guided by ongoing insights and feedback. As a result, Agile principles contribute to a more responsive, adaptable, and customer-centric development approach, ensuring that the delivered features are not just timely but impactful for the organisation and its end-users.

Example: In an e-commerce platform development, Agile prioritisation may focus on implementing essential checkout or incomplete basket functionalities as opposed to other UX requirements as this will impact commercial revenue.

Agile Ceremonies: Utilise Backlog Refinement sessions to collaboratively prioritise backlog items based on their business value.

Online Resources: Platforms like ***Scrum Inc.*** and ***Scaled Agile*** provide insights into Agile prioritisation techniques.

Rapid Time-to-Market:

Iterative development and Agile methodologies play a pivotal role in expediting the time-to-market for technology solutions. By adopting these approaches, organisations embrace a dynamic and flexible development process that prioritises delivering functional increments in short cycles. This iterative approach allows teams to release valuable features incrementally and regularly, ensuring that tangible progress is consistently made. Consequently, organisations can respond promptly to changing market demands and user needs, positioning themselves to gain a competitive edge in fast-paced and dynamic markets.

The iterative development model's emphasis on regular releases facilitates a continuous feedback loop between development teams and end-users. This ongoing interaction enables organisations to receive real-time input on the evolving product, allowing for quick adjustments and refinements. Agile methodologies, with their focus on collaboration, transparency, and adaptability, create an environment where teams can swiftly incorporate feedback, address emerging challenges, and enhance features. This iterative feedback loop not only ensures the relevance and alignment of the technology solution with market demands but also accelerates the overall development process.

Furthermore, the incremental delivery of features in short cycles enables organisations to establish an early presence in the market with a minimum viable product (MVP). This early market entry provides a tangible product for users to engage with, creating opportunities for organisations to gather insights, build user loyalty, and capture market share ahead of competitors. The ability to quickly adapt and release improvements based on user feedback becomes a strategic advantage, allowing organisations to stay ahead in the competitive landscape and deliver technology solutions that resonate with evolving customer expectations.

Example: Continuous delivery practices in Agile enable software updates and releases to occur more frequently, responding to market demands swiftly.
Agile Ceremonies: Embrace Sprint Reviews to showcase completed increments, ensuring that valuable features are delivered regularly.
Online Resources: Platforms like ***Agile Testing Alliance*** and ***Agile Testing Days*** provide insights into Agile delivery practices.

Improved Team Collaboration and Enhanced Morale:

Adopting Agile principles proves instrumental in fostering improved team collaboration and enhancing employee morale. Agile principles actively encourage collaboration among cross-functional teams, breaking down silos and promoting collective ownership. This collaborative culture ensures that the technology roadmap is shaped by diverse perspectives, ultimately leading to more innovative and comprehensive solutions.

For instance, daily stand-up meetings in Agile provide a forum for team members to share progress, discuss challenges, and align on goals, promoting collaborative problem-solving. To further enhance team communication, collaboration, and alignment, it is recommended to implement regular Daily Stand-ups as part of Agile ceremonies.

Additionally, Agile's iterative development practices contribute significantly to creating a positive work environment. These practices provide teams with a sense of accomplishment at the end of each iteration, fostering continual progress that, in turn, enhances employee morale and motivation. Celebrating small victories at the end of each sprint, a common Agile practice, contributes to a culture of achievement and continuous improvement.

To embrace and formalise these positive practices, incorporating Agile ceremonies such as Sprint Retrospectives is recommended. These ceremonies provide a structured opportunity to celebrate achievements, reflect on improvements, and boost overall team morale.

For further insights into fostering collaboration within Agile teams and cultivating a positive Agile culture, online resources like the Agile Coaching Institute, Agile Project Management, Agile For All, and The Agile Manifesto can be explored. These platforms offer valuable guidance and strategies to promote a collaborative and morale-boosting environment within Agile teams.

Examples: Daily stand-up meetings in Agile provide a forum for team members to share progress, discuss challenges, and align on goals, promoting collaborative problem-solving. The Agile practice of celebrating small victories at the end of each sprint contributes to a culture of achievement and continuous improvement.

Agile Ceremonies: Implement regular Daily Stand-ups to enhance team communication, collaboration, and alignment. Embrace Sprint Retrospectives to celebrate achievements, reflect on improvements, and boost team morale.

Online Resources: Explore _**Agile Coaching Institute**_ and _**Agile Project Management**_ for insights into fostering collaboration within Agile teams. Platforms like _**Agile For All**_ and _**The Agile Manifesto**_ provide insights into fostering a positive Agile culture.

Data-Driven Decision Making:

Iterative development allows for data-driven decision-making by incorporating metrics and feedback into the development process. Data-driven insights help CTOs refine the roadmap based on the actual performance of features and user engagement.

This approach seamlessly integrates metrics and user feedback into the development process, offering valuable insights that empower CTOs to refine the roadmap based on the actual performance of features and user engagement. Rather than relying solely on theoretical projections, CTOs can leverage real-time data obtained at each phase of the development process to make informed decisions.

By adopting an iterative development model, organisations gain the ability to collect empirical data on user interactions and system performance with every deliverable. Key performance indicators (KPIs) and user feedback mechanisms are strategically incorporated, allowing CTOs to assess the success of specific features and identify areas for improvement. This data-driven approach enables CTOs to make agile adjustments to the technology roadmap, ensuring it remains responsive to evolving user needs and market trends.

The adaptability inherent in iterative development facilitates swift decision-making based on concrete insights, allowing CTOs to make informed adjustments without the need for a continuous feedback loop. This dynamic approach ensures that the technology roadmap aligns effectively with user expectations and strategic objectives. By grounding decisions in empirical data, CTOs can enhance the efficiency of the development process, contributing to the overall success of technology initiatives and the organisation's strategic goals.

Example: A/B testing and analytics tools enable teams to gather data on user interactions, informing decisions about feature improvements or adjustments in the technology roadmap.

Agile Ceremonies: Integrate data reviews into Sprint Reviews, leveraging metrics to guide future iterations and strategic decisions.

Online Resources: Platforms like **_Agile Metrics_** and **_Lean Agile Intelligence_** offer resources on Agile metrics and data-driven decision-making.

These tools may also be of interest.

1. **UsabilityHub:** This platform enables quick and easy user testing. You can gather feedback on designs, prototypes, or interfaces to make informed decisions based on user preferences.
2. **New Relic:** A performance monitoring tool that provides insights into the performance of web applications. It helps in identifying and resolving issues that may impact user experience.
3. **Optimizely:** A platform for experimentation and A/B testing. It enables you to test different versions of your website or app to determine which performs better based on user engagement metrics.
4. **Qualtrics:** A comprehensive experience management platform that allows you to collect and analyse feedback from various sources, including surveys, social media, and direct customer interactions.
5. **Crazy Egg:** This tool provides visualisations like heatmaps and scrollmaps to help you understand how users interact with your website. It aids in optimising page layouts and content based on user behaviour.

The incorporation of iterative development and Agile principles into the creation of a technology roadmap is not just a strategic choice; it's recognised as a good approach and currently best practice. This approach ensures adaptability, aligns with user expectations, fosters collaboration, and empowers teams to deliver value consistently. By leveraging Agile methodologies and iterative practices, organisations can navigate the complexities of the technological landscape with agility and resilience. For in-depth exploration and guidance, the mentioned online resources serve as valuable references on Agile methodologies, ceremonies, and best practices.

Benefits of Data Driven decision Making

Continual Growth

Valuable Insights

Improved Program Outcomes

Prediction of future trends

Actionable Insights

Optimised Operations

A 10-Step Action Plan for Introducing Agile Principles

Creating a technology roadmap infused with iterative development and Agile principles requires a structured and practical approach. This 10-step plan outlines actionable strategies for CTOs aiming to leverage these methodologies, embracing adaptability and fostering innovation.

1. Initiate Agile Training Sessions:

- **Implementation:** Kickstart the journey by organising Agile training sessions for the technology teams. Ensure a comprehensive understanding of Agile principles, ceremonies, and collaborative practices.
- **Example:** Engage external Agile coaches or trainers to conduct workshops on Scrum, Kanban, and Agile best practices.

2. Assess Current Development Processes:

- **Implementation:** Conduct a thorough assessment of existing development processes to identify areas that can benefit from iterative development and Agile methodologies.
- **Example:** Use retrospectives or surveys to gather feedback from development teams on existing workflows and potential pain points.

3. Establish Agile Ceremonies:

- **Implementation:** Introduce key Agile ceremonies such as Sprint Planning, Daily Stand-ups, Sprint Reviews, and Retrospectives. Clearly communicate the purpose and expectations for each ceremony.
- **Example:** Begin with a well-structured Sprint Planning session to set the agenda for the upcoming development cycle.

4. Integrate Feedback Loops:

- **Implementation:** Create mechanisms for continuous feedback loops. Implement tools and practices that enable stakeholders, users, and team members to provide feedback regularly.
- **Example:** Introduce user feedback sessions during Sprint Reviews and encourage open discussions on improvements.

5. Incorporate Risk Management Practices:

- **Implementation:** Integrate risk management discussions into Agile ceremonies, particularly Sprint Planning. Identify potential risks and collaboratively strategise risk mitigation.
- **Example:** Use risk matrices to assess the likelihood and impact of potential risks, developing proactive mitigation plans.

6. Prioritise Value-Based Development:

- **Implementation:** Collaboratively prioritise backlog items based on their business value. Emphasise the importance of delivering high-value features early in the development process.
- **Example:** Implement story mapping sessions to visualise and prioritise features according to user and business value.

7. Implement Continuous Delivery Practices:

- **Implementation:** Embrace continuous delivery practices to accelerate time-to-market. Foster a culture of delivering functional increments in short cycles.
- **Example:** Utilise automated testing and deployment pipelines to streamline the release process and reduce time-to-market.

8. Foster Cross-Functional Collaboration:

- **Implementation:** Encourage cross-functional collaboration by breaking down silos between development, testing, and other relevant teams. Promote a shared sense of ownership.
- **Example:** Implement cross-functional teams where members from different departments work together on specific projects.

9. Integrate Data-Driven Decision-Making:

- **Implementation:** Implement data-driven decision-making by incorporating metrics and analytics into the development process. Leverage insights to guide future iterations and strategic decisions.
- **Example:** Integrate A/B testing tools to gather real-time data on user interactions and preferences.

10. Utilise Agile Tools for Visibility:

- **Implementation:** Introduce Agile tools such as Jira, Trello, or Kanban boards to visualise work progress. Leverage metrics like velocity and burn-down charts for insights into team performance.
- **Example:** Implement Jira to manage user stories, sprints, and track progress visually, enhancing communication and planning.

This 10-step practical plan provides a structured approach for CTOs to infuse iterative development and Agile principles into the creation of a dynamic technology roadmap. By embracing these methodologies, organisations can navigate the complexities of the technological landscape with adaptability, collaboration, and a relentless focus on delivering value.

Summary of Agile Tools: Jira, Trello, and Kanban Boards

Agile tools play a crucial role in facilitating and managing Agile methodologies, providing teams with the necessary frameworks and features to enhance collaboration, visibility, and efficiency. Here is a brief summary of three popular Agile tools: Jira, Trello, and Kanban boards.

1. Jira:

Overview:
- **Type:** Project and issue tracking tool.
- **Key Features:**
  - **Scrum and Kanban Boards:** Jira supports both Scrum and Kanban boards, providing flexibility for different Agile methodologies.
  - **Customisable Workflows:** Users can tailor workflows to match their specific development processes.
  - **Advanced Reporting:** Jira offers robust reporting and analytics tools for tracking team performance and project progress.
  - **Integration Capabilities:** Integrates seamlessly with various third-party tools and plugins.

Use Cases:
- **Project Management:** Jira is widely used for planning, tracking, and releasing software.
- **Issue Tracking:** It allows teams to create, assign, and prioritise tasks or issues.
- **Sprint Planning:** Effective for Scrum teams in planning and executing sprints.

2. Trello:

Overview:
- **Type:** Visual project management tool.
- **Key Features:**
  - **Boards, Lists, and Cards:** Trello organises projects into boards, lists, and cards for visual task management.
  - **Collaboration:** Supports team collaboration through comments, attachments, and real-time updates.
  - **Customisable:** Users can adapt Trello to fit their preferred workflow.
  - **Integration Options:** Integrates with various applications like Slack, Google Drive, and more.

Use Cases:
- **Task Management:** Trello is ideal for tracking individual tasks within a project.
- **Project Planning:** Provides a visual overview of project progress.
- **Team Collaboration:** Enhances communication and collaboration among team members.

3. Kanban Boards:

Overview:
- **Type:** Visual project management method.
- **Key Features:**
  - **Visual Workflow:** Kanban boards use cards and columns to represent tasks and their progress through different stages.
  - **Work-in-Progress (WIP) Limits:** Helps teams manage and control the number of tasks in progress simultaneously.
  - **Continuous Improvement:** Encourages a focus on process improvement based on real-time feedback.
  - **Flexibility:** Adaptable to various project management styles and industries.

Use Cases:
- **Workflow Visualisation:** Ideal for visualising work and optimising processes.
- **Efficiency Improvement:** Helps teams identify bottlenecks and optimise their workflow.
- **Limiting Work-in-Progress:** Encourages a steady flow of tasks and prevents overloading teams.

Each of these Agile tools brings unique strengths to project management and development processes. The choice between Jira, Trello, Kanban boards, or a combination depends on the specific needs and preferences of the team. These tools contribute to creating an Agile and collaborative environment, promoting transparency, and ultimately enhancing the success of Agile projects.

4.6 Communicating the Roadmap Effectively:

Creating a technology roadmap is a visionary exercise, but its success lies not just in the formulation but in the effective communication of the roadmap's intricacies, objectives, and expected outcomes. The ability to convey this roadmap with clarity and impact is crucial for gaining buy-in, alignment, and support from all stakeholders.

Transparency and Trust:

Transparent communication plays a pivotal role in establishing trust within an organisation. Articulating the rationale behind strategic decisions and offering a transparent view of the roadmap fosters confidence among stakeholders. For instance, providing stakeholders with access to a detailed and regularly updated online dashboard that visually represents the progress of the roadmap, along with key milestones, enhances transparency. This approach ensures that stakeholders are well-informed and have a clear understanding of the organisation's strategic direction, reinforcing trust in the decision-making processes.

Customer and Stakeholder Confidence:
Effective communication is paramount in instilling confidence not only within the internal team but also among customers and external stakeholders. When these external entities are well-informed about the technology roadmap, they gain confidence in the organisation's capability to meet their needs. For instance, sharing a simplified version of the roadmap with key clients during strategic meetings can go a long way in demonstrating the organisation's commitment to understanding and addressing their specific needs and concerns. This transparent approach builds a strong foundation of trust and confidence, enhancing the overall relationship between the organisation and its external stakeholders.

Enhancing Organisational Credibility:
Beyond transparency and customer confidence, clear communication also contributes to the overall credibility of the organisation. By openly sharing the reasoning behind strategic decisions and providing stakeholders with a clear view of the roadmap, the organisation establishes itself as trustworthy and capable. This credibility extends to various facets, including partnerships, collaborations, and industry reputation. Effectively communicating the strategic vision through transparent channels reinforces the organisation's position as a reliable and forward-thinking entity, strengthening its standing within the industry and among stakeholders.

Metrics for Progress Evaluation:

Effective communication establishes key metrics for evaluating progress. Stakeholders can track milestones and key performance indicators (KPIs) to gauge the success of the roadmap.
Example: Implementing data visualisation tools like **_Tableau_** to create interactive dashboards allows stakeholders to monitor progress and evaluate the impact of the roadmap.

C-Suite Communication:

Effectively communicating the Technology Roadmap to the C-suite is a crucial step in ensuring alignment with organisational goals and garnering support from top-level executives. The process involves a strategic approach to presenting intricate technical information in a clear and business-oriented manner.

The communication process typically begins with an executive summary that provides a high-level overview of the Technology Roadmap. This summary should succinctly highlight key objectives, anticipated benefits, and potential risks. Using non-technical language is crucial at this stage to ensure that executives from diverse backgrounds can grasp the strategic significance of the roadmap. Visual aids such as infographics, charts, and timelines can enhance the clarity of the message.

Following the executive summary, a detailed presentation delves into the specific components of the Technology Roadmap. This includes outlining the timeline for major milestones, key technological initiatives, and the expected impact on the organisation's overall strategy. Utilising visual tools like Gantt charts or interactive presentations can aid in breaking down complex technical details into digestible segments. Emphasising how the roadmap aligns with broader business objectives and contributes to the company's competitive advantage is crucial for executive buy-in.

In addition to formal presentations, interactive sessions and workshops can be organised to facilitate discussions and address any questions or concerns from the C-suite. This allows executives to actively engage with the roadmap, provide input, and ensure that their perspectives are considered in the strategic planning process. Collaboration tools such as virtual whiteboards or collaborative platforms can be employed during these sessions to encourage real-time participation and feedback.

To enhance ongoing communication, regular updates and progress reports are vital.

Implementing a structured reporting mechanism, such as monthly or quarterly briefings, ensures that the C-suite remains informed about the roadmap's implementation and any adjustments made based on evolving circumstances. These updates should focus on key performance indicators, achievements, and challenges, providing a comprehensive picture of the roadmap's impact on the organisation. Implementing a dedicated communication platform, possibly utilising project management or collaborative tools, can facilitate seamless information sharing and iterative loops between the technology team and the C-suite.

Example: Provide a concise executive summary of the roadmap that highlights key milestones, anticipated outcomes, and the alignment of technology initiatives with business growth.

Communicating the technology roadmap effectively is not just a means of informing stakeholders; it's a strategic imperative for aligning, motivating, and inspiring teams while gaining trust and support from all levels of the organisation. Utilising examples and data sources enhances the clarity and impact of the communication, making the roadmap a dynamic and shared vision for success.

Communicating the Roadmap: A 10-Step Action Plan

Effectively communicating the technology roadmap is essential for gaining support and alignment from various stakeholders. This 10-step practical plan ensures a clear and impactful communication strategy:

1. **Define Your Audience:**

   - **Action:** Identify all stakeholders, including internal teams, executives, clients, and external partners. Tailor your communication approach based on their level of technical understanding and specific interests.

2. **Craft a Clear Message:**

   - **Action:** Develop a concise and compelling narrative that highlights the purpose, key milestones, and benefits of the roadmap. Avoid jargon and focus on the broader impact on the organisation.

3. **Utilise Visual Aids:**

   - **Action:** Create visual representations of the roadmap using charts, graphs, or timelines. Visual aids enhance understanding and retention, especially for stakeholders with varying levels of technical expertise.
   - **Example:** Use tools like *Microsoft PowerPoint, Lucidchart,* or specialised roadmap software to create visually engaging presentations.

4. **Host Interactive Workshops:**

   - **Action:** Organise workshops where stakeholders can actively participate in discussions, ask questions, and provide input. This fosters engagement and ensures a shared understanding.
   - **Example:** Conduct virtual or in-person workshops using collaboration tools like *Zoom* or *Microsoft Teams*.

5. **Establish a Communication Cadence:**

   - **Action:** Define a regular communication schedule to provide updates on the roadmap's progress, changes, and achievements. Consistency builds trust and keeps stakeholders informed.
   - **Example:** Set up monthly or quarterly town hall meetings, webinars, or newsletters for comprehensive updates.

6. Tailor Messages for Different Audiences:

- **Action:** Customise communication materials for different stakeholders. Emphasise specific aspects that resonate with each group, addressing their unique concerns and interests.
- **Example:** Highlight cost efficiencies and business impact when communicating with executives, and technical details for development teams.

7. Leverage Collaboration Platforms:

- **Action:** Use collaboration tools like Slack, Microsoft Teams, or project management platforms to share real-time updates, engage in discussions, and gather feedback.
- **Example:** Create dedicated channels for roadmap discussions, ensuring easy access and participation from relevant stakeholders.

8. Provide Accessible Documentation:

- **Action:** Develop comprehensive documentation that details the roadmap, its objectives, and key milestones. Make these documents easily accessible to all stakeholders.
- **Example:** Use platforms like *Confluence* or *Google Docs* for creating and sharing detailed documentation.

9. Facilitate Q&A Sessions:

- **Action:** Host regular Q&A sessions where stakeholders can ask questions, seek clarifications, and express concerns. This promotes an open and transparent communication culture.
- **Example:** Use video conferencing tools for virtual Q&A sessions or dedicate time during team meetings for questions.

10. Solicit Feedback Actively:

- **Action:** Encourage stakeholders to provide feedback on the roadmap. Actively seek input, suggestions, and concerns to make adjustments and ensure continuous improvement.
- **Example:** Deploy surveys, feedback forms, or conduct one-on-one sessions to gather valuable insights.

By following this 10-step practical plan, you'll create a communication strategy that engages all stakeholders effectively, ensuring a shared understanding and commitment to the technology roadmap.

4.7 Monitoring and Key Performance Indicators (KPIs) in Creating a Technology Roadmap

The creation of a robust Technology Roadmap is not a one-time endeavour but a continuous journey that demands vigilant monitoring and assessment. The integration of Key Performance Indicators (KPIs) is instrumental in providing a quantitative measure of the roadmap's success and aligning technological initiatives with organisational goals.

Establishing Relevant KPIs:

Begin by defining KPIs that directly align with the objectives outlined in the Technology Roadmap. For instance, if one of the roadmap goals is to enhance cybersecurity, a relevant KPI could be the percentage reduction in security incidents over time.
Example: *KPI:* Cybersecurity Incident Reduction Rate
Data Source: Security incident logs and reports from the cybersecurity team.

Utilising Data-Driven KPI's:

We covered data driven decision making in the previous section, but we can additionally utilise data to help create the KPI's to track progress and make informed decisions throughout the implementation of the Technology Roadmap. Regularly analyse KPI data to identify trends, areas for improvement, and successes.

To create effective Key Performance Indicators (KPIs) for tracking progress during the implementation of the Technology Roadmap, organisations should begin by leveraging available data. Conduct a thorough analysis of historical performance data, user engagement metrics, and relevant industry benchmarks. By understanding past trends and performance indicators, organisations can identify the most relevant and impactful KPIs aligned with the strategic goals outlined in the roadmap.

Strategic KPI Selection:
Once the data has been analysed, the next step involves selecting KPIs that directly align with the objectives outlined in the Technology Roadmap. These KPIs should be specific, measurable, achievable, relevant, and time-bound (SMART). For example, if a roadmap objective is to enhance customer satisfaction, KPIs may include customer retention rates, Net Promoter Score (NPS), or response times to customer queries. By grounding KPIs in data-driven insights, organisations ensure that they are tracking metrics that truly reflect progress and success in line with the roadmap.

Continuous Monitoring and Analysis:

The implementation of the Technology Roadmap is an ongoing process, and regular monitoring and analysis of KPI data are imperative. Establish a routine for collecting and reviewing KPI data at predefined intervals. Identify trends, anomalies, and areas for improvement. Continuous analysis allows organisations to adapt their strategies in real-time, making informed decisions to address challenges or capitalise on emerging opportunities. Regular KPI reviews ensure that the organisation remains agile and responsive throughout the roadmap implementation.

Driving Informed Decision-Making:

Data-driven KPIs serve as a compass for informed decision-making. Through consistent analysis, organisations gain insights into the effectiveness of implemented strategies and can make timely adjustments. For instance, if KPIs indicate a decline in user engagement after the introduction of a new feature, the organisation can investigate the cause and implement corrective measures swiftly. By integrating data-driven decision-making into the roadmap implementation process, organisations enhance their ability to navigate challenges and optimise outcomes, ultimately contributing to the success of the technology initiatives outlined in the roadmap.

Examples of Data Driven KPI's for the Technology Roadmap
User Adoption Rate:

Percentage of users or stakeholders who have adopted new technologies or features outlined in the Technology Roadmap. Indicates how well the organisation is succeeding in getting users to embrace and use the new technologies, providing insights into the roadmap's impact.

Technology Integration Efficiency:

Measures the time and resources required to integrate new technologies into existing systems or processes. Highlights the efficiency of the integration process, ensuring that the roadmap's implementation aligns with planned timelines and resource allocations.

System Downtime and Stability:

Tracks the amount of downtime experienced due to technology implementations or updates outlined in the roadmap. Provides insights into the reliability and stability of the systems during the implementation phase, ensuring minimal disruptions to operations.

Cost of Technology Implementation:

Calculates the actual costs incurred during the implementation of technologies compared to the budgeted costs in the roadmap. Helps in evaluating the financial efficiency of the roadmap, ensuring that cost projections align with the actual expenses incurred.

Customer Satisfaction Score (CSAT):
Measures customer satisfaction with the implemented technologies or changes outlined in the roadmap. Gauges how well the roadmap aligns with customer expectations, ensuring that technological advancements resonate positively with end-users.

Return on Investment (ROI) from Technological Initiatives:
Evaluates the financial returns generated by the implemented technologies in comparison to the initial investment outlined in the roadmap. Assesses the economic impact of the roadmap, ensuring that technological initiatives contribute positively to the organisation's bottom line.

Employee Productivity Metrics:
Tracks changes in employee productivity, efficiency, or task completion rates after the implementation of new technologies. Provides insights into how well the technological changes outlined in the roadmap positively influence workforce productivity.

Data Security and Compliance Metrics:
Monitors adherence to data security measures and compliance standards post-implementation of new technologies. Ensures that the roadmap aligns with regulatory requirements and maintains the security of sensitive information.

Innovation Velocity:
Measures the speed at which new technologies or innovative features are introduced as per the roadmap. Reflects the organisation's agility in adapting to technological advancements and staying ahead of the competition.

Scalability Metrics:
Assesses the scalability of the implemented technologies to accommodate growth and increased demands as outlined in the roadmap. Ensures that the roadmap anticipates and supports future organisational growth without major disruptions.

Measuring Operational Efficiency:

Include KPIs that measure the efficiency of operational processes impacted by the roadmap. This could involve tracking the time taken for key processes before and after the implementation of new technologies.

A practical approach involves monitoring the time taken for key processes before and after the implementation of new technologies. For example, if a roadmap involves integrating a new software system, KPIs can be established to observe the time required to complete specific tasks within that system. This may include measuring the time taken to process customer orders, handle inventory updates, or generate financial reports. Comparing these metrics before and after implementation provides a tangible measure of the impact on operational efficiency.

Another pertinent KPI is the reduction in error rates. Operational processes often encompass intricate workflows where errors can result in inefficiencies and disruptions. By monitoring error rates before and after the roadmap implementation, organisations can assess the effectiveness of the changes made. For instance, if a roadmap involves streamlining communication channels through a unified platform, a KPI could be the decrease in communication errors or misinterpretations. This not only measures efficiency but also reflects improvements in accuracy and overall process reliability.

Cost-effectiveness is a crucial aspect of operational efficiency, and KPIs related to resource utilisation can offer valuable insights. Organisations can track resource allocation metrics, such as staff hours, equipment usage, or material consumption, before and after the roadmap implementation. If the roadmap involves the introduction of automation or optimisation tools, the corresponding KPIs would indicate whether these changes have led to more efficient resource utilisation. This could include tracking the reduction in overtime hours, decrease in material waste, or improved equipment downtime. Such KPIs help organisations gauge the financial impact and return on investment resulting from the operational changes outlined in the strategic roadmap.

Examples of Operational efficiency KPI's

Mean Time to Resolve (MTTR): Measure the average time taken to resolve technical issues. A lower MTTR indicates quicker problem resolution and improved operational efficiency.

System Uptime: Track the percentage of time that critical systems and services are available and operational. Higher system uptime reflects better operational efficiency and reliability.

Deployment Frequency: Monitor how often new updates, features, or releases are deployed. A higher deployment frequency suggests a streamlined development and release process.

Automated Test Coverage: Assess the percentage of tests automated in the testing process. Higher automation coverage reduces manual effort, accelerates testing, and improves overall efficiency.

Incident Response Time: Evaluate the time it takes to respond to and address incidents or service disruptions. Faster incident response contributes to better operational efficiency.

Infrastructure Utilisation: Measure the efficiency of resource utilisation within the infrastructure. Optimising resource usage ensures cost-effectiveness and operational efficiency.

Energy Consumption per Transaction: Evaluate the energy consumption associated with each transaction or operation. Lower energy consumption per transaction indicates energy efficiency and cost savings.

Code Churn: Track the rate of change in the source code. Higher code churn may suggest inefficiencies, while lower churn indicates stable and efficient development processes.

Release Cycle Time: Assess the time taken from code commit to actual deployment. A shorter release cycle time signifies an agile and efficient development and release pipeline.

Customer Satisfaction Index (CSI): Gauge end-user satisfaction with technology solutions. A higher CSI indicates that the technology roadmap aligns with user needs, leading to enhanced operational efficiency.

Aligning with User Experience Metrics:

Incorporate KPIs related to user experience for technology-driven solutions. This ensures that improvements made align with the end-users' needs and expectations.

One effective approach is to employ quantitative metrics that provide a clear, numerical representation of UX elements. Metrics such as page load times, response times, and error rates can be tracked using tools like Google Analytics or specialised UX monitoring platforms. These quantitative indicators offer tangible insights into the efficiency and reliability of the technology, allowing teams to identify areas for improvement and gauge the impact of implemented changes.

In addition to quantitative metrics, qualitative data is equally important in understanding the user experience. Surveys, user interviews, and feedback forms can help capture subjective insights from users, providing valuable information about their perceptions, preferences, and pain points. Employing sentiment analysis tools on user feedback can help convert qualitative data into measurable KPIs, allowing for a more holistic evaluation of the user experience. This combination of quantitative and qualitative approaches ensures a comprehensive understanding of the technology's impact on users, enabling teams to prioritise improvements that address both objective performance metrics and subjective user sentiments.

A/B testing and usability studies can provide valuable insights into the effectiveness of specific changes, allowing teams to fine-tune their approach and deliver an optimal user experience over time.

Examples of User Experience KPI's

User Satisfaction Score (USS): Assessing overall user satisfaction through surveys or feedback forms can provide a quantitative metric, indicating how well the technology aligns with user expectations.

Task Success Rate (TSR): TSR evaluates the percentage of successfully completed tasks within the technology solution, helping to gauge its efficiency and user-friendliness.

Average Session Duration: The average time users spend interacting with the technology can indicate engagement levels and whether the solution meets their needs effectively.

Error Rate: Tracking the frequency and types of errors users encounter provides insights into the system's reliability and identifies areas for improvement in the user experience.

Conversion Rate: For technology solutions with specific conversion goals, tracking the conversion rate helps evaluate how well the design and functionality lead users to desired actions.

Onboarding Completion Rate: Assessing the percentage of users who successfully complete the onboarding process measures the effectiveness of the initial user experience and ease of adoption.

User Retention Rate: Retention rate measures the percentage of users who continue to engage with the technology over time, reflecting its ability to maintain user interest and satisfaction.

Loading Time (Page Load Speed): Monitoring the time it takes for pages or features to load ensures a smooth and responsive user experience, directly impacting user satisfaction.

User Engagement Metrics (e.g., clicks, scrolls): Analysing user engagement metrics provides insights into how users interact with different elements, helping to refine the user interface and overall experience.

Mobile Responsiveness: With an increasing number of users accessing technology on mobile devices, ensuring a seamless and responsive experience on various screen sizes is crucial for overall satisfaction.

Suggested Sentiment Analysis Tools

TextBlob: TextBlob is a Python library that offers simple and effective sentiment analysis capabilities. It's open-source and easy to use, making it a cost-effective choice for smaller projects or those on a budget.

MonkeyLearn: MonkeyLearn provides a user-friendly platform for sentiment analysis. While it offers more advanced features, its pricing is designed to accommodate smaller businesses and startups, making it a cost-effective solution.

Google Cloud Natural Language API: Google's Natural Language API is a powerful tool that offers sentiment analysis features. It can analyse text and provide sentiment scores, helping businesses understand how their users feel about specific topics or products.

Microsoft Azure Text Analytics: Azure Text Analytics, part of the Microsoft Azure cloud services, includes sentiment analysis as one of its features. It allows users to assess the sentiment of text data, enabling businesses to gauge customer opinions and make data-driven decisions.

VADER (Valence Aware Dictionary and sEntiment Reasoner): VADER is a Python-based sentiment analysis tool that is specifically designed for social media text. It comes with a pre-built sentiment lexicon and is known for its ability to handle nuances in sentiment expressed in short texts.

Amazon Comprehend: Amazon Comprehend is a natural language processing service that includes sentiment analysis among its features. It can determine the sentiment of text data, making it easier for businesses to understand customer opinions and feedback.

Tracking Financial Performance:

Tracking Financial KPIs in the Technology Roadmap is crucial for several reasons. Firstly, it provides a clear understanding of the financial health of technology initiatives, allowing stakeholders to make informed decisions. By aligning financial metrics with the roadmap, organisations can allocate resources more strategically, ensuring that investments contribute effectively to business objectives.

Secondly, monitoring financial KPIs helps identify potential cost overruns or inefficiencies early in the process, enabling timely adjustments and preventing financial setbacks.

Lastly, it enhances accountability and transparency, as stakeholders can assess the return on investment (ROI) and understand the economic impact of technological advancements.

Key Financial KPIs that should be considered in a Technology Roadmap include but are not limited to:

Budget Variance: Tracking the variance between planned and actual expenditures helps to identify financial discrepancies and ensures that the project remains within budget constraints.

Return on Investment (ROI): Evaluating the ROI provides insights into the profitability and effectiveness of technology investments, guiding future decision-making.

Cost per User/Transaction: Understanding the cost associated with each user or transaction provides a granular view of expenses, aiding in resource allocation and optimisation.

Time to Market: This KPI measures the speed at which technology solutions are delivered. Shorter timeframes often correlate with reduced costs and increased revenue.

Resource Utilisation: Assessing how efficiently resources are utilised throughout the project lifecycle helps identify areas for improvement and cost-saving opportunities.

Presenting Financial KPIs in the Technology Roadmap should be clear, concise, and easily digestible for stakeholders. Visual representations such as charts and graphs can simplify complex financial data, providing a quick overview of the project's financial performance. Regular reporting and updates during key project milestones or reviews help keep stakeholders informed, fostering a transparent and collaborative environment. Additionally, incorporating narrative explanations alongside the KPIs can provide context and insights into the financial implications of various technology initiatives, enabling a more comprehensive understanding for all stakeholders involved.

Monitoring Cybersecurity Resilience:

Given the rising importance of cybersecurity, integrate KPIs that gauge the organisation's resilience against cyber threats.

These metrics serve as vital tools for gauging the effectiveness of an organisation's cybersecurity posture and its ability to withstand and recover from potential cyberattacks. Tracking the frequency of security audits is crucial, as it provides insights into the proactive measures taken to identify vulnerabilities and rectify them before malicious actors exploit them. Regular security audits not only help in maintaining compliance but also contribute to a proactive cybersecurity strategy, ensuring that the organisation remains vigilant and prepared for evolving cyber threats.

Additionally, KPIs such as incident response times play a pivotal role in measuring an organisation's ability to react swiftly and effectively when a cyber incident occurs. A shorter incident response time indicates a more agile and resilient cybersecurity infrastructure, reducing the potential impact of security breaches. Monitoring the effectiveness of implemented security measures, whether through the analysis of successful threat mitigations or the reduction of security incidents, provides valuable feedback on the overall strength of the organisation's cyber defences. By integrating and consistently evaluating these KPIs, organisations can not only identify areas for improvement but also demonstrate a commitment to cybersecurity best practices, instilling confidence among stakeholders and customers in an increasingly interconnected digital ecosystem.

Example Cyber KPI's

Incident Response Time: Measure the time it takes for the organisation to respond to a cybersecurity incident, demonstrating the efficiency of the incident response plan and minimising potential damage.

Patch Management Effectiveness: Track the percentage of systems and software applications that are up-to-date with the latest security patches, indicating the organisation's resilience against known vulnerabilities.

Phishing Resilience Rate: Assess the ability of employees to identify and report phishing attempts. A higher resilience rate indicates the effectiveness of cybersecurity awareness training.

Vulnerability Remediation Rate: Monitor the speed at which identified vulnerabilities are addressed and mitigated, ensuring a proactive approach to reducing the attack surface.

User Access Monitoring: Keep track of user access and privilege changes to identify unusual or unauthorised activities, reducing the risk of insider threats and unauthorised access.

Network Traffic Anomalies: Detect and analyse unusual patterns in network traffic, helping to identify potential threats or malicious activities that might go unnoticed through standard security measures.

Security Audit Frequency: Evaluate how often security audits are conducted to ensure that the organisation remains compliant with industry standards and regulations, and that security controls are regularly tested.

Endpoint Security Effectiveness: Measure the success rate of endpoint security solutions in detecting and preventing malware or other malicious activities on end-user devices.

Data Encryption Compliance: Ensure that sensitive data is encrypted according to established security policies, providing an additional layer of protection against data breaches.

Security Awareness Training Completion Rate: Track the percentage of employees who complete cybersecurity awareness training, ensuring that the workforce is educated and informed about security best practices, reducing the risk of human error.

The integration of Monitoring and Key Performance Indicators into the creation and execution of a Technology Roadmap is pivotal for steering the organisation towards technological excellence. The continuous evaluation of KPIs allows for informed decision-making, ensures the roadmap's alignment with organisational objectives, and provides the necessary agility to thrive.

Example Cyber KPIs

KPI	Description	Target	Benefits of Tracking
Mean Time to Detect (MTTD)	Measures the average time it takes to identify a security incident.	Reduce MTTD to hours or even minutes.	Faster incident identification minimizes potential damage and allows for quicker response.
Mean Time to Remediate (MTTR)	Tracks the average time it takes to resolve a detected security incident.	Strive for MTTR within hours or a day.	Quicker resolution limits the impact of incidents and restores normal operations.
False Positive Rate	Measures the percentage of security alerts triggered that are not actual threats.	Aim for a low false positive rate (ideally below 10%).	Minimizing false positives reduces alert fatigue and improves security analyst efficiency.
Phishing Email Click-Through Rate	Tracks the percentage of employees who click on phishing links in emails.	Reduce click-through rate to below 5%.	Identifying susceptible employees allows for targeted training and awareness programs.
Patch Deployment Rate	Measures the percentage of devices within your network that have received critical security patches.	Maintain a patch deployment rate exceeding 95%.	Timely patching closes vulnerabilities and minimises the risk of exploitation.
Security Awareness Training Completion Rate	Tracks the percentage of employees who have completed mandatory security awareness training.	Achieve a training completion rate of 100%.	Educated employees are better equipped to recognise and report security threats.
Vulnerability Scan Findings	Identifies the number and severity of vulnerabilities discovered during security scans.	Reduce the number and severity of vulnerabilities over time.	Proactive vulnerability identification allows for prioritisation and timely patching.
Security Incident Count	Tracks the total number of security incidents detected and reported.	Minimise the overall number of security incidents.	Lower incident count indicates a stronger overall security posture.

4.8 Specialised Roadmap Tools

There are several specialised roadmap software tools designed to create visually engaging presentations for communicating project plans, strategies, and technology roadmaps. Here are a few notable options:

1. *Roadmunk*:
    1. **Features:**
        1. Intuitive drag-and-drop interface.
        2. Customisable roadmap templates.
        3. Collaboration features for team input.
        4. Data visualisation for key milestones.
        5. Scenario planning and "what-if" analysis.

    2. **Use Case:** Roadmunk is suitable for creating high-level technology roadmaps and strategic plans.

2. *Aha!:*

    1. **Features:**
        1. Customisable visual roadmaps.
        2. Goal tracking and progress reporting.
        3. Integration with project management tools.
        4. Collaboration and feedback features.
        5. Presentation-ready exports.

    2. **Use Case:** Aha! is well-suited for comprehensive product and technology roadmaps, linking strategies with features and initiatives.

3. *ProductPlan: (My personal recommendation)*

    1. **Features:**
        1. Drag-and-drop roadmap builder.
        2. Visual timeline with color-coded items.
        3. Collaboration tools for team input.
        4. Integration with popular project management tools.
        5. Export options for presentations.

    2. **Use Case:** ProductPlan is ideal for creating product-focused roadmaps, aligning product development with business goals.

4. *Monday.com*:

   1. **Features:**
      1. Visual timeline view for roadmaps.
      2. Customisable templates.
      3. Collaboration and communication tools.
      4. Integration with various third-party apps.
      5. Real-time updates and progress tracking.

   2. **Use Case:** Monday.com is a versatile work operating system, suitable for creating project roadmaps and managing team workflows.

5. *Smartsheet*:

   1. **Features:**
      1. Gantt chart functionality.
      2. Collaboration and sharing capabilities.
      3. Resource management tools.
      4. Integration with project management platforms.
      5. Customisable reporting and dashboards.

   2. **Use Case:** Smartsheet is a flexible platform suitable for creating project roadmaps with a focus on task and timeline management.

6. *Wrike:*

   1. **Features:**
      1. Interactive Gantt charts for timelines.
      2. Collaboration and communication tools.
      3. Customisable dashboards.
      4. Integration with various apps.
      5. Reporting and analytics.

   2. **Use Case:** Wrike is a versatile project management tool that can be adapted for creating technology roadmaps with a focus on project execution.

Before selecting a tool, consider factors such as the specific needs of your team, collaboration requirements, integration capabilities, and the level of customisation you require. Each of these tools offers a trial period, so you can experiment and choose the one that aligns best with your communication and presentation needs.

4.9 Detailed Technology Roadmap Outline Structure

1. Executive Summary:
- **Purpose:** Provide a concise overview for executive stakeholders.
- **Content:**
    - Brief introduction to the Technology Roadmap.
    - Summary of key strategic objectives.
    - High-level timeline highlighting major milestones.

2. Introduction:
- **Purpose:** Establish the context and rationale for the roadmap.
- **Content:**
    - Articulate the business goals driving technology initiatives.
    - Analyse the current technology landscape.
    - Explain the need for and benefits of the roadmap.

3. Current State Assessment:
- **Purpose:** Evaluate the existing technological infrastructure comprehensively.
- **Content:**
    - Detailed examination of current systems and technologies.
    - SWOT analysis encompassing strengths, weaknesses, opportunities, and threats.
    - In-depth overview of key performance indicators (KPIs) reflecting the current state.

4. Strategic Objectives:
- **Purpose:** Define the overarching goals guiding the roadmap with clarity.
- **Content:**
    - Clear articulation of strategic objectives aligned with organisational goals and mission.
    - Explanation of how these objectives contribute to the overall business strategy.

5. Technology Initiatives:
- **Purpose:** Outline specific projects and initiatives in detail.
- **Content:**
    - Comprehensive description of each technology initiative.
    - Clearly defined objectives, scope, and expected outcomes.
    - Resource requirements including budget, personnel, and other necessities.

6. Timeline and Milestones:

- **Purpose:** Provide a visual representation of the roadmap schedule.
- **Content:**
    - Gantt chart or timeline illustrating project phases.
    - Specific milestones for key achievements and deliverables.
    - Dependencies and critical paths.

7. Resource Allocation:

- **Purpose:** Detail the allocation of resources for transparency.
- **Content:**
    - Explicit breakdown of budget allocation for each initiative.
    - Personnel requirements, including skill sets and roles.
    - Investment plans for technology and infrastructure.

8. Risk Analysis and Mitigation:

- **Purpose:** Identify potential challenges and develop robust mitigation strategies.
- **Content:**
    - Identification and categorisation of potential risks.
    - In-depth analysis of the impact of each risk.
    - Mitigation plans and contingency measures.

9. Performance Metrics and KPIs:

- **Purpose:** Establish metrics for measuring success and progress.
- **Content:**
    - Key performance indicators aligned with strategic objectives.
    - Clear targets for each KPI.
    - Comprehensive details on data sources for ongoing monitoring.

10. Communication Plan:

- **Purpose:** Define how progress and updates will be communicated across the organisation.
- **Content:**
    - Stakeholder communication plan outlining target audiences.
    - Reporting frequency, methods, and channels.
    - Protocols for addressing issues, changes, and unexpected developments.

11. Technology Stack and Architecture:

- **Purpose:** Define the technological foundation for initiatives.

- **Content:**
  - Detailed overview of the technology stack.
  - Infrastructure and architecture considerations.
  - Integration points between systems and technologies.

12. Training and Change Management:

- **Purpose:** Plan for user training and manage organisational change effectively.
- **Content:**
  - Comprehensive training programs for end-users.
  - Strategic change management strategies.
  - Communication plan for managing expectations and facilitating a smooth transition.

13. Governance and Oversight:

- **Purpose:** Establish clear roles and responsibilities for oversight.
- **Content:**
  - Governance structure with defined roles.
  - Responsibilities of key stakeholders.
  - Escalation procedures for issue resolution.

14. Post-Implementation Review:

- **Purpose:** Plan for evaluating the success and impact of initiatives.
- **Content:**
  - Criteria for determining the success of initiatives.
  - Detailed post-implementation review process.
  - Capture lessons learned and areas for continuous improvement.

15. Budget and Financial Planning:

- **Purpose:** Provide detailed financial planning for transparency and accountability.
- **Content:**
  - Detailed breakdown of budget allocations for each initiative.
  - Forecasted costs and potential cost savings.
  - Financial impact analysis.

16. Resource Development and Skill Enhancement:

- **Purpose:** Outline plans for enhancing the skills of the team and acquiring necessary resources.
- **Content:**
  - Training and development programs for team members.

- Recruitment plans for acquiring new skills.
- Strategies for ensuring the team is adequately equipped for the roadmap.

17. Cybersecurity Measures:
- **Purpose:** Detail plans to enhance cybersecurity in line with technological initiatives.
- **Content:**
    - Cybersecurity strategy and measures.
    - Risk assessment and mitigation plans related to cybersecurity.
    - Continuous monitoring and improvement strategies.

18. Stakeholder Engagement:
- **Purpose:** Outline plans for engaging and involving stakeholders throughout the roadmap.
- **Content:**
    - Stakeholder engagement strategy.
    - Inclusive communication plans for various stakeholder groups.
    - Mechanisms for feedback and input.

19. Environmental Impact Considerations:
- **Purpose:** Address environmental impact considerations in technological decisions.
- **Content:**
    - Evaluation of the environmental impact of technological choices.
    - Strategies for sustainable technology adoption.
    - Compliance with environmental regulations.

20. Appendix:
- **Purpose:** Include supplementary information for reference.
- **Content:**
    - Supporting documentation for key elements.
    - Detailed technical specifications.
    - Additional resources, charts, and graphs.

Chapter 5.

Creating The Technology Strategy

5.1 Navigating the Intersection: Technology Strategy vs. Digital Strategy

This chapter explores the intersection of technology and digital strategy, emphasising the importance of crafting a comprehensive digital strategy that extends beyond technical implementation. It also delves into effective communication and measurement of strategy success, drawing insights from industry leaders.

Technology or Digital Strategy?

Before people start getting disgruntled, I do understand there are two camps on this topic. One that Digital Strategy is the same as Technology Strategy and the other that they constitute very different things. My view is that it doesn't matter providing you understand what your organisation expects. For the purposes of this book I'll give different views. The table below isn't comprehensive but might provide a flavour of the differences. Some organisations I have worked for have had a Chief Technology Officer and a Chief Digital Officer, other organisations have had either one or the other, some had neither and had a Chief Information Officer and my most recent role was as Chief Product and Technology Officer. So to be honest regardless of you title, your organisation may have a Digital strategy, a Technology Strategy, one strategy that combines them both, or it may have neither but the strategy is one and the same as the organisation strategy.

Digital Strategy	Technology Strategy
Business Model Transformation	Infrastructure Management
Customer Experience Enhancement	Cyber Security & Data Management
Digital Marketing & Branding	Application Development & Integration
Data Driven Decision Making	IT Governance and Compliance
Cultural Transformation	Technology Infrastructure Roadmap
Multi-Channel Customer Engagement	IT Service Management
E-Commerce & Online Presence	Cloud Strategy & Adoption
Innovation & Digital Products	Enterprise Architecture
Data Analytics & Business Intelligence	Technology Standards & Policies
Social Media Engagement	IT Resource Planning & Allocation

Technology Strategy: Crafting a Holistic Tech Roadmap

- **Purpose:** The Technology Strategy serves as the overarching blueprint that guides an organisation's technology-related decisions and investments. It goes beyond the immediate operational concerns, encompassing a broad spectrum to ensure that technological initiatives align with and contribute to the achievement of overall business objectives.

- **Distinguishing Elements:**

    - **Scope:** Encompasses a wide array of technology-related aspects, including infrastructure, systems, processes, and talent management. This comprehensive scope ensures a holistic approach to technology planning.

    - **Focus:** Oriented towards maximising the efficiency and effectiveness of all technology-driven facets within the organisation, from the backend infrastructure to the front-end user interfaces.

- **Influence:** Shapes the overall technological direction and investments, impacting various business functions such as operations, finance, and strategic planning.

- **Components of a Technology Strategy may include:**

  - **Infrastructure Optimisation:** Rigorous assessment and enhancement of the existing technological infrastructure to support business objectives, considering aspects such as scalability, reliability, and performance.

  - **Talent Management:** Strategic planning for attracting, retaining, and developing the right talent for technology-related roles, ensuring a skilled workforce aligned with business goals.

  - **Process Improvement:** Streamlining and optimising business processes through technology integration, with a focus on enhancing operational efficiency and reducing redundancies.

  - **Innovation Framework:** Establishing a framework for continuous innovation in technology adoption, fostering a culture that embraces emerging technologies and adapts to industry trends.

Digital Strategy: A Blueprint for Digital Transformation

- **Purpose:** In contrast, the Digital Strategy is a specific subset of the broader Technology Strategy, focusing explicitly on leveraging digital technologies to achieve strategic business goals. It hones in on the transformative power of digitisation and user-centric enhancements.

- **Distinguishing Elements:**

  - **Scope:** Centred around the use of digital tools, technologies, and platforms to enhance customer experiences, operational processes, and overall competitiveness in the digital realm.

  - **Focus:** Prioritises digitisation, user experience, and innovation in digital technologies, recognising the pivotal role of digital advancements in modern business operations.

  - **Influence:** Directly impacts customer-facing aspects, internal processes, and the organisation's competitive position in the digital landscape.

- **Components of a Digital Strategy may include:**

  - **User-Centric Design:** Prioritising user needs and experiences in the design and deployment of digital solutions, ensuring a seamless and satisfying user journey.

- **Data-Driven Decision Making:** Leveraging data analytics for informed decision-making and continuous improvement, harnessing the power of data to drive strategic initiatives.

- **Cybersecurity Measures:** Ensuring robust security protocols are integrated into digital initiatives, safeguarding sensitive information and mitigating cyber threats. Of course this can equally sit in a Technology Strategy.

- **Scalability and Flexibility:** Designing digital solutions with scalability and adaptability for future growth, acknowledging the dynamic nature of digital technologies.

Bridging the Gap: Synergies and Collaboration

- **Synergies:** While some consider distinct, Technology Strategy and Digital Strategy are not mutually exclusive. Their synergistic integration is crucial for ensuring a holistic approach that addresses both the broader technological landscape and the specific challenges and opportunities presented by digitisation.

- **Collaboration:** Effective collaboration between technology and digital teams is vital for a cohesive strategy. It facilitates a unified approach that considers both the broad technological landscape and the specific nuances of digital transformation.

Data Sources for Informed Strategies:

- **Importance:** Both strategies heavily rely on data-driven insights for formulation and continuous improvement. The integration of relevant data enriches the strategies, providing actionable insights for informed decision-making.

- **Examples:**

  - **Technology Strategy:** Utilises data on system performance, talent capabilities, and infrastructure effectiveness. This includes metrics on system uptime, talent acquisition and retention, and infrastructure utilisation.

  - **Digital Strategy:** Incorporates customer feedback, market trends, and user behaviour analytics. Examples include customer satisfaction scores, market research insights, and data on user interactions with digital platforms.

5.2 The Strategy Presentation: Navigating the Digital and Technology Future

No matter how brilliant your strategy is, if you fail to communicate it, it will fail before you even begin. This is a good outline structure for a strategy presentation.

Introduction:

- **Objective:** Set the stage by highlighting the transformative role of digital technologies in achieving the organisation's strategic vision.

- **Content:**

    - Thorough exploration of the Digital Strategy's pivotal role in enhancing organisational agility, elevating customer experiences, and bolstering overall competitiveness.

    - Introduction to key components with an emphasis on their direct alignment with broader business objectives.

1. Aligning with Business Objectives:

- **Objective:** Ensure precise alignment between the Digital Strategy and the overarching goals of the organisation.

- **Content:**

    - Detailed articulation of how each digital initiative directly supports specific business objectives.

    - Illustration of the Digital Strategy's impact on revenue growth, cost efficiency, and market expansion through real-world examples.

2. User-Centric Design:

- **Objective:** Prioritise user needs and experiences for an unparalleled level of customer satisfaction.

- **Content:**

    - In-depth presentation of meticulously crafted user personas and journey maps guiding the development of digital solutions.

    - Showcasing exemplary instances of user-centric design principles in leading digital products within the industry.

3. Data-Driven Decision Making:

- **Objective:** Leverage the power of data analytics for nuanced decision-making and continuous improvement.

- **Content:**

  - Overview of key data sources that profoundly influence decision-making processes (customer feedback, market research, user analytics).

  - Case studies meticulously detailing the tangible impact of data-driven decisions on various digital initiatives.

4. Integration with Existing Systems:

- **Objective:** Ensure seamless integration of new digital solutions with current systems for optimal operational efficiency.

- **Content:**

  - Strategic frameworks for integrating state-of-the-art digital solutions with existing infrastructure.

  - Detailed examples of successful integration strategies that resulted in improved operational effectiveness.

5. Innovation and Emerging Technologies:

- **Objective:** Embrace innovation through the exploration and integration of cutting-edge technologies.

- **Content:**

  - Comprehensive identification of relevant emerging technologies with a profound potential impact on organisational operations.

  - Showcasing organisations that successfully adopted innovative technologies, highlighting resultant improvements.

6. Digital Security and Compliance:

- **Objective:** Prioritise digital security measures and compliance strategies to mitigate potential risks.

- **Content:**

  - In-depth overview of robust security protocols seamlessly integrated into digital solutions.

- Explicit demonstration of compliance measures with data protection laws and industry-specific regulations.

7. Scalability and Flexibility:

- **Objective:** Design digital solutions with scalability and adaptability to accommodate future growth.

- **Content:**

  - Presentation of comprehensive strategies for scalable architecture and adaptable solutions.

  - Engaging case studies illustrating how scalable solutions effectively accommodated business growth.

8. Employee Training, Talent and Change Management:

- **Objective:** Facilitate effective employee training and streamline organisational change management.

- **Content:**

  - Present what skills are required to make the business successful, and the plan to acquire/bridge any skill gaps in the organisation.

  - Outline of the professional training programmes for employees, ensuring seamless integration with new digital tools.

  - Communication strategies for navigating the cultural shift associated with widespread digital adoption.

  - Real-world case studies spotlighting successful change management within digital transformation.

9. Monitoring and Reengineering:

- **Objective:** Establish robust metrics for monitoring and a well-defined framework for reengineering.

- **Content:**

  - Comprehensive list of key performance indicators (KPIs) designed to gauge the success of digital/technology solutions.

  - Transparent processes for regular evaluation and strategic enhancement based on performance data.

- Identification of crucial data sources pivotal for ongoing monitoring and iterative improvement.

Conclusion: Orchestrating Digital Excellence:

- Recapitulation of key points, underscoring the transformative impact of the Digital Strategy on organisational success.

- A compelling call to action, seeking executive support and collaboration for the effective implementation and continuous refinement of the Digital Strategy.

Q&A Session:

- Invitation for open dialogue, questions, discussions, and feedback from the CEO and other key stakeholders.

This presentation outline provides a thorough overview of the Digital Strategy, effectively communicating its nuanced components, overarching objectives, and potential impact on the organisation's overarching success.

5.3 Measuring Success in the Digital Landscape

How do you know your Strategy is a success and how do you evidence that to your stakeholders. Success is an evolving narrative, woven through the threads of continuous refinement and adaptation. This section delves into the pivotal facets of gauging success, key performance indicators (KPIs), Completion Criteria, methodologies of evaluation, and the strategic infusion of data-driven insights.

Also success, or how your CEO or board perceives success, may not <u>only</u> be measured by tangible results. They may say that is the benchmark, and for sure it's a primary one, but from experience success is also about how you make them feel. Yes, your CFO, CEO and Chairman are likely to be human too, well probably! It's about confidence in the future and in you. You can have a "pig's ear" of a project, and we will all have one, but the transparency of how you handle that can create confidence with your board that they are fully aware of the challenges and risks. The biggest fear of a board or CEO is the unknown or feeling they don't have control. Shit happens, if they are experienced board members or C Suite then they will have seen it before, and possibly sat in your shoes. It just shouldn't be a surprise, as that makes it look like you are not in control. Your board will also be looking at the horizon as well as your feet, so give them something to look at. The opportunities even future ones you create on the horizon can also be seen as success.

As a Chief Technology Officer (CTO), distinguishing between Key Performance Indicators (KPIs) and project completion criteria is crucial for effective oversight of technology initiative and strategy. KPIs in a technological context often revolve around quantifiable metrics that reflect the ongoing health and efficiency of technical processes. For instance, in a software development project, KPIs could include code deployment frequency, system uptime, or response times. These metrics provide real-time insights into the technology team's performance, enabling data-driven decisions to enhance efficiency and productivity.

On the other hand, project completion criteria or the launch of a new product or penetration into a new market, are fundamental benchmarks that denote a successful conclusion. These criteria are typically set at the strategy outset, providing clear expectations for the final deliverables and conditions that signify success. As a CTO, understanding and articulating these criteria is vital for ensuring that technology initiatives align with broader organisational goals. Completion criteria might encompass the successful implementation of a new technology system, meeting security standards, or achieving specific performance targets or it could be a new product in a new geography, or the building of a world class team.

For a CTO, striking a balance between monitoring ongoing performance through KPIs and adhering to project completion criteria is pivotal. KPIs serve as early warning systems, allowing the CTO to identify and address issues promptly during the project lifecycle. Completion criteria, on the other hand, provide a definitive measure of project success, helping the CTO and the technology team focus on delivering tangible, impactful outcomes aligned with strategic objectives. Together, these perspectives empower the CTO to steer technology initiatives towards success, ensuring that both the process and the final results are optimally aligned with the organisation's technological vision.

Aligning with Organisational Goals:

The true litmus test of digital success lies in its alignment with overarching organisational goals. A successful digital or technology strategy is one that seamlessly integrates with and propels the broader business objectives. Stakeholders should witness a harmonious relationship between the digital initiatives and the strategic vision of the organisation.

In the previous section we covered the importance and examples of utilising KPI's to track and measure success as tangible metrics are the storytellers of success. I won't repeat those in this section, but this is a high level re-cap of some areas you may consider for showcasing success are as follows.

Defining Success Metrics: To embark on the journey of measuring success in the digital domain, we must first delineate what success entails within the context of our Digital Strategy. This involves aligning overarching business goals with the objectives of our digital initiatives, crafting a narrative that is both specific and measurable, and setting the stage for a trajectory guided by SMART criteria.

Key Performance Indicators (KPIs): Within this digital narrative, KPIs become the compass, offering directional insights into the performance of our digital initiatives. The canvas of our evaluation is painted with metrics such as conversion rates, customer satisfaction scores, and user engagement indicators. Each brushstroke tells a story of how well our digital strategy resonates with the pulse of our goals.

Identify and meticulously track KPIs that directly relate to the goals of the digital strategy. Whether it's conversion rates, user engagement, or revenue growth, these indicators should vividly portray the positive impact of the digital journey.

Return on Investment (ROI): Within the digital narrative, ROI emerges as a financial subplot, evaluating the economic impact of our digital ventures in relation to our investments. We calculate the return, not just in terms of revenue generation but also in the efficiencies gained, crafting a story of sustainable growth and prudent investment.

Conveying success to stakeholders necessitates demonstrating a return on their investment. Showcase the financial gains, cost efficiencies, and revenue increases attributed to the digital initiatives. Make the connection explicit, showcasing that every pound or dollar invested has contributed to tangible organisational growth.

Adaptability and Scalability: As the narrative unfolds, we examine the chapters dedicated to adaptability and scalability. Here, we assess how well our digital solutions weather the winds of change and scale with the crescendo of business growth. The narrative is one of resilience, ease of updates, and seamless integration with emerging technologies.

Cybersecurity Effectiveness: Woven into the fabric of our narrative is the theme of cybersecurity. This subplot explores the measures we take to safeguard our digital initiatives, acknowledging the role it plays in the overall success story. We scrutinise breach incidents, response times, and weave a tale of robust cybersecurity contributing to our digital triumphs.

Employee Adoption and Satisfaction: Within the human dimension of our narrative, we delve into the chapter dedicated to employee adoption and satisfaction. Through surveys and testimonials, we capture the emotional resonance of our digital tools, ensuring our narrative is one where employees are not just participants but champions of our digital journey.

Continuous Improvement and Innovation: The narrative embraces the theme of continuous improvement and innovation, chronicling how our strategies adapt and enhance over time. We measure success through iteration cycles and the transformative impact of innovations responding to the cadence of market trends.

Stakeholders appreciate not just success but a commitment to ongoing improvement. Illustrate how the organisation, guided by digital insights, is continuously evolving and staying ahead of market trends. This narrative of constant refinement signals to stakeholders that success is not a destination but a journey of perpetual enhancement.

Stakeholder Feedback and Alignment: As the narrative concludes, we shift our focus to the collective voice of our stakeholders. Through their feedback, we gauge alignment with expectations and gather insights that shape our future chapters. This segment is a dialogue, a reflection on how our narrative resonates with those who share in our digital odyssey.

Case Studies and Success Stories:

Humanise success through compelling narratives. Develop case studies and success stories that narrate the journey, challenges overcome, and the transformative impact of digital initiatives. Stakeholders resonate with real-world examples that bring the abstract concept of success into a relatable and tangible context.

Employee Engagement and Productivity:

A successful digital strategy is one that empowers and engages employees. Highlight improvements in employee productivity, satisfaction, and collaboration facilitated by digital tools. Stakeholders find assurance in knowing that the digital strategy is not just about external success but also internal empowerment.

Benchmarking Against Competitors:

Benchmarking against industry peers adds an external layer of evidence. Demonstrate how the organisation stands out or competes favourably in the market due to its successful digital or technology strategy. Comparative data can be a compelling tool to evidence success.

Future Roadmap:

Stakeholders are reassured by a vision for the future. Clearly articulate the next steps and the future roadmap of the strategy. Whether it's further innovations, expansions, or new initiatives, a forward-looking vision signals that success is not static but an ongoing commitment to excellence.

Don't Forget the Emotional Success

Success in the eyes of a CEO or board often extends beyond tangible results. While concrete achievements may be articulated as benchmarks, the emotional resonance of success is equally crucial. The way we as Chief Technology Officers make the executive team feel about the future of technology within the organisation can significantly shape the perception of success. It involves instilling confidence, fostering a sense of optimism, and showcasing adaptability in the face of challenges. For instance, if a project encounters setbacks, the CTO's ability to transparently communicate the issues, present viable solutions, and maintain a positive outlook can be perceived as a form of emotional success. This demonstrates resilience and leadership qualities that contribute to the overall confidence the board or CEO may have in the CTO's strategic direction.

Moreover, creating opportunities on the horizon, even amidst project challenges, can be seen as an emotionally successful approach. This involves the CTO identifying innovative pathways and new prospects that align with the organisation's goals. For example, introducing a strategic pivot in response to changing market dynamics or identifying emerging technologies that can enhance the company's competitive edge can be perceived as success beyond traditional project metrics. This forward-thinking approach can generate a positive sentiment among leadership, showcasing the CTO as a visionary and adaptive leader who can navigate the organisation through evolving technological landscapes.

Ultimately, emotional success for a CTO involves not only delivering tangible outcomes but also cultivating a positive and forward-looking atmosphere. This may include fostering a culture of innovation, highlighting opportunities for growth, and ensuring the technology strategy aligns with the broader aspirations of the organisation. By consistently promoting a sense of confidence, enthusiasm, and strategic vision, the CTO can enhance their impact and contribute to a holistic perception of success within the leadership team.

The success of a digital strategy, as perceived by stakeholders, involves weaving together a narrative that integrates quantitative data, human stories, and a vision for the future. By aligning digital success with organisational goals, maintaining transparency in communication, and showcasing innovation, stakeholders not only observe success but actively engage in and contribute to the ongoing digital journey. Success is not a static point but a journey – a narrative crafted through measurement, adaptation, and the integration of insights. This ensures that our strategy not only meets current objectives but also lays the foundation for future advancements.

5.4 Examples of Digital and Technology Strategies worthy of further study

Amazon - Seamless Customer Experience:

Netflix - Data-Driven Content Recommendations:

The BBC – Navigating Digital Frontiers:

British Aerospace – Digital Innovation in Aviation:

Tesla - Direct-to-Consumer Sales Model:

Starbucks - Mobile App and Loyalty Programme:

All these strategies are within the public domain.

Amazon - Seamless Customer Experience:

1. Customer-Centric DNA:

- Philosophy of Obsession: Amazon's unwavering commitment to customer satisfaction is embedded in its DNA. The digital strategy pivots around understanding customer needs, preferences, and pain points to provide solutions that go beyond mere transactions.

2. Personalised Recommendations:

- Algorithmic Magic: At the heart of Amazon's success is its recommendation engine. Leveraging machine learning algorithms, the platform analyses past behaviours and preferences, offering personalised product recommendations that enhance user engagement and satisfaction.

3. One-Click Purchase:

- Frictionless Checkout: Amazon revolutionised the online shopping experience with its patented one-click purchase feature. Reducing the checkout process to a single click minimises friction, streamlining the buying journey and encouraging impulse purchases.

4. Prime Membership Ecosystem:

- Beyond Free Shipping: Amazon Prime exemplifies a holistic approach to customer loyalty. Offering perks like expedited shipping, access to streaming services, and exclusive deals, it fosters a symbiotic relationship where customers receive value beyond the transaction.

5. Amazon Go - Checkout-Free Stores:

- Future of Retail: The Amazon Go concept redefines in-store shopping. Using advanced technologies such as computer vision and sensors, customers can enter, shop, and leave without the need for traditional checkouts, marking a paradigm shift in the brick-and-mortar experience.

6. Alexa - Voice-Powered Convenience:

- Voice-Activated Commerce: Amazon's foray into voice-activated technology with Alexa aligns with the ethos of convenience. Users can effortlessly place orders, track packages, and access a myriad of services through simple voice commands, further integrating Amazon into users' daily lives.

7. Data-Driven Innovation:

- Insights for Evolution: Amazon's digital strategy thrives on data. Through continuous analysis of user behaviour and market trends, Amazon identifies opportunities for innovation. Data-driven decision-making informs everything from product offerings to the enhancement of user interfaces.

8. Prime Day - Fostering Excitement:

- Retail Holiday: Amazon's Prime Day has evolved into a global retail phenomenon. The digital strategy leverages this event to create excitement, boost Prime memberships, and drive sales through exclusive deals and promotions.

9. Community Engagement:

- Reviews and Ratings: Amazon's platform encourages a community-driven approach to product discovery. Customer reviews and ratings play a pivotal role, offering social proof and aiding other users in making informed purchasing decisions.

10. Continuous Evolution:

- Agility and Adaptability: Amazon's digital strategy is not static. It evolves in response to market dynamics and technological advancements. The company embraces change, from implementing new technologies to entering diverse business verticals, ensuring its strategy remains at the forefront of innovation.

Studying Amazon's Digital Strategy:

- Annual Reports: Explore Amazon's annual reports for insights into their overarching strategy and key performance indicators.

- Jeff Bezos' Shareholder Letters: Bezos' letters to shareholders provide a deep dive into Amazon's long-term thinking and strategic priorities.

- Case Studies: Numerous business journals and academic publications feature case studies on Amazon's digital strategy evolution.

- Technology and Retail Analyses: Industry analyses from reputable sources shed light on the impact of Amazon's digital strategies on the technology and retail sectors.

Amazon's seamless customer experience is a testament to the power of a customer-centric digital strategy. From the frictionless one-click purchase to the innovative Prime ecosystem, each element contributes to a holistic and evolving approach that positions Amazon at the pinnacle of digital retail innovation.

Netflix - Data-Driven Content Recommendations:

1. Algorithmic Precision:

Netflix's digital strategy focusses on advanced algorithms designed to scrutinise user behaviour with unparalleled precision. The platform meticulously dissects what users watch, how long they engage with content, and even when they hit pause. This granular data forms the bedrock for the tailored content recommendations that define the Netflix experience.

2. Personalisation Paradigm:

At the core of Netflix's strategy lies an unwavering commitment to personalisation. The platform transcends traditional genre preferences, delving into the subtleties of individual viewing habits, time-of-day preferences, and the devices used for streaming. This personalisation paradigm creates a bespoke content catalogue for each user, fostering an intimate and curated viewing journey.

3. Machine Learning Mastery:

Netflix's data scientists harness the power of machine learning, deploying models that evolve continually. These models, driven by algorithms, predict user preferences based on historical data, ensuring that recommendations remain dynamic and adaptive to changing viewing patterns.

4. Dynamic Content Tags:

Every piece of content on Netflix is meticulously tagged with a plethora of descriptors. These dynamic tags span genres, moods, and intricate details like character attributes or plot twists. This meticulous tagging enhances the precision of recommendations, enabling the algorithm to grasp nuanced viewer preferences with exceptional accuracy.

5. A/B Testing Innovation:

In the pursuit of excellence, Netflix employs a robust A/B testing methodology. Different algorithms undergo rigorous testing on user subsets to evaluate their effectiveness. This iterative process ensures that the recommendation engine is not static but perpetually refined for optimal content discovery.

6. Original Content Synergy:

Netflix's strategy seamlessly integrates its expansive library of original content into the recommendation engine. Original series and films are strategically recommended based on user affinities for specific genres or themes, creating a symbiotic relationship that promotes engagement with exclusive content.

7. User Ratings Reinforcement:

While some platforms have moved away from user ratings, Netflix strategically harnesses them. User ratings play a pivotal role in fine-tuning the recommendation algorithm, offering direct feedback on content satisfaction. This approach underscores the platform's commitment to prioritising user preferences.

8. Global Adaptability:

Netflix's data-driven strategy transcends cultural and geographic boundaries. The recommendation engine's adaptability to diverse viewer preferences worldwide showcases its global scalability. This global adaptability underscores the platform's dedication to providing a personalised experience to users across the globe.

9. Transparency and User Control:

Netflix places a premium on transparency and user control within its recommendation strategy. Users are empowered to view and adjust their viewing history, refine content preferences, and even reset recommendations. This user-centric approach ensures that individuals have granular control over the content discovery process.

10. Continuous Innovation:

Netflix's data-driven content recommendation strategy is a testament to the company's commitment to relentless innovation. The platform continually explores new avenues, such as interactive storytelling and experimental features, keeping the digital experience at the forefront of innovation.

Studying Netflix's Digital Strategy:

- Netflix Technology Blog: Delve into the ***Netflix Technology Blog*** for in-depth articles and technical insights on their recommendation algorithms, machine learning initiatives, and data-driven strategies.

- Academic Journals: Explore journals such as the "Journal of Data Science" and "International Journal of Artificial Intelligence and Applications" for academic perspectives on Netflix's content recommendation strategies.

- Industry Publications: Regularly peruse tech-focused magazines like Wired, TechCrunch, and industry publications such as Variety and The Hollywood Reporter. These sources cover the latest trends in streaming services and provide insights into the evolution of Netflix's digital strategy.

The BBC – Navigating Digital Frontiers

1. User-Centric Design:

At the heart of the BBC's digital strategy is a commitment to user-centric design. The BBC's digital platforms are intuitively designed to enhance user experience, ensuring easy navigation and accessibility. Features such as personalised content recommendations and customisable interfaces embody this commitment to putting the audience first.

2. Content Personalisation:

The BBC harnesses data analytics and user insights to curate personalised content recommendations. Through a combination of user viewing habits, preferences, and demographic data, the BBC tailors content suggestions, creating a bespoke viewing experience for each user.

3. Digital-First Content Creation:

The BBC has embraced a digital-first approach to content creation. Original series, documentaries, and news features are often conceptualised with a digital audience in mind. This strategy allows the BBC to engage with a diverse global audience through digital channels.

4. Cross-Platform Accessibility:

Recognising the changing media consumption landscape, the BBC ensures cross-platform accessibility. Audiences can seamlessly transition from traditional broadcasting to online platforms, with the BBC iPlayer serving as a flagship example of how the broadcaster has embraced streaming services.

5. Data-Driven Journalism:

In the realm of news reporting, the BBC employs data-driven journalism to enhance storytelling. Infographics, interactive features, and immersive multimedia experiences enrich the news consumption journey, catering to the evolving preferences of digital audiences.

6. Global Digital Presence:

The BBC's digital strategy extends globally, reaching audiences beyond national borders. Through international versions of its website and dedicated digital content, the BBC establishes a global footprint, delivering news, entertainment, and educational content to a diverse audience.

7. Innovation in Broadcasting:

Innovation is embedded in the BBC's digital DNA. From pioneering live streaming to experimenting with virtual reality (VR) experiences, the broadcaster continually explores new technologies to enrich the audience experience and stay ahead in the digital landscape.

8. Inclusive Digital Accessibility:

The BBC is committed to digital inclusivity. The development of accessible technologies, subtitles, and sign language interpretation for digital content ensures that a wide audience, including those with disabilities, can engage with the BBC's diverse range of programming.

9. Audience Engagement Through Social Media:

Social media platforms play a pivotal role in the BBC's digital strategy. Engaging with audiences on platforms such as Twitter, Facebook, and Instagram allows the broadcaster to build communities, share content, and gather real-time feedback.

10. Transparency and Trust:

Central to the BBC's digital strategy is a commitment to transparency and trust. Providing clear information on how user data is handled, ensuring impartial reporting, and maintaining editorial standards contribute to building and retaining audience trust in the digital space.

Studying The BBC's Digital Strategy:

- BBC Research & Development Blog: Explore the **_BBC R&D Blog_** for insights into the technological innovations and experiments conducted by the BBC's Research & Development team.

- Academic Journals: Investigate academic journals in media studies, such as "Media, Culture & Society" and "Journal of Broadcasting & Electronic Media," for scholarly perspectives on the BBC's digital strategy.

- BBC Annual Reports: Delve into the BBC's annual reports, available on their corporate website, for comprehensive overviews of their digital initiatives and strategies.

- Media Industry Conferences: Attendance or exploration of conferences like the International Broadcasting Convention (IBC) offers insights into the broader trends shaping the digital strategies of media organisations, including the BBC.

British Aerospace – Digital Innovation in Aviation

1. Digital Twin Technology:

At the heart of BAE's digital strategy is the integration of digital twin technology, creating virtual replicas of physical assets. This enables real-time monitoring, predictive maintenance, and enhanced operational efficiency for complex aerospace systems, from aircraft to defence equipment.

2. Advanced Data Analytics:

BAE harnesses the power of advanced data analytics to extract meaningful insights from vast datasets. This includes analysing flight data, equipment performance, and supply chain information. The insights gleaned fuel decision-making processes, optimise operations, and contribute to the development of cutting-edge aerospace solutions.

3. Cybersecurity Vigilance:

In an era where cybersecurity threats loom large, BAE's digital strategy places paramount importance on cybersecurity vigilance. Robust cybersecurity measures safeguard critical aviation systems, ensuring the integrity of data, communication channels, and the overall security of aerospace operations.

4. Smart Manufacturing Processes:

BAE embraces smart manufacturing processes, leveraging the Internet of Things (IoT) and automation to enhance production efficiency. The integration of sensors and connectivity in manufacturing equipment facilitates real-time monitoring and optimisation, reducing downtime and enhancing overall productivity.

5. AI-Driven Decision Support:

Artificial Intelligence (AI) plays a pivotal role in BAE's digital strategy, providing decision support across various facets of operations. AI algorithms aid in route planning, predictive maintenance, and even in the design and development of innovative aerospace solutions.

6. Collaborative Ecosystems:

BAE actively participates in collaborative ecosystems, partnering with technology innovators, research institutions, and governmental bodies. This collaborative approach fosters the exchange of ideas, accelerates innovation cycles, and positions BAE at the forefront of emerging technologies in aviation.

7. Digital Skills Development:

Recognising the transformative power of digital technologies, BAE invests in developing digital skills within its workforce. Continuous training programmes ensure that employees are equipped to navigate the digital landscape, fostering a culture of innovation and adaptability.

8. Sustainable Aviation Initiatives:

BAE's digital strategy extends to addressing sustainability challenges in aviation. The company explores innovative solutions, such as fuel-efficient aircraft designs and optimised flight routes through data analytics, contributing to the broader industry goal of sustainable aviation.

9. Customer-Centric Digital Platforms:

In the digital era, customer-centricity remains a focal point for BAE. Digital platforms facilitate seamless communication with clients, providing real-time updates on projects, maintenance schedules, and fostering a transparent and collaborative relationship.

10. Continuous Digital Evolution:

BAE's digital strategy is not static; it is a continuous evolution. The company stays abreast of emerging technologies, anticipates industry trends, and remains agile in adopting new digital solutions. This commitment to continuous evolution ensures that BAE remains a pioneer in digital innovation within the aerospace sector.

Studying BAE's Digital Strategy:

- BAE Systems Insights: Explore BAE's official website for insights, case studies, and whitepapers on their digital initiatives in aerospace.

- Aerospace Journals: Investigate academic journals such as the "Journal of Aircraft" and "AIAA Journal of Aerospace Information Systems" for scholarly perspectives on digital advancements in aerospace.

- Industry Conferences: Attend aerospace and technology conferences where BAE might showcase its digital innovations. These events provide opportunities for networking and firsthand insights into the company's digital strategy.

- Technology News Outlets: Stay updated through technology-focused publications like Aviation Week and Technology Review for news and analyses on BAE's digital initiatives and their impact on the aerospace industry.

Tesla - Direct-to-Consumer Sales Model:

1. Online Sales Platform Prowess:

Central to Tesla's digital strategy is its proficiency in online sales. By predominantly selling its electric vehicles through its official website, Tesla circumvents the conventional dealership model. Customers can seamlessly explore models, customise features, and place orders—all with the click of a button.

2. Over-the-Air Software Updates:

A hallmark of Tesla's digital strategy is its commitment to ongoing customer engagement. Through over-the-air software updates, Tesla enhances vehicle features and performance remotely. This strategy not only improves customer satisfaction but also ensures that Tesla vehicles evolve and remain cutting-edge long after the initial purchase.

3. Customer-Centric Customisation:

Tesla's direct-to-consumer approach places a premium on customer experience. The online platform allows customers to customise their vehicles according to their preferences, from performance options to aesthetic details. This level of customer-centric customisation contributes to a sense of ownership and personalisation.

4. Transparent Pricing Structure:

Tesla's digital strategy embraces transparency in pricing. The online platform provides clear and detailed information about the cost breakdown, including potential savings from incentives and fuel savings. This transparency builds trust with customers, a key element in the success of the direct-to-consumer sales model.

5. Virtual Showrooms and Test Drives:

Tesla complements its online platform with virtual showrooms and test drives. These experiences, often located in high-footfall areas, allow potential customers to interact with Tesla vehicles and experience the brand firsthand. The combination of online accessibility and physical experiences contributes to a holistic customer journey.

6. Bypassing Traditional Dealerships:

Tesla's strategy disrupts the traditional dealership model, eliminating intermediaries between the company and the customer. This direct relationship not only streamlines the sales process but also enables Tesla to maintain control over the entire customer experience, from product education to after-sales service.

7. Global Accessibility:

The digital nature of Tesla's sales model enables global accessibility. Customers from various parts of the world can engage with the online platform, place orders, and receive vehicles, creating a borderless market. This global accessibility aligns with Tesla's vision of accelerating the world's transition to sustainable energy.

8. Enhanced Customer Education:

Tesla's digital strategy includes robust customer education initiatives. The online platform provides detailed information about electric vehicles, sustainability benefits, and the unique features of Tesla models. This educational approach helps demystify electric vehicles and fosters a deeper understanding among consumers.

9. Streamlined Purchase Process:

The direct-to-consumer sales model streamlines the purchase process. Customers can complete the entire transaction online, from configuring their vehicle to arranging financing. This streamlined process enhances efficiency, reduces friction in the buying journey, and aligns with the digital expectations of modern consumers.

10. Community Building through Online Forums:

Beyond the transactional aspect, Tesla fosters a sense of community through online forums and social media. Customers share experiences, tips, and insights, creating a virtual community of Tesla enthusiasts. This community-building aspect enhances brand loyalty and contributes to the overall success of the digital strategy.

Studying Tesla's Digital Strategy:

- Tesla Official Website: Explore Tesla's <u>official website</u> for insights into the online sales platform, vehicle customisation options, and the overall customer journey.

- Tesla Blog and Updates: Delve into Tesla's official blog for announcements, updates on software releases, and insights into the company's approach to customer engagement.

- Automotive Industry Publications: Industry publications such as Automotive News and Electric Vehicle (EV) forums provide analyses of Tesla's direct-to-consumer sales model and its impact on the automotive industry.

Starbucks - Mobile App and Loyalty Programme:

1. Seamless Mobile Ordering and Payment:

At the heart of Starbucks' digital strategy lies the seamless integration of mobile ordering and payment through its app. Customers can browse the menu, customise orders, and pay ahead, reducing wait times and enhancing the overall convenience of the Starbucks experience.

2. Rewards Programme Integration:

The cornerstone of Starbucks' digital engagement is its sophisticated Rewards Programme. The app seamlessly integrates with the loyalty programme, allowing customers to earn points for every purchase. These points translate into personalised promotions, discounts, and even free beverages, creating a symbiotic relationship between customer loyalty and digital innovation.

3. Geolocation-Powered Store Locator:

Starbucks leverages geolocation technology within its app to provide users with a store locator feature. This functionality not only helps customers find the nearest Starbucks but also enables the app to offer location-specific promotions and updates, enhancing the overall user experience.

4. Personalised Recommendations and Offers:

Through data analytics and machine learning, the Starbucks app analyses customer preferences and purchase history. This information fuels the app's ability to provide personalised recommendations and targeted offers, creating a tailored and enticing digital journey for each user.

5. In-App Engagement and Content:

Starbucks elevates user engagement by embedding rich content within its app. From featuring the stories behind coffee blends to providing exclusive access to limited-time promotions, the app becomes a hub for immersive and interactive content, extending the Starbucks brand beyond the physical store.

6. Mobile Payment Innovation:

Starbucks pioneered mobile payments within its app, allowing customers to link their payment methods seamlessly. The app's innovative approach to mobile payments not only facilitates transactions but also enhances security through features like touch ID and facial recognition.

7. Gamification of Loyalty:

The Starbucks app gamifies the loyalty experience, turning the act of collecting stars and reaching new membership levels into a rewarding and enjoyable journey. Gamification elements encourage continued app usage and foster a sense of achievement among users.

8. Integration with Social Responsibility Initiatives:

Starbucks weaves its social responsibility initiatives into the app experience. Users can learn about Starbucks' sustainability efforts, charitable partnerships, and community projects, fostering a connection between customers and the brand's values.

9. Real-Time Order Tracking:

Enhancing the transparency of the ordering process, Starbucks' app features real-time order tracking. Customers can monitor the status of their orders, from preparation to pick up, providing a sense of control and assurance.

10. Continuous App Evolution:

Starbucks maintains a commitment to continuous app evolution. Regular updates introduce new features, refine existing functionalities, and respond to user feedback. This dedication ensures that the Starbucks app remains a cutting-edge tool for enhancing customer experiences.

Studying Starbucks Digital Strategy:

- Starbucks Investor Relations: Explore the ***Starbucks Investor Relations*** website for annual reports and updates on the performance and impact of the digital strategy.

- Marketing Journals: Dive into marketing journals such as the "Journal of Marketing" and "Journal of Interactive Marketing" for academic analyses on Starbucks' mobile app and loyalty programme strategy.

- Business News Outlets: Keep an eye on business news outlets like Bloomberg, Forbes, and CNBC for industry analyses and insights into Starbucks' digital innovations.

Apple - Ecosystem Integration:

1. Seamless Device Integration:

At the core of Apple's digital strategy lies the seamless integration of its diverse range of devices. From the iPhone and iPad to Mac, Apple Watch, and Apple TV, these devices operate cohesively within the ecosystem. The connectivity allows users to effortlessly transition between devices, creating a unified and continuous user experience.

2. App Store Monetisation:

A key pillar of Apple's digital strategy is the App Store, a marketplace where users can access a vast array of applications. The App Store not only serves as a distribution platform but also as a Monetisation avenue for developers. Apple's revenue model, with a percentage cut from app sales, in-app purchases, and subscriptions, is integral to the sustainability of the ecosystem.

3. iCloud Synchronisation:

Apple's ecosystem extends into the cloud with iCloud, a service that synchronises data across devices. From photos and documents to app preferences, iCloud ensures a seamless experience, allowing users to access their content irrespective of the device they are using. This synchronisation creates a unified digital environment.

4. Continuity and Handoff:

The continuity feature allows users to start an activity on one Apple device and seamlessly pick it up on another. Whether drafting an email, browsing Safari, or taking a call, the handoff between devices is a testament to Apple's commitment to a frictionless user experience within its ecosystem.

5. Apple Music and Media Integration:

Apple strategically integrates its media services, such as Apple Music, Apple TV+, and the iTunes Store, within the ecosystem. This integration ensures that users can access their music, movies, and TV shows seamlessly across devices, reinforcing the notion of a comprehensive and interconnected digital environment.

6. HomeKit and Smart Home Integration:

With the HomeKit framework, Apple extends its ecosystem into the realm of smart homes. Users can control compatible smart home devices through the Home app on their Apple devices, creating a unified hub for home automation within the ecosystem.

7. Wearables Integration:

Apple's foray into wearables, including the Apple Watch and AirPods, is strategically woven into the ecosystem. These wearables not only complement existing devices but also enhance user experience with features like health tracking, seamless audio connectivity, and Siri integration.

8. Developer Ecosystem:

Apple fosters a thriving ecosystem for developers through tools like Xcode and SwiftUI. This support enables developers to create apps that seamlessly integrate with Apple's devices and services, contributing to the vibrancy and innovation within the ecosystem.

9. Security and Privacy Emphasis:

Security and privacy are paramount in Apple's digital strategy. The end-to-end encryption of iMessages, Face ID, and the App Tracking Transparency feature exemplify Apple's commitment to safeguarding user data and fostering trust within the ecosystem.

10. Continuous Innovation:

Apple's digital strategy is marked by a culture of continuous innovation. The introduction of new features, devices, and services ensures that the ecosystem remains dynamic and aligned with evolving technological trends.

Studying Apple's Digital Strategy:

- Apple Developer Documentation: Explore the ***Apple Developer*** portal for in-depth technical insights, documentation, and tools for developers within the Apple ecosystem.

- Academic Journals: Delve into academic journals such as the "International Journal of Human-Computer Interaction" and "Journal of Computer Science & Technology" for scholarly analyses of Apple's ecosystem integration strategies.

- Industry Publications: Stay informed through tech-focused publications like Macworld, 9to5Mac, and industry analyses featured in publications like Forbes and Bloomberg.

Book 2 – Leadership and Innovation

This book offers a thorough exploration of technology leadership and challenges confronting CTOs today. It covers navigating leadership styles, building strategic relationships, fostering innovation, driving digital transformation, embracing ethics, continuous learning, managing remote/hybrid teams, diversity, inclusion, and navigating mergers and acquisitions.

Chapter 6.

Introduction: Leadership & The CTO

In this chapter, we delve into the diverse landscape of technology leadership, aiming to define the essence of a CTO's role in the broader context of organisational success.

Here is a summary of sections in this chapter.

6.1 Defining Leadership in Technology

This section delves into the intricacies of technology leadership, outlining how a CTO not only navigates the technical aspects but also orchestrates a vision that aligns with organisational goals. From crafting innovative strategies to leading technological initiatives, the definition of leadership is tailored to the unique demands of the tech landscape.

6.2 Navigating Leadership Styles in the Tech Industry

Leadership in the tech industry is a nuanced art. Here, we explore diverse leadership models and their applications within technology environments. Understanding these nuances is crucial for CTOs seeking alignment with the dynamic and ever-evolving tech landscape. From visionary leaders to collaborative facilitators, each style contributes uniquely to the success of technology initiatives.

6.3 Building Strategic Relationships with Executive Leadership & C-Suite

Effective technology leadership extends beyond technical proficiency. This section illuminates the significance of building robust relationships with executive leadership and the C-Suite. As a CTO, understanding and navigating the intricate web of organisational dynamics is paramount for strategic alignment, fostering collaboration, and ensuring the successful implementation of technological strategies.

6.4 Leading High-Performance Teams

Cultivating a culture of excellence is at the heart of effective technology leadership. This section delves into the art of building and leading high-performance teams, exploring practices that elevate teams to achieve technological brilliance. Topics include fostering collaboration, cultivating innovation, and navigating the challenges inherent in the tech landscape to achieve optimal team performance.

6.5 Fostering Innovation Leadership

Innovation is the cornerstone of technological progress. This section explores how technology leaders foster a culture of innovation. It delves into the strategies for leading technological change and adapting to the ever-evolving landscape of technological advancements. Case studies and practical examples highlight successful innovation leadership in action.

6.6 Navigating Digital Transformation Leadership

Digital transformation is a strategic imperative in the contemporary tech landscape. This section provides insights into strategies for driving digital transformation. It explores overcoming the challenges that arise during the transformative journey, offering practical guidance on successfully steering organisations through the digital transition.

6.7 Embracing Ethics and Responsible Technology Leadership

Ethical considerations are integral to decision-making in technology leadership. This section examines the ethical dimensions of technology choices. It guides leaders in ensuring responsible technology innovation, taking into account societal impacts and ethical implications. Case studies illustrate the importance of ethical leadership in maintaining trust and reputation.

6.8 Continuous Learning, Professional Development, and Leading Through Challenges

In a landscape where change is constant, leaders must embrace continuous learning. This segment outlines strategies for lifelong learning, leadership in personal and team development, and navigating technological challenges and crises with resilience and foresight. Practical tips and resources empower CTOs to foster a culture of continuous improvement and adaptability.

6.9 Pioneering Thought Leadership in the Tech Industry

Establishing thought leadership is not just about individual recognition; it's about shaping the discourse of the entire industry. This section explores the significance of personal and organisational thought leadership. It provides insights into influencing industry trends and discourse. Practical guidance helps CTOs position themselves and their organisations as leaders in shaping the technological narrative.

6.10 Exploring Environmental Sustainability & ESG

Leadership in the modern era extends to broader responsibilities. This section delves into the imperative of integrating sustainability into technological strategies and leading green technology initiatives. Practical examples showcase successful initiatives, emphasising the role of technology leaders in contributing to environmental sustainability and corporate social responsibility.

6.11 Remote & Hybrid Team Management

In an era of evolving work structures, effective leadership also encompasses the management of remote teams. This section explores the nuances of leading distributed technology teams and strategies for effective leadership in remote work environments. Best practices, tools, and case studies offer insights into successfully managing remote teams and maintaining high levels of productivity and collaboration.

6.12 Strategies for Diversity and Inclusion

Diversity and inclusion are essential components of effective leadership. This section focuses on fostering inclusive leadership practices and leading diversity initiatives in technology. Practical strategies, case studies, and success stories highlight the positive impact of diversity in driving innovation, enhancing team dynamics, and ensuring a more inclusive and equitable tech industry.

6.13 Leadership in Mergers and Acquisitions

Navigating the complexities of mergers and acquisitions requires unique leadership skills. This section provides strategies for leading technology integration in M&A and navigating technological transitions seamlessly. Practical insights, case studies, and key considerations equip CTOs to play a pivotal role in ensuring the success of technological transitions during organisational mergers and acquisitions.

6.1 Defining Leadership in Technology:

At the core of effective technology leadership lies a recognition that a CTO's responsibilities extends further than technical competency. Or in other words, it's not just about understanding the intricacies of algorithms or the latest programming languages; it's about being a versatile leader. A successful CTO must embrace a role that includes strategic thinking, effective communication, team collaboration, and foresighted decision-making.

Having a clear and compelling vision is paramount. A CTO is, or should be, the visionary architect, seamlessly blending technological possibilities with broader organisational objectives. This involves not just crafting a vision but also communicating it effectively. A compelling vision guides technological strategies, fosters innovation, and ensures alignment with the overarching goals of the organisation.

Innovation is the heartbeat of technology, and the CTO needs to be at the forefront of crafting strategies that nurture and harness it.

Beyond conceptualisation, effective technology leadership involves the practical implementation of initiatives. A CTO acts as the catalyst for change, leading technological initiatives that propel the organisation forward. This encompasses everything from choosing the right project management methodologies to fostering a culture of adaptability and continuous innovation.

As we delve into the CTO's role, it becomes evident that effective technology leadership is a balance between technical prowess and visionary acumen. The CTO's ability to navigate this journey determines not only the success of technological initiatives but also the organisation's resilience and relevance in the ever-evolving world of technology.

6.2 Navigating Leadership Styles in the Tech Industry: A Spectrum of Strategies

Technology leadership manifests in various nuanced forms. As the guiding force, a Chief Technology Officer (CTO) employs diverse leadership styles to adeptly navigate the complex landscape of technological innovation, team dynamics, and organisational objectives. Let's delve into the spectrum of leadership styles that a CTO may adopt in the ever-evolving world of technology.

The Visionary Pathfinder

Some CTOs wholeheartedly embrace the role of a visionary pathfinder. Beyond merely articulating a clear vision, they meticulously lay the groundwork for its realisation. These leaders not only inspire innovation but actively foster an environment that encourages risk-taking and experimentation. They champion the pursuit of uncharted territories, fostering a culture of creativity and forward-thinking that permeates every facet of the tech team.

The Agile Adaptor

Agility is not merely an advantage; it's a necessity. Agile adaptors among CTOs demonstrate a proactive approach to change, swiftly responding to shifts in technology, market trends, and customer needs. These leaders champion iterative development, allowing their teams the flexibility to pivot swiftly when required. The result is an organisation that remains responsive and resilient, adapting seamlessly amidst constant change.

The Collaborative Facilitator

Effective tech projects often hinge on seamless cross-functional collaboration. A CTO adopting the role of a collaborative facilitator excels at building bridges between departments. They actively encourage open communication channels, fostering a culture of teamwork that transcends traditional silos. By ensuring that the tech team integrates seamlessly with other facets of the organisation, this leadership style promotes a holistic approach to problem-solving, leveraging the collective intelligence of the entire enterprise.

The Technological Evangelist

For some CTOs, leadership extends beyond technical expertise; it involves being a passionate advocate for technology's potential. These technological evangelists actively engage with emerging trends, not only within the industry but also across broader technological landscapes. Their leadership is marked by a commitment to driving innovation and ensuring that the organisation stays at the forefront of technological advancements. This infectious enthusiasm motivates the entire team to strive for continuous innovation and explore the full spectrum of technological possibilities.

The Data-Driven Strategist

In the era of big data, some CTOs gravitate towards a data-driven leadership style.

Beyond the mere utilisation of data for decision-making, these leaders immerse themselves in comprehensive analytics. They leverage data to not only forecast trends but also identify potential opportunities and mitigate risks strategically. This approach ensures that every technological initiative is not just visionary but is also grounded in empirical evidence, contributing to informed and strategic decision-making with a tangible impact on organisational outcomes.

The Empathetic Mentor

Leadership in the tech industry is inherently people-centric. The empathetic mentor, as a leadership style, goes beyond traditional roles. These CTOs focus on the personal and professional development of their team members. They are keenly attuned to individual strengths and aspirations, creating a supportive environment that not only nurtures growth and innovation but also fosters a sense of belonging and purpose among team members.

The Pragmatic Problem-Solver

In the face of technological challenges, some CTOs adopt a pragmatic problem-solving approach. Beyond merely identifying issues, they thrive on dissecting complex problems and developing practical solutions. This leadership style ensures that the team remains focused on achievable goals while navigating obstacles effectively. The pragmatic problem-solver excels in maintaining the smooth functioning of the tech engine, ensuring that day-to-day operations align seamlessly with broader strategic objectives.

The Inclusive Innovator

Innovation thrives in diverse and inclusive environments. The inclusive innovator, as a leadership style, actively fosters diversity within the tech team. These CTOs not only welcome but actively seek varied perspectives and ideas. They recognise the strength in differences, creating a culture where innovation can truly thrive. Groundbreaking solutions emerge from a collaborative and diverse team, reflecting the depth and richness of varied experiences and insights.

Which leadership style are you?

CTO Leadership Style Identification Questionnaire

By permission or the Marlow Business school

Keep a record of which statement, a, b. c, d, e, sounds most like you. Answer the questions truthfully of how you think you are perceived by your colleagues. Then retake the quiz in how you would like to be perceived.

Section 1: Vision and Direction

1. **How do you typically approach setting a vision for your team or projects?**

a) Clearly articulate a visionary roadmap.

b) Proactively respond to shifts and changes.

c) Build bridges between departments and encourage teamwork.

d) Actively engage with emerging tech trends and advocate for innovation.

e) Leverage data extensively for decision-making.

2. **How do you handle challenges and uncertainties in your projects?**

a) Embrace risk-taking and experimentation.

b) Adapt swiftly and embrace change.

c) Encourage cross-functional collaboration.

d) Explore technological advancements and innovation as a solution

e) Rely on data-driven insights for strategic decision-making.

3. **To what extent do you involve your team in the vision-setting process?**

a) Encourage active participation and creativity.

b) Solicit input for agile responses to changes.

c) Promote collaborative vision-building across departments.

d) Inspire the team with the latest tech trends and possibilities.

e) Incorporate data insights to shape the vision collaboratively.

Section 2: Collaboration and Flexibility

4. **How do you promote teamwork and collaboration within your team?**

a) Encourage creativity and forward-thinking.

b) Champion iterative development and flexibility.

c) Build bridges between departments for seamless collaboration.

d) Foster an environment that thrives on continuous innovation.

e) Promote a holistic approach to problem-solving.

5. **How do you encourage adaptability in response to market trends and tech advancements?**

a) Emphasise the importance of a long-term vision.

b) Pivot swiftly and adapt to changes.

c) Collaborate with various departments for a well-rounded response.

d) Embrace emerging trends to stay at the forefront of technology.

e) Utilise data analytics to forecast trends and plan strategically.

6. **How do you foster a culture of continuous learning within your team?**

a) Encourage exploration of new ideas and technologies.

b) Support skill development for agile responses.

c) Facilitate cross-departmental knowledge sharing.

d) Promote awareness of cutting-edge tech advancements.

e) Integrate data-driven insights into the learning process.

Section 3: Reflection and Aspiration

7. **When reflecting on past successes, which trait was most prominent in your leadership approach?**

a) Visionary and innovative thinking.

b) Agility and adaptability to change.

c) Collaboration and teamwork.

d) Passionate advocacy for technology.

e) Data-driven decision-making.

8. **Which CTO role aligns most closely with the impact you aspire to make?**

a) Visionary Pathfinder

b) Agile Adaptor

c) Collaborative Facilitator

d) Technological Evangelist

e) Data-Driven Strategist

9. **How do you approach balancing short-term goals with long-term vision in your projects?**

a) Prioritise visionary goals for long-term impact.

b) Adjust goals based on short-term changes and needs.

c) Ensure that short-term goals align with long-term objectives.

d) Balance tech advancements for immediate impact.

e) Use data insights to inform the balance between short-term and long-term goals.

Section 4: Continuous Learning and Adaptability

10. **How open are you to adopting new leadership styles and learning from different approaches within the CTO spectrum?**

a) Highly receptive to new ideas and approaches.

b) Willing to adapt to changing circumstances.

c) Open to collaboration and diverse perspectives.

d) Eager to explore emerging technological landscapes.

e) Committed to integrating data insights for informed decision-making.

11. How do you stay informed about technological trends and innovations?

a) Regularly engage with industry thought leaders.

b) Stay agile and adapt to new technologies.

c) Collaborate with diverse teams for varied insights.

d) Actively participate in industry events and forums.

e) Utilise data analytics tools for trend analysis.

12. How do you foster a culture of innovation and experimentation within your team?

a) Encourage brainstorming and creative thinking.

b) Support iterative approaches and risk-taking.

c) Facilitate cross-functional innovation workshops.

d) Promote exploration of cutting-edge technologies.

e) Incorporate data insights to guide innovative initiatives.

Results:

Add up the scores for each section (a-e). The section with the highest total corresponds to the CTO role that aligns most closely with your current leadership tendencies. Try taking the quiz again this time selecting the answer you wish was more like you.

a) Visionary Pathfinder, b) Agile Adaptor, c) Collaborative Facilitator, d) Technological Evangelist, e) Data-Driven Strategist

A strategic CTO should be able to play all these roles depending on the environment, culture or organisation requirement. If you would like to explore personal development options to enhance the number of roles you play review chapter 6.7

Understanding and Influencing CTO Leadership Styles

Understanding and influencing a CTO's leadership style requires a multifaceted approach, combining quantitative and qualitative methods. The tools and methodologies outlined in this chapter provide a comprehensive framework for organisations seeking to unravel the complexities of tech leadership and guide it towards sustained success.

1. Leadership Style Assessments

Utilise established leadership assessments to gauge a CTO's predominant style. Instruments like the Myers-Briggs Type Indicator (MBTI), DISC assessment, Marlow Business School Assessment (MBSA), or Leadership Grid can provide valuable insights into communication preferences, decision-making approaches, and overall leadership tendencies.

Example: The Myers-Briggs Type Indicator (MBTI) can offer insights into whether a CTO leans towards being more visionary, analytical, or people-oriented.

2. 360-Degree Feedback Surveys

Implementing 360-degree feedback surveys involves collecting input from peers, subordinates, and superiors. This comprehensive approach provides a holistic view of a CTO's leadership style, highlighting strengths, weaknesses, and areas for improvement.

Example: Tools like SurveyMonkey or Qualtrics can facilitate the creation and deployment of 360-degree feedback surveys.

3. Behavioural Interviews

Conducting behavioural interviews delves into past experiences, revealing patterns in decision-making and problem-solving. By examining how a CTO handled specific situations, one can discern their preferred leadership style under various circumstances.

Example: The STAR (Situation, Task, Action, Result) method can structure behavioural interviews, eliciting detailed responses about a CTO's actions and thought processes.

4. Psychometric Assessments

Leverage psychometric assessments tailored to leadership traits. Instruments such as the Leadership Versatility Index (LVI) or Hogan Assessments can provide nuanced insights into a CTO's personality, potential derailers, and areas for development.

Example: Hogan Assessments, with its suite of tools, can offer in-depth insights into leadership competencies and potential challenges.

5. Observational Analysis

Apply observational analysis by closely observing a CTO's day-to-day interactions, decision-making processes, and communication style. This hands-on approach allows for real-time understanding of how a leader operates within the organisational context.

Example: Developing a structured observation checklist to note behaviours, communication patterns, and responses in different situations.

6. Leadership Development Programmes

Engage CTOs in targeted leadership development programmes that encourage self-reflection and skill enhancement. These programmes, often facilitated by external experts, can guide leaders in recognising and refining their existing styles.

Example: Harvard Business Review's Leadership Development Programmes or executive education courses at top-tier business schools.

7. Stakeholder Interviews

Conduct one-on-one interviews with key stakeholders, including team members, peers, and executives. Direct conversations provide qualitative data on a CTO's leadership impact and effectiveness.

Example: Structured interviews with predefined questions can gather qualitative insights into a CTO's leadership style.

8. Cultural and Industry Benchmarks

Benchmark a CTO's leadership style against industry and cultural norms. Understanding the prevalent leadership models in the tech industry and aligning them with organisational culture can provide context for assessing effectiveness.

Example: Industry reports, articles, and case studies on leadership in technology-driven organisations.

9. Continuous Feedback Mechanisms

Establish regular feedback mechanisms, creating an environment where ongoing discussions about leadership styles are encouraged. This approach fosters adaptability and evolution in response to changing organisational dynamics.

Example: Implementing regular check-ins or feedback sessions as part of the organisational culture.

10. Mentorship and Coaching

Encourage mentorship and coaching relationships for CTOs. Having an external mentor or coach can offer an objective perspective, guiding leaders in refining their styles and addressing specific challenges.

Example: Engaging executive coaches with expertise in technology leadership.

6.3 Building Strategic Relationships with Executive Leadership & C-Suite

As a CTO, understanding organisational dynamics is paramount for strategic alignment, fostering collaboration, and ensuring the successful implementation of technological strategies.

Nurturing Strategic Alliances: A CTO's Guide to Executive Relationships and Collaborative Ecosystems

Confucious said that "Proficiency transcends technical prowess". Actually he didn't say that but as a million quotes are attributed to him, let's give him another!

Collaboration within the C Suite and Senior Management Team is essential if innovation is to be successful in an organisation. From regular strategic meetings to cross-functional workshops, building an environment that encourages the exchange of ideas and perspectives is fundamental for harnessing the collective intelligence of the executive team and creating a solution that has buy-in.

At its core, collaboration has to be more than a buzzword. The CTO, as a catalyst for technological advancement needs to recognise the transformative potential of collaborative ecosystems.

Regular strategic meetings within the C-Suite provide the stage for the exchange of ideas and insights. These exchanges may not always be harmonious, but they can be constructive. The CTO, in a leadership role, takes charge of creating an environment where executives feel empowered to share their perspectives. An important lesson for any CTO, especially someone new to the role, is that their core team is not the technology or digital team that they lead, but the C Suite itself.

Cross-Functional Workshops: Bridging Disciplinary Divides

Instigating cross-functional workshops are an important innovation step, where the CTO facilitates collaborative thinking across different executive functions.

These workshops, serve as a think tank or creative space for different executive functions to work or brainstorm together.

Purposeful Design: Crafting Collaborative Experiences

The success of cross-functional workshops lay in their design which need to align with the strategic objectives but also foster an environment where executives from various disciplines feel empowered to contribute. For example, the workshop may be themed around a specific organisational challenge or a technological opportunity, ensuring relevance to each participating function.

Inclusivity and Diversity: Fostering a Confluence of Perspectives

One strength of cross-functional workshops is in the diversity of perspectives they bring to the table and how they actively encourage inclusivity, ensuring representation from all relevant executive functions. By bringing together individuals with varied expertise, ranging from technology and marketing to finance and operations, the workshops become melting pots of ideas, challenging traditional boundaries and encouraging innovative thinking.

Facilitation Mastery: Guiding the Collaborative Journey

The CTO, as facilitator, needs to play a pivotal role in guiding the collaborative journey within these workshops. This involves skillfully managing discussions, encouraging active participation, and ensuring that every voice is heard. Techniques such as design thinking workshops or agile frameworks may be incorporated to stimulate creative thinking and problem-solving. The aim is to channel the collective intelligence of the participants towards actionable outcomes.

Actionable Outcomes: Transforming Ideas into Initiatives

The true measure of a successful cross-functional workshop lies in its ability to translate ideas into actionable outcomes. These outcomes are not abstract concepts but concrete plans that align with the organisation's strategic goals. For instance, a workshop focused on digital transformation may culminate in a roadmap for implementing specific technologies across different business functions.

Post-Workshop Integration: Ensuring Continuity of Collaboration

The impact of cross-functional workshops extends beyond the workshop room. It is a good idea to employ post-workshop strategies to ensure the continuity of collaboration. This may involve establishing task forces or project teams that bring together individuals from different functions to execute the initiatives developed during the workshop. The CTO takes a proactive stance in overseeing the implementation phase, reinforcing the collaborative spirit ignited in the workshop.

Success Stories: Showcasing the Impact of Collaboration

To inspire future workshops and reinforce the value of collaboration, it can be positive to showcase success stories emanating from these collaborative endeavours. This involves highlighting how initiatives born in workshops have positively influenced the organisation, whether through enhanced efficiency, improved customer experiences, or innovative product developments. These success stories serve as testimonials to the power of cross-functional collaboration.

Effective Communication Strategies

To navigate executive relationships successfully, it's crucial to employ adept communication strategies when explaining intricate technical concepts to non-technical stakeholders. This ensures that the C-Suite is not only informed but actively engaged in the technological journey. From delivering concise presentations to adopting tailored communication styles, effective communication serves as a strategic tool for garnering support and fostering a shared vision.

Understanding the Audience: Understanding the correct communication style can be important to land important messaging. Different executives may have varying levels of technical acumen and distinct preferences in communication. Adapting the communication approach to resonate with each executive ensures that the message is not only heard but also understood.

Concise Presentations: Operating where time is a precious commodity, the C-Suite values concise presentations as a means to cut through complexity and deliver essential information efficiently. A CTO's ability to articulate intricate technical concepts succinctly not only saves time but also guarantees that critical information is communicated effectively.

8 tips for explaining technical concepts to non-technical C Suite members.

1. **Use Analogies and Metaphors:** Employ relatable analogies and metaphors to translate complex technical concepts into familiar terms. Drawing parallels to everyday experiences helps non-tech-savvy C-Suite members grasp the essence of intricate ideas.

2. **Storytelling Approach:** Frame technical details within a narrative or story. Narratives create a context that makes the information more accessible and memorable, aiding comprehension for individuals who may not be familiar with technical intricacies. Review *Vyond* at a tool for helping tell your story.

3. **Visual Aids and Infographics:** Leverage visual aids, infographics, and diagrams to represent complex concepts in a visual format. Visual tools provide a clearer understanding and serve as effective communication aids for those who may not be accustomed to technical terminology.

4. **Eliminate Jargon:** Avoid excessive technical jargon and acronyms. Instead, opt for plain language to convey ideas. Clear and straightforward communication helps prevent confusion and ensures that non-tech-savvy executives can follow the discussion.

5. **Interactive Demonstrations:** Incorporate hands-on or interactive demonstrations whenever possible. Practical examples and real-time demonstrations can significantly enhance understanding, making complex technical concepts more tangible for non-tech-savvy members.

6. **Provide Context and Relevance:** Establish the relevance of the technical concept within the broader business context. Clearly articulate how the technology aligns with organisational goals and impacts the overall strategy, demonstrating the practical significance of the discussed concepts.

7. **Encourage Questions and Dialogue:** Foster an environment where non-tech-savvy C-Suite members feel comfortable asking questions. Encouraging dialogue creates a space for clarification, ensuring that everyone is on the same page and can contribute to the discussion.

8. **Create Summary Documents:** Develop concise summary documents that encapsulate key points in a non-technical language. These documents can serve as reference materials, allowing C-Suite members to revisit and reinforce their understanding of complex technical concepts at their own pace.

Reference Tools for Concise Presentations

1. **Presentation Software:** Everyone knows tools like Microsoft PowerPoint, Google Slides, or Keynote, but don't over rely on these. They generally turn out to be less exciting than actually demonstrating concepts. If you have data to show, try not to use a set of numbers in a static chart, use Power BI to show drill down of numbers and answer questions with the data in real time.

2. **Data Visualisation Tools:** Platforms like Tableau or Microsoft Power BI can be employed for creating dynamic visualisations that distil complex data into comprehensible graphics, aiding in concise presentations.

3. **Infographic Creation Tools:** Canva, Piktochart, or Venngage offer user-friendly interfaces for designing infographics that condense information into visually appealing formats, suitable for concise presentations.

4. **Slide Note Annotations:** It's amazing how few people use the notes section of presentation software. It doesn't mean that you have to read the script, although it can be a valuable pointer if needed, but more so when you revisit presentations, which I end up doing regularly, I can see my thinking at the time.

20 seconds of Courage

Rooted in the idea that transformative decisions often hinge on brief moments of boldness, this methodology offers a compelling approach to moving business decisions forward. Embracing the essence of seizing opportunities with a burst of courage, it becomes a catalyst for innovation, growth, and resilience in the face of challenges.

Seizing the Moment with 20 Seconds of Courage:

In the realm of business, where risks and rewards intertwine, decisions can define the trajectory of an organisation. The "20 seconds of courage" methodology encourages business leaders to embrace moments of uncertainty and make decisions with conviction. This approach acknowledges that profound change often requires overcoming initial hesitations, pushing boundaries, and taking bold steps forward. By instilling a sense of urgency and decisiveness, this methodology empowers individuals to confront challenges head-on, fostering a culture of innovation and adaptability within the business ecosystem.

The Power of Decisive Action:

At its core, the "20 seconds of courage" methodology recognises that pivotal decisions are not always made through lengthy deliberations but rather in those brief windows of audacity. In the fast-paced business environment, where opportunities can be fleeting, the ability to act decisively becomes a competitive advantage. By embracing this approach, organisations can navigate uncertainties, break through stagnation, and propel themselves towards success. This methodology challenges traditional paradigms, urging business leaders to embrace the transformative potential within brief moments of bold decision-making.

The "20 seconds of courage" is not just used in business decisions but also in personal decisions. In film it was highlighted in We Bought a Zoo starring Matt Damon and Scarlett Johansson. The book "We Bought a Zoo" on which the film is based is a memoir by Benjamin Mee, recounting his real-life experiences of purchasing a rundown zoo in Dartmoor, England. Published in 2008, the memoir delves into Mee's decision to make a drastic life change following the death of his wife. The concept of "20 seconds of courage" serves as a transformative force, propelling the characters into uncharted territories of personal growth and resilience. Matt Damon's character, Benjamin Mee, embodies this philosophy when he decides to purchase and revive the struggling zoo. The impact of "20 seconds of courage" is palpable as Benjamin confronts his grief, steps out of his comfort zone, and takes on the monumental challenge of not only managing a zoo but also providing his family with a fresh start. This brief yet powerful burst of courage becomes the driving force behind the Mee family's journey, inspiring them to overcome obstacles, forge new connections, and rediscover the joy in living. The film beautifully illustrates how a moment of bold decision-making, fueled by courage, can lead to profound transformations.

Storytelling Techniques: Weaving a Compelling Narrative

The ability to convey complex technical concepts through storytelling is a potent skill that transforms information into engaging narratives. This section explores storytelling techniques that empower a Chief Technology Officer (CTO) to connect with non-technical stakeholders within the C-Suite.

The Narrative Arc Methodology

Effective storytelling follows a narrative arc, providing structure and coherence to communication. In other fields we successfully use tried and tested methodologies to navigate, so why do we, factual and data centric CTO's, find it so hard to use The Narrative Arc Methodology? It creates successful movies and books of course, and when you understand the formula you will see it clearly every day. I challenge you to re look at some of your favourite movies once you have mastered it. Try Star Wars and you will see it in a different light.   The storytelling elements include exposition, rising action, climax, falling action, and resolution. For instance, when presenting a new technology implementation, the exposition may introduce the current challenges, the rising action builds tension, the climax reveals the innovative solution, and the resolution outlines the positive impact on the organisation. This structured approach captivates the audience and enhances understanding. Human culture, neuroscience and psychology over centuries has been programmed to understand information through stories and there is a formula for success. I recommend reading The Science of Storytelling by Will Storr.

Components of the Narrative Arc Methodology

Not all components need to be utilised in a presentation or report, but it's a good checklist to the audience wanting to know the outcome and being engaged.

Personifying Technology: The Hero's Journey

An effective storytelling technique involves personifying technology as the hero on a journey. By casting technological initiatives as protagonists overcoming challenges, a CTO makes the narrative relatable and engaging. For example, the implementation of a cutting-edge cybersecurity system can be portrayed as the hero safeguarding the organisation against cyber threats. This technique humanises technology, fostering a deeper connection with the C-Suite.

Establish Business Context (Exposition):

Commence the storytelling journey by delving into the roots of an organisation, providing a comprehensive background that immerses stakeholders in its evolution. Introduce the organisation's founding principles, key milestones, and the individuals behind its inception. Detail the prevailing market conditions at the time of the organisation's establishment, offering insights into the industry landscape and challenges faced by businesses in that era. Illuminate the historical journey, highlighting pivotal moments that shaped the organisation's identity and positioned it within the broader market.

Present a Key Business Challenge (Call to Adventure):

Introduce a pivotal moment in the organisation's timeline, where external factors either disrupt its stability and demand a transformative response or new opportunity arises. Articulate the challenge or opportunity, setting the stage for a compelling narrative. Whether it's a seismic shift in consumer preferences, technological advancements, or market dynamics, vividly portray the catalyst that propels the organisation into uncharted territory. This is the call to adventure that signals a departure from the status quo and beckons the organisation to embark on a new, transformative journey.

Acknowledge Initial Reservations (Refusal of the Call):

Navigate the nuanced landscape of change within the organisation by addressing the initial hesitations and resistance among stakeholders. Capture the doubts and concerns that naturally arise in the face of disruption. Illuminate the human side of the business, portraying the emotional responses of employees, leadership, and other key players. By acknowledging and exploring these reservations, the story gains authenticity, making it relatable to the audience and setting the stage for a compelling transformation.

Introduce Expert Guidance (Meeting the Mentor):

Highlight a pivotal turning point where direction from a strategic advisor or industry expert guides direction or where new evidence or information appears that gives encouragement or direction. Introduce the mentor figure who will guide this new adventure, it may be the CEO or a visionary within the business or the CTO. The individual must be seen as a guiding force, offering invaluable insights and steering the organisation through the complexities of the transformative journey. Illustrate the impact of this mentorship on decision-making and strategic direction. This introduces a layer of mentorship and wisdom that contributes to the organisation's resilience and adaptability.

Define Strategic Approach (Crossing the Threshold):

Outline the strategic initiatives undertaken by the organisation in response to the identified challenge. Portray the organisation's commitment to navigating uncharted territory with a clear and defined approach. Provide details on the strategies implemented, whether it involves product innovation, market expansion, or organisational restructuring. This phase marks the organisation's commitment to crossing the threshold into the transformative journey, symbolising a collective determination to achieve sustainable growth.

Navigate Challenges Collaboratively (Tests, Allies, Enemies):

Embark on a dynamic rising action by describing collaborative efforts, alliances formed, and challenges faced during the organisation's strategic shift. Showcase how the organisation navigates complexities through teamwork, innovation, and adaptability. Introduce allies who contribute to the collective effort and potential adversaries that add tension to the narrative. This collaborative navigation becomes a testament to the organisation's resilience and ability to thrive amidst uncertainty.

Approach Core Business Challenge (Approach to the Inmost Cave):

Shift the focus to the critical phase where the organisation approaches the core business challenge. Emphasise the gravity of this moment, escalating tension in the narrative. Illuminate the strategic decisions made or will need to be made during this phase, highlighting the pivotal choices that will shape the organisation's destiny. The approach to the inmost cave symbolises reaching the climax-building phase

Overcome Challenges (Ordeal):

Navigate the climax of the story by narrating how the organisation faces and triumphs over significant challenges. Illustrate the moments of intense adversity, portraying the organisation's resilience and collective will to overcome obstacles. Detail the strategies employed, innovative solutions developed, and transformative decisions made during this ordeal. The climax becomes the high-stakes moment where the organisation emerges victorious against all odds.

Achieve Milestones and Gains (Reward - Seizing the Sword):

Transition into the falling action by showcasing the rewards and gains derived from the organisation's strategic efforts. Highlight milestones, breakthrough innovations, or market successes achieved during the transformative journey. This phase serves as the moment of seizing the sword, where the organisation reaps the rewards of its resilience and strategic prowess.

Strategise Business Return (The Road Back):

Outline the transition from the challenging phase to the next business objective. Describe how the organisation strategises its return, setting the stage for sustained growth and future objectives. This phase builds the narrative towards a resolution, guiding stakeholders through the roadmap for the organisation's ongoing journey.

Navigate Final Climactic Challenge (Climax):

Illustrate the ultimate obstacle or transformative turning point where the organisation faces a final climactic challenge. Highlight the strategies, innovations, or decisions that contribute to overcoming this pinnacle challenge. This moment becomes the climax of the business story, a pivotal turning point that solidifies the organisation's position as a resilient industry player.

Conclude with Business Achievements (Falling Action):

Summarise the business journey, emphasising the lessons learned and showcasing the positive impact on market positioning. Reflect on the organisation's achievements during the transformative journey, instilling confidence in stakeholders. As the story enters the falling action, provide closure by narrating how the organisation's narrative arc has positively shaped its future.

Resolution and Reflection (Resolution):

Offer a satisfying conclusion that addresses challenges and aligns with the organisation's vision. Reflect on the profound transformation, leaving stakeholders with a sense of closure and anticipation for the company's future. Highlight the enduring impact of the organisation's journey, emphasising how the company's narrative aligns with its long-term goals. This resolution becomes a powerful conclusion that resonates with stakeholders and positions the organisation for continued success in the evolving business landscape.

Visual Storytelling with Vyond

Vyond, an animation tool, offers a dynamic approach to visual storytelling. CTOs can use Vyond to create animated videos that simplify complex technical concepts. For instance, a CTO can use animation to illustrate the journey of data through a secure network, making it accessible and visually appealing. This tool adds a layer of creativity to storytelling, ensuring that the narrative is not only informative but also visually compelling.

Recommended reading on Storytelling

- **The Storytelling Animal: How Stories Make Us Human" by Jonathan Gottschall**
  Jonathan Gottschall explores the evolutionary and psychological aspects of storytelling, offering a unique perspective on why humans are inherently drawn to stories. This book provides insights into the innate human connection to narrative and its impact on culture and society.

- **The Science of Storytelling" by Will Storr: Highly Recommended**
  Will Storr delves into the neuroscience and psychology behind storytelling, exploring the science that makes stories resonate with audiences. This book provides a unique perspective on the cognitive and emotional aspects of effective storytelling.

Showcasing Real-World Impact: Case Studies and Analogies

Tangible examples and relatable analogies enrich storytelling by providing real-world context. CTOs can share case studies showcasing how similar technological implementations have positively impacted other organisations. Analogies, such as comparing a sophisticated algorithm to a well-oiled machine, bridge the knowledge gap for non-technical stakeholders. This technique grounds the narrative in reality and demonstrates the potential impact of the proposed technology.

Utilising Metaphors and Imagery

Metaphors and imagery add depth to storytelling, making abstract technical concepts more tangible. A CTO can use metaphors, like describing cloud computing as a virtual storage warehouse, to simplify intricate ideas. Incorporating vivid imagery through descriptive language enhances the audience's ability to visualise complex scenarios. This technique aids in creating a mental picture that resonates with the C-Suite.

Strategic Sequencing for Impactful Delivery

Storytelling is not only about what is said but also how it is delivered. CTOs can strategically sequence information to maintain suspense and reveal key points at pivotal moments. By aligning the narrative with the audience's curiosity, a CTO can ensure that each revelation builds on the previous one, creating a compelling and memorable storytelling experience.

In conclusion, storytelling techniques offer CTOs a powerful means to convey technical information with resonance and impact. By crafting narratives with a well-defined arc, personifying technology, incorporating visual storytelling with tools like Vyond, showcasing real-world impact through case studies, utilising metaphors and imagery, and strategically sequencing information, a CTO can effectively engage the C-Suite and weave a compelling narrative around technological initiatives.

Interactive Communication Platforms: Fostering Engagement

Beyond traditional communication channels, interactive platforms provide further engagement opportunities to create an environment conducive to active involvement of executive leadership in the technological conversation.

Workshops for Collaborative Innovation

Organising workshops tailored to specific technological initiatives can be a powerful strategy. Platforms like *Miro* or **MURAL** provide virtual whiteboards where C-Suite members can collaboratively brainstorm, outline strategies, and visualise complex concepts. Workshops not only facilitate idea generation but also create a space for executive leaders to actively contribute to the development of technological solutions. For instance, conducting design thinking workshops can be effective in fostering a shared vision and encouraging collaborative problem-solving. Even with the growth of virtual technologies like Miro, my personal preference when it comes to brainstorming, collecting requirements, innovation sessions and project planning is still to do this face to face. Leave the working from home for operational or administrative process, but ideas and creativity need people together with a pen and whiteboard – are more successful in my opinion.

Q&A Sessions for Direct Engagement

Interactive question-and-answer (Q&A) sessions offer a direct avenue for executive leaders to engage with the CTO and deepen their understanding of technological initiatives. Platforms like **_Slido_** or **_Mentimeter_** allow participants to submit and upvote questions, ensuring that the discussion addresses the most pertinent concerns. This approach not only fosters transparency but also encourages a culture of openness, where the C-Suite feels empowered to seek clarifications and provide insights.

Collaborative Discussions through Enterprise Social Networks

Enterprise social networks, such as **_Slack_**, serve as virtual spaces for ongoing collaborative discussions. These platforms enable real-time communication, idea sharing, and quick updates. Creating dedicated channels for specific technological projects allows executive leaders to actively participate in discussions, share insights, and stay informed about progress. The asynchronous nature of these platforms accommodates busy schedules, promoting continuous engagement.

Virtual Reality (VR) Platforms for Immersive Experiences

Innovative technologies like virtual reality (VR) can elevate the engagement level within the C-Suite. Platforms like **_ChilloutVR_**, **_VRChat_** or **_Helios_** offer virtual environments where executive leaders can attend immersive meetings, presentations, or workshops. While this may be more cutting-edge, VR provides a unique opportunity for a CTO to showcase prototypes, demonstrate technological concepts, and create a more immersive understanding of complex initiatives.

Collaboration Tools for Documented Contribution

Everyone uses tools like Google Workspace or Microsoft 365, but still few use them as collaborative documents where many people can edit the same document simultaneously. This enables executive leaders to actively collaborate on documents, proposals, or project plans, not just see them at the end. The ability to see and track contributions in real time enhances engagement and ensures that the C-Suite has a direct impact on shaping technological strategies and outcomes.

Examples in Action:

- **Workshop Success with _MURAL_:**
    - A CTO organises a virtual workshop using MURAL to collaboratively map out the technological roadmap for an upcoming project. Executive leaders actively contribute ideas, visually identify challenges, and

collectively strategise, fostering a shared understanding and commitment to the initiative.

- **Q&A Session via *Slido*:**

    - During a town hall meeting, the CTO utilises Slido to host a Q&A session. Executive leaders submit questions in real time, and the most upvoted ones are addressed. This interactive format encourages direct engagement and ensures that concerns or inquiries from the C-Suite are directly addressed.

- **Immersive VR Meeting with *VRChat*:**

    - To discuss a groundbreaking technological innovation, the CTO arranges a virtual meeting using VRChat. Executive leaders put on VR headsets, immersing themselves in a 3D environment where they can interact with prototypes and simulations, fostering a deeper understanding and engagement.

- **Document Collaboration with *Google Workspace*:**

    - The CTO initiates a collaborative document on Google Workspace, where the C-Suite collectively contributes to a technology strategy document. This real-time collaboration ensures that each executive leader has a documented impact on the final output, reinforcing a sense of ownership and engagement.

Leveraging interactive communication platforms is pivotal for fostering engagement within the C-Suite. The examples and reference tools outlined above showcase practical approaches a CTO can adopt to create an environment where executive leaders actively contribute to and shape technological discourse, ultimately fostering a shared vision and commitment.

Influencing Decision-Making Processes

Understanding how decisions are made at the executive level is a key aspect of effective leadership. By leveraging data-driven presentations, compelling business cases, strategic advocacy, scenario planning, technology roadmaps, expert panels, and risk assessment tools, a CTO can guide decisions effectively. These methods not only empower decision-makers with the necessary information but also position technology as an integral driver for organisational progress.

1. Data-Driven Presentations: Shaping Decisions with Insights

One powerful way to influence decision-making is through data-driven presentations. By leveraging comprehensive data analytics tools, such as **_Tableau_** or **_Power BI_**, a CTO can visually articulate the impact and benefits of proposed technological initiatives. Real-time dashboards and visually compelling reports help convey complex information in a digestible manner, empowering decision-makers to make informed choices aligned with technological goals. Ditching PowerPoint and using these tools to drill down into the data in a presentation is much more powerful.

2. Compelling Business Cases: Aligning Technology with Strategic Objectives

Crafting compelling business cases is an art that involves aligning technological proposals with strategic business objectives. The CTO can use tools like business case templates, financial models, and ROI calculators to present a comprehensive rationale. By clearly illustrating the anticipated return on investment and strategic alignment, a CTO can sway decision-makers towards embracing technology as a driver of organisational success. Investigate **Prophix** a comprehensive financial modeling and performance management software. It enables businesses to create detailed financial models, scenario planning, and budgeting. Its features include forecasting tools, financial consolidation, and reporting functionalities.

3. Strategic Advocacy: Building Consensus for Technological Initiatives

Influence often stems from strategic advocacy within the C-Suite. A CTO can employ tools like stakeholder analysis and influence maps to identify key decision-makers and build alliances. By understanding the motivations and concerns of each stakeholder, strategic advocacy becomes a targeted effort to garner support for technological initiatives, creating a consensus that paves the way for smooth decision-making processes. Investigate https://www.stakeholdermap.com/ StakeholderMap is specifically designed for stakeholder analysis. It provides tools to identify, analyse, and manage stakeholders effectively. The platform allows you to create stakeholder maps, assess their interests and influence, and develop strategies for engagement.

4. Scenario Planning: Anticipating Decision Outcomes

Scenario planning tools, such as Monte Carlo simulations or decision trees, provide a proactive approach to decision-making. A CTO can use these tools to outline potential outcomes based on different decision paths. By presenting decision-makers with a clear understanding of the potential risks and rewards associated with each choice, the CTO guides the decision-making process towards options that align with technological objectives. Explore **_palisade.com_**

5. Technology Roadmaps: Communicating Long-Term Vision

A well-crafted technology roadmap serves as a visual guide for decision-makers. Using tools like Gantt charts or project management software, a CTO can illustrate the phased implementation of technological initiatives. This visual representation not only communicates the long-term vision but also showcases the strategic planning behind each phase, facilitating decision-making that aligns with the overall technology strategy. Explore *ProductPlan*

6. Expert Panels and Advisory Boards: Leveraging External Insights

To augment internal perspectives, a CTO can establish expert panels or advisory boards comprising external industry experts. Tools for managing expert opinions, such as Delphi method applications or collaborative platforms, enable the collection and synthesis of diverse insights. This external input adds credibility to technological proposals, influencing decision-makers by incorporating a broader industry perspective.

7. Risk Assessment Tools: Mitigating Concerns

Decision-makers often grapple with risk considerations. A CTO can use risk assessment tools, such as risk matrices or SWOT analysis, to systematically evaluate potential challenges associated with technological decisions. By presenting comprehensive risk mitigation strategies, the CTO instils confidence in decision-makers, addressing concerns and facilitating a more affirmative approach towards technology adoption. Explore *Metricstream*

Building Trust and Credibility

Building trust and credibility as a CTO doesn't have an exact formula but it does require work. You can positively influence credibility by having a plan. It does encompass consistent results, transparent communication, reliable decision-making, accountability, collaborative relationships, technical expertise, customer satisfaction, ethical leadership, continuous learning, and feedback mechanisms. Leveraging recognised tools, frameworks, and methodologies can also enhance the effectiveness of these strategies, establishing the CTO as a trusted leader in guiding the organisation towards technological excellence.

How to build trust may be obvious to many but if that trust is missing; this list as a reminder may help.

1. Consistent Delivery of Results:

Trust is earned through the consistent delivery of tangible results. A CTO can build trust by consistently meeting or exceeding performance expectations. For instance, if a technology initiative consistently achieves its goals and contributes to the overall success of the organisation, it fosters trust among the executive leadership.

2. Transparent Communication:

Transparent communication is a vital element in building trust. A CTO should communicate openly about technology strategies, challenges, and successes. Utilising communication tools such as regular executive briefings, project dashboards, and collaborative platforms ensures that the executive team is well-informed, contributing to a culture of transparency.

3. Reliability in Decision-Making:

Reliability in decision-making enhances trust. A CTO should make well-informed, strategic decisions that align with organisational goals. Using decision-making frameworks, data analytics tools, and risk assessment methodologies can bolster the reliability of decisions, instilling confidence in the executive team.

4. Accountability and Ownership:

Taking accountability and ownership of technology initiatives fosters trust. When faced with challenges or setbacks, a CTO should openly acknowledge them, outline corrective actions, and take responsibility for the outcomes. Utilising project management tools and accountability frameworks aids in demonstrating ownership.

5. Building Relationships Through Collaboration:

Trust is nurtured through relationship-building. A CTO can foster collaborative relationships within the executive team by actively participating in cross-functional projects, seeking input from colleagues, and utilising relationship management tools to understand individual perspectives and needs.

6. Demonstrating Expertise:

Demonstrating technical expertise contributes significantly to building trust. By staying updated on industry trends, certifications, and leveraging knowledge-sharing platforms, a CTO showcases a commitment to professional development and expertise, reinforcing credibility within the executive ranks.

7. Customer and Stakeholder Satisfaction:

Customer and stakeholder satisfaction are indicators of the success of technology initiatives. Utilising customer feedback tools, satisfaction surveys, and engagement metrics, a CTO can measure and communicate the positive impact of technology on end-users and stakeholders, strengthening trust.

8. Ethical Leadership:

Trust is closely tied to ethical leadership. Upholding ethical standards in decision-making, data handling, and technology implementations builds credibility. Utilising ethical frameworks, compliance tools, and industry best practices demonstrates a commitment to responsible and trustworthy leadership. Having a reputation for doing the "right thing" is a powerful part of trust.

9. Continuous Learning and Adaptation:

Demonstrating a commitment to continuous learning and adaptation is a trust-building strategy. A CTO should stay abreast of emerging technologies, industry best practices, and invest in professional development. Utilising online courses, industry conferences, and mentorship programmes showcases a dedication to staying current and adaptable.

10. Feedback Mechanisms:

Implementing feedback mechanisms is crucial for understanding and addressing concerns. By utilising employee feedback tools, conducting regular check-ins, and actively seeking input from the executive team, a CTO can demonstrate a willingness to listen and adapt, fostering an environment of trust.

Navigating Organisational Politics

Organisational politics is an inevitable facet of sitting in senior positions, where your head is above the battlements. By leveraging tools, examples, and reference materials, it is possible to adeptly maneuver through power structures whilst building strong relationships, understanding cultural nuances, foster transparent decision-making, and implement effective conflict resolution strategies. Ultimately, these skills contribute to creating a technology-centric environment that aligns with organisational objectives and thrives amidst the dynamics of corporate politics.

Interpersonal Relationship Building

Building strong interpersonal relationships is a key component of successful leadership amidst organisational politics. Case studies and examples of successful relationship-building within the C-Suite can serve as valuable references for a CTO. Establishing connections through mentorship programs, cross-functional projects, and collaborative initiatives contributes to a positive and supportive environment, fostering a collective commitment to technological goals.

One key technique is the cultivation of open and transparent communication. Establishing a culture where senior leaders feel comfortable sharing ideas, concerns, and feedback creates an environment of trust and mutual respect. Regular communication channels, such as team meetings, one-on-one sessions, and collaborative platforms, enable executives to understand each other's perspectives, align on strategic goals, and address challenges collectively. This open dialogue also promotes a sense of unity and shared responsibility within the senior management team.

Another effective technique involves team-building activities designed specifically for senior management. These activities can range from strategic planning retreats to team-building exercises that encourage collaboration and camaraderie. Engaging in activities outside the usual work setting provides opportunities for executives to connect on a personal level, fostering a deeper understanding of each other's strengths, working styles, and values. This shared experience can translate into improved communication and cooperation within the C-suite. If you are considering off-site "team-building" events, it is important to know your audience. For some - trekking through wilderness might be a nightmare rather than a positive team building experience. Also not all medical issues are known, so for some it may highlight or prevent them from participating, causing stress rather than collaboration.

Mentorship and leadership development programmes represent another technique for enhancing interpersonal relationships within senior management. Establishing mentorship relationships among executives facilitates knowledge transfer, skill development, and the exchange of valuable insights. Leadership development programmes, workshops, and training sessions offer a structured approach to skill enhancement and encourage ongoing personal and professional growth. These initiatives not only strengthen individual capabilities but also create a supportive network within the senior management team, reinforcing bonds and promoting a collaborative leadership culture.

Cultural Nuances and Sensitivity

Every organisation possesses its own distinct culture, and for a Chief Technology Officer (CTO), comprehending cultural nuances is paramount for successfully navigating through organisational dynamics. Surveys, feedback mechanisms, and cultural assessment tools can prove invaluable in helping a CTO gauge the pulse of the organisation. For instance, tools like the ***Organisational Culture Assessment Instrument (OCAI)*** or the ***Cultural Intelligence (CQ) Assessment*** can provide valuable insights into the prevailing cultural values and norms. This insight enables the CTO to discern the organisational culture, offering a foundation for aligning technological strategies accordingly. By doing so, the CTO can not only reduce potential friction but also foster an environment that encourages collaboration.

One good example to consider is the context of First Nation cultures such as Australia and New Zealand, with rich traditions and distinct approaches to technology, and where cultural sensitivity becomes even more pronounced. For example, many First Nation communities place a strong emphasis on communal decision-making processes, where consensus and collective input play a central role. This contrasts with the more individualistic decision-making often found in Western cultures. Understanding and respecting these communal dynamics can significantly impact how technological strategies are introduced and accepted within First Nation organisations.

Another notable difference lies in the perception of time. First Nation cultures often embrace a more cyclical and holistic view of time, emphasising the interconnectedness of past, present, and future. In contrast, Western cultures often adhere to a linear and segmented concept of time. A culturally sensitive CTO should recognise these differences when implementing technological initiatives, acknowledging that time considerations may vary and adapting strategies accordingly.

Moreover, the spiritual and symbolic significance attached to technology may differ. In some First Nation cultures, technology is seen not just as a practical tool but also as a reflection of cultural values and connections to the land. Understanding and incorporating these symbolic dimensions in your storytelling can greatly enhance the success of technological initiatives in a way that respects and aligns with the cultural fabric of First Nation communities. If interested in this topic review Lera Boroditsky's TED talk on How Language shapes the way we think:
https://www.youtube.com/watch?v=RKK7wGAYP6k

Example: Cultural intelligence assessments, combined with regular pulse surveys, can provide data-driven insights into the cultural landscape of the organisation.

Transparent Decision-Making Processes

Transparent decision-making processes are an antidote to suspicion and speculation that often accompany organisational politics. Leveraging project management tools, communication platforms, and documentation systems can help a CTO streamline decision-making processes and make them accessible to relevant stakeholders. This transparency fosters trust and mitigates political manoeuvring around technological initiatives.

Conflict Resolution Strategies

Inevitably, conflicts arise within any organisational setting. A CTO equipped with effective conflict resolution strategies can turn potential disruptions into opportunities for growth. Reference materials such as conflict resolution models, case studies, and best practices in conflict management enable a CTO to approach conflicts with a proactive and constructive mindset. When training with IBM we had management training in conflict resolution, but these days investment in good base management training is seen less often.

I would recommend The Center for Conflict Resolution (CCR). Enrolling in professional conflict resolution training programs, such as those offered by CCR, provides individuals with comprehensive skills in mediation, negotiation, and conflict management. These programs often incorporate theoretical frameworks, role-playing scenarios, and practical exercises to simulate real-world conflict situations.

One fundamental technique employed by mediators is active listening. This involves providing individuals involved in the conflict with a dedicated platform to express their concerns, ensuring that each party feels heard and understood. Skilled mediators use empathetic listening to grasp the underlying emotions and perspectives driving the conflict, creating a foundation for constructive dialogue. By acknowledging the feelings and concerns of all parties involved, mediation aims to uncover shared interests and common ground, paving the way for mutually agreeable solutions.

Another effective mediation technique is reframing. Mediators work to reframe the issues at hand in a manner that encourages a more positive and collaborative mindset. This involves steering the conversation away from blame and focusing on shared goals and interests. By reframing the narrative, mediators guide participants towards a solution-oriented mindset, helping them to explore alternative perspectives and potential resolutions. This technique not only shifts the focus from individual positions to collective problem-solving but also promotes a culture of understanding and cooperation within the workplace.

Reference: Utilising the **Thomas-Kilmann Conflict Mode Instrument** to assess individual conflict-handling styles and tailor resolution strategies accordingly.

6.4 Leading High-Performance Teams

Cultivating a culture of excellence is at the heart of effective technology leadership. By integrating these practices, leveraging relevant tools, and drawing insights from exemplary companies, a CTO can cultivate a culture of excellence, steer teams towards high performance, and ensure technological brilliance within the evolving landscape of technology leadership.

1. Fostering Collaborative Environments

Cultivating Collaboration: Creating an environment that fosters collaboration is foundational to high-performance teams. Tools like Slack, Microsoft Teams, or Asana can facilitate seamless communication and project collaboration. A real-world example is how Atlassian's collaboration tools have empowered teams worldwide to work cohesively on complex projects.

Cross-Functional Collaboration: Encouraging collaboration across diverse skill sets enhances problem-solving. Platforms like Miro or Trello facilitate visual collaboration, allowing cross-functional teams to ideate and innovate collectively. The success of cross-functional collaboration at Google, where engineers, designers, and product managers work collaboratively, exemplifies the potential impact.

2. Cultivating Innovation and Creativity

Innovation Workshops: Organising innovation workshops can spark creativity. Tools like Stormboard or MURAL enable virtual brainstorming, ensuring remote teams actively contribute ideas. Companies like Apple are renowned for their innovation workshops, fostering a culture where employees are encouraged to think outside the box.

Hackathons and Innovation Challenges: Events like hackathons provide a platform for intense, focused collaboration. GitHub, a platform popular for hosting code repositories, often sponsors hackathons that stimulate innovation. Facebook's tradition of "hackathons" played a pivotal role in the creation of features like the "Like" button.

3. Navigating Tech Landscape Challenges

Agile Methodologies: Adopting agile methodologies like Scrum or Kanban enhances adaptability to tech challenges. Tools such as Jira or Monday.com assist in implementing agile workflows. Spotify's successful use of the Spotify Model, an agile framework, has set a benchmark for tech companies seeking efficient collaboration and adaptability.

Remote Work Challenges: Navigating challenges posed by remote work requires tailored strategies. Tools like Zoom or Microsoft Teams facilitate virtual team interactions. GitHub, a leading platform for version control, has effectively embraced remote work, maintaining high team productivity.

4. Measuring Team Performance

Key Performance Indicators (KPIs): Defining and tracking KPIs is crucial for assessing team performance. Tools like Google Analytics or GitHub Insights provide quantitative insights. Amazon's "two-pizza teams" model, where teams are small enough to be fed by two pizzas, is a testament to the effectiveness of measuring team performance.

Employee Feedback and Surveys: Regularly seeking feedback through tools like SurveyMonkey or Officevibe ensures continuous improvement. Netflix, renowned for its high-performance culture, emphasises the importance of candid feedback in its "360-degree feedback" approach.

Data Sources and Reference Tools:

1. **State of DevOps Reports:** The annual State of DevOps Report provides valuable insights into high-performance team practices.

2. **Harvard Business Review:** Articles on team dynamics and leadership provide academic perspectives on cultivating high-performance teams.

3. **Project Management Institute (PMI):** PMI's resources offer insights into effective project management practices for high-performance teams.

4. **Leaders Eat Last by Simon Sinek:** This book explores leadership principles that contribute to creating high-performing and fulfilled teams.

Building a High-Performing Team: A Practical 10-Step Action Plan

Creating a high-performing team involves a strategic and intentional approach. Here's a practical 10-point plan for CTOs to build and nurture teams that excel:

1. Define Clear Goals and Expectations:

- Clearly articulate team goals, both short-term and long-term.
- Set specific, measurable, achievable, relevant, and time-bound (SMART) expectations for each team member.
- Ensure alignment between individual objectives and the broader organisational mission.

2. Foster a Collaborative Culture:

- Encourage open communication and idea-sharing within the team.
- Promote a culture of trust and mutual respect.
- Facilitate team-building activities to strengthen interpersonal relationships.

3. Invest in Team Development:

- Identify the skills and competencies needed for success.
- Provide ongoing training and development opportunities.
- Foster a learning culture where team members can continuously upskill.

4. Empower and Delegate:

- Delegate responsibilities based on individual strengths and expertise.
- Empower team members to make decisions within their areas of responsibility.
- Create a sense of ownership by involving team members in decision-making processes.

5. Provide Regular Feedback:

- Conduct regular performance reviews and offer constructive feedback.
- Recognise and celebrate individual and team achievements.
- Establish a feedback loop for continuous improvement.

6. Encourage Innovation and Creativity:

- Foster an environment where team members feel empowered to suggest and implement innovative ideas.

- Allocate time for creative thinking and experimentation.

- Recognise and reward creative solutions to challenges.

7. Foster Inclusivity and Diversity:

- Build a diverse team that brings different perspectives and experiences.

- Create an inclusive environment where everyone feels valued.

- Ensure that decision-making processes consider a variety of viewpoints.

8. Promote Work-Life Balance:

- Encourage a healthy work-life balance to prevent burnout.

- Provide flexible work arrangements where feasible.

- Support well-being initiatives to ensure the overall health and happiness of the team.

9. Utilise Technology for Efficiency:

- Implement tools and technologies that enhance collaboration and productivity.

- Leverage project management tools for streamlined workflows.

- Ensure that the team has access to the latest and most effective technologies.

10. Lead by Example:

- Demonstrate the values and work ethic expected from the team.

- Model effective communication and problem-solving.

- Be transparent and accountable, creating a leadership style others can emulate.

Building a high-performing team is an ongoing process that requires a combination of strategic planning, continuous development, and a supportive work culture. This 10-point plan can lay the foundation for a team that not only meets but exceeds expectations, driving innovation and success within the organisation.

6.5 Fostering Innovation Leadership

Innovation is the cornerstone of technological progress. This section explores how technology leaders foster a culture of innovation. It delves into the strategies for leading technological change and adapting to the ever-evolving landscape of technological advancements. Case studies and practical examples highlight successful innovation leadership in action. Review Book 3 for an-depth look at innovation.

1. Understanding the Innovation Landscape

Successful innovation leadership begins with a deep understanding of the innovation landscape. Technology leaders must stay abreast of emerging trends, disruptive technologies, and market dynamics. Tools such as **_Gartner's Hype Cycle_** and **_CB Insights_** provide valuable insights into emerging technologies and their potential impact. By comprehending the innovation terrain, leaders can proactively identify opportunities for technological advancement.

2. Fostering a Culture of Innovation

Creating a culture that fosters innovation is a cornerstone of effective leadership. Examples from companies like Google, known for its "20% time" policy allowing employees to pursue passion projects, showcase how a supportive environment stimulates creativity. Tools like idea management platforms such as **_Spigit/ Idea Place_** or **_Ideawake_** can facilitate the generation and evaluation of innovative ideas within the organisation.

3. Leading Technological Change

Innovation often requires embracing technological change. Leaders must guide their teams through transitions, whether adopting new tools, methodologies, or digital transformation. Case studies from companies like Netflix, which successfully navigated the shift from DVD rentals to streaming, offer insights into effective leadership during transformative periods. Change management tools such as Prosci's ADKAR model can aid in planning and executing technological transitions seamlessly.

4. Adapting to Technological Advancements

The technological landscape is ever-evolving, requiring leaders to adapt continually. Case studies from Amazon's AWS evolution demonstrate how embracing advancements in cloud computing can redefine an organisation. Tools like **_TechCrunch_** and **_Wired_** provide real-time updates on technological trends, aiding leaders in making informed decisions about adopting cutting-edge technologies.

5. Encouraging Cross-Functional Collaboration

Innovation thrives when diverse perspectives converge. Leaders should encourage cross-functional collaboration to break down silos and stimulate creative thinking. Collaboration tools like Slack or Microsoft Teams facilitate real-time communication and idea exchange. Case studies from innovative companies like Pixar emphasise the power of interdisciplinary collaboration in fostering groundbreaking technological solutions.

6. Metrics for Innovation Success

Measuring the success of innovation initiatives is crucial. Key performance indicators (KPIs) such as the number of implemented ideas, time-to-market for new products, or customer satisfaction after the introduction of innovations provide tangible metrics. Tools like Innovation Cloud or Innovation Management Software can assist in tracking and evaluating the impact of innovation efforts.

7. Continuous Learning and Adaptation

Innovation leadership is an ongoing process that requires a commitment to continuous learning. Leaders should actively seek opportunities for professional development and stay informed about emerging technologies. Platforms like Coursera and MIT Sloan Executive Education offer courses on innovation leadership, providing valuable insights and frameworks.

8. Collaborative Innovation Platforms

Utilise collaborative innovation platforms to harness the collective intelligence of your team. Tools like Brightidea or InnoCentive enable organisations to crowdsource innovative ideas from employees, fostering a culture of collective innovation.

Harnessing Innovation Acceleration Through Hackathons

In the dynamic landscape of technology, organisations seek innovative ways to accelerate the development of groundbreaking solutions.

Hackathons represent a dynamic and effective strategy for accelerating innovation within organisations. By fostering collaboration, encouraging experimentation, aligning with organisational goals, and tapping into the collective creativity of participants, hackathons can serve as powerful catalysts for driving technological advancements and shaping the future of innovation within the organisation. Successful case studies and practical examples illuminate the transformative potential of hackathons as a cornerstone of innovation acceleration.

1. Understanding the Essence of Hackathons

Hackathons, short for "hacking marathons," are intensive, time-bound events where cross-functional teams collaborate to solve specific challenges or create innovative solutions. Understanding the core concept of hackathons is crucial for leveraging their potential as innovation accelerators.

2. Catalyst for Cross-Functional Collaboration

Hackathons break down silos and encourage cross-functional collaboration. Engineers, designers, marketers, and other specialists collaborate intensively, bringing diverse perspectives to the table. This collaborative synergy often sparks creative solutions that might not emerge in traditional work settings.

3. Rapid Prototyping and Solution Validation

The time-bound nature of hackathons prompts teams to focus on rapid prototyping. Participants develop tangible prototypes or minimum viable products (MVPs) within a short timeframe, allowing organisations to quickly validate ideas and assess their potential viability. This accelerates the innovation cycle and reduces time-to-market for new concepts.

4. Fostering a Culture of Innovation

Hackathons play a pivotal role in fostering a culture of innovation within an organisation. They provide a platform for employees to explore unconventional ideas, experiment with emerging technologies, and think outside the box. Successful hackathons contribute to a culture that values continuous learning, experimentation, and creative problem-solving.

5. Tapping into Employee Creativity

Employees are often reservoirs of untapped creativity. Hackathons provide an avenue for individuals at all levels of the organisation to unleash their creative potential. Recognising and valuing employees' innovative ideas can lead to transformative solutions that benefit the entire organisation.

6. Aligning Hackathons with Organisational Goals

Strategic alignment is crucial for the success of hackathons. Organisations should design hackathon challenges that align with their overarching goals and priorities. This ensures that the innovative solutions generated during these events contribute directly to the organisation's strategic objectives.

7. Encouraging Experimentation and Risk-Taking

Innovation often involves a degree of experimentation and risk-taking. Hackathons provide a safe space for teams to experiment with unconventional ideas without the fear of failure. This fosters a culture where calculated risks are embraced, and failure is seen as a valuable learning opportunity.

8. Incorporating External Perspectives

Hackathons can extend beyond internal teams. Collaborating with external participants, such as startups, industry experts, or even customers, introduces fresh perspectives and diverse skill sets. This external collaboration can inject new ideas and approaches into the innovation process.

9. Providing Resources and Support

Successful hackathons require adequate resources and support. Providing access to cutting-edge technologies, mentorship from subject matter experts, and a supportive organisational infrastructure are essential components for maximising the impact of hackathons.

10. Evaluating and Implementing Results

The ultimate success of a hackathon lies in the evaluation and implementation of results. Organisations should establish clear criteria for assessing the outcomes of the hackathon and create a structured process for transitioning viable solutions into actual projects or products.

Conclusion: Shaping the Future through Innovation Leadership

Fostering innovation leadership is not merely about embracing change but actively shaping the future of technology. By understanding the innovation landscape, fostering a culture of innovation, leading technological change, adapting to advancements, encouraging collaboration, measuring success, committing to continuous learning, and leveraging collaborative platforms, technology leaders can navigate the dynamic landscape with vision and resilience. Case studies and practical examples illustrate that innovation leadership is not a theoretical concept but a tangible and transformative force driving technological progress.

6.6 Embracing Ethics and Responsible Technology Leadership

Ethical considerations are integral to decision-making in technology leadership. This section examines the ethical dimensions of technology choices. It guides leaders in ensuring responsible technology innovation, taking into account societal impacts and ethical implications. Case studies illustrate the importance of ethical leadership in maintaining trust and reputation.

Embracing Ethics and Responsible Technology Leadership

In the ever-evolving landscape of technology leadership, the role of ethics and responsible decision-making has become paramount. This section delves into the ethical dimensions of technology choices, providing a comprehensive guide for leaders to navigate the complex intersection of innovation, societal impacts, and ethical considerations.

1. Understanding the Ethical Landscape

Before embarking on ethical decision-making, it's crucial for technology leaders to understand the broader ethical landscape. Concepts such as privacy, data security, artificial intelligence ethics, and inclusivity are central. Tools like the Ethical Matrix and Ethical AI Guidelines can serve as compasses, helping leaders navigate the multifaceted ethical considerations inherent in technological choices.

2. Societal Impact Assessment

Responsible technology leadership necessitates a meticulous assessment of societal impacts. Leaders must consider how their technological decisions might affect different segments of society. Case studies, such as the impact of algorithmic bias on marginalised communities, offer tangible examples. The UN's Sustainable Development Goals (SDGs) can serve as a framework, guiding leaders to align technological initiatives with broader societal well-being.

3. Balancing Innovation and Ethical Considerations

Innovation and ethics are not mutually exclusive. This section guides leaders in finding a delicate balance between pushing technological boundaries and ensuring ethical considerations are at the forefront. Tools like the ***Innovation Ambition Matrix*** can assist in aligning innovation goals with ethical principles, ensuring that technology is a force for positive change.

4. Responsible Data Management

Data is a cornerstone of technological advancements, but its responsible management is equally critical. Leaders must establish robust data governance frameworks, ensuring compliance with regulations such as GDPR. Tools like **_DataRobot_** and **_Collibra_** facilitate ethical data management, providing transparency, accountability, and safeguards against misuse.

5. Case Studies in Ethical Leadership

Drawing from real-world examples is a powerful way to illustrate the impact of ethical leadership. Case studies, such as the Cambridge Analytica scandal and the challenges faced by companies like Google in balancing user privacy with data-driven innovation, shed light on the consequences of ethical lapses. These cases serve as cautionary tales and sources of valuable lessons for leaders.

6. Incorporating Ethical Considerations into Decision-Making Frameworks

Ethics should not be an afterthought but an integral part of decision-making frameworks. This section provides guidance on incorporating ethical considerations into existing frameworks. Tools like the Ethical Decision-Making Framework offer a structured approach, helping leaders systematically evaluate the ethical implications of their choices.

7. Building a Culture of Ethical Innovation

Leadership extends beyond individual decisions; it shapes organisational culture. Leaders must foster a culture that values ethical innovation. Initiatives such as ethics training, whistleblower protection, and the implementation of ethical guidelines contribute to building a culture where ethical considerations are woven into the fabric of technological advancement.

8. Leveraging Ethical Certification and Standards

Ethical certifications and standards, such as B Corp Certification and IEEE Global Initiative on Ethics of Autonomous and Intelligent Systems, provide external validation of an organisation's commitment to ethical practices. This section explores how leaders can leverage such certifications to build trust with stakeholders and demonstrate a dedication to responsible technology leadership.

9. Continuous Monitoring and Adaptation

Ethical considerations are dynamic, requiring continuous monitoring and adaptation. Leaders should stay abreast of evolving ethical norms, industry best practices, and regulatory updates. Subscription to ethical tech newsletters, participation in industry forums, and engagement with ethics-focused organisations contribute to ongoing awareness.

10. Stakeholder Engagement and Transparency

Transparent communication with stakeholders is fundamental to ethical leadership. Leaders should engage with stakeholders, including customers, employees, and the broader community, to understand their concerns and expectations. Tools like Transparency-One and Ethical Explorer facilitate transparency, allowing organisations to communicate their ethical commitments effectively.

Case Study 1: NHS Digital and the *Care.data* Initiative

Background: The **Care.data** initiative was a project by NHS Digital in the UK aimed at collecting and sharing patient data from GP records to improve healthcare research and services.

Ethical Dilemma: The initiative faced significant ethical challenges related to patient privacy and data security. Concerns were raised about the potential misuse of sensitive medical information and the lack of clear consent from patients.

Response: Due to widespread concerns and public outcry, the **Care.data** initiative was eventually scrapped in 2016. The decision reflected the importance of respecting patient privacy and the need for transparent communication about data-sharing initiatives in the healthcare sector.

Lessons Learned:

- The critical importance of obtaining informed consent for data-sharing initiatives.
- The need for transparency in communicating the goals and potential impacts of large-scale data projects, especially in sensitive areas like healthcare.

This case study from the UK healthcare sector highlights the ongoing challenges and ethical considerations associated with leveraging technology for data-driven improvements in public services. It underscores the necessity of robust ethical frameworks and transparent communication when handling sensitive information for societal benefit.

Case Study 2: Facebook and the Cambridge Analytica Scandal

Background: In 2018, the **Cambridge Analytica** scandal unfolded, revealing that the personal data of millions of Facebook users had been harvested without their consent for political purposes.

Ethical Dilemma: The scandal raised significant ethical concerns regarding data privacy, user consent, and the responsible management of user information.

Response: Facebook faced immense public backlash and regulatory scrutiny. The company implemented various measures to enhance data privacy, including stricter data access policies, increased transparency, and changes to its platform to give users more control over their data.

Lessons Learned:

- The importance of proactive data governance and protection measures.
- The enduring impact of ethical lapses on user trust and organisational reputation.

Case Study 3: Google's Project Maven

Background: In 2018, Google faced ethical challenges when it was revealed that the company had been working on Project Maven, a military AI initiative with the U.S. Department of Defence. The project aimed to use artificial intelligence for analysing drone footage.

Ethical Dilemma: Google employees and external stakeholders raised concerns about the use of AI for military purposes, citing potential risks to human rights and the blurred lines between civilian and military applications.

Response: In response to the ethical concerns raised by employees and the public, Google decided not to renew its contract for Project Maven. This case highlights the importance of considering societal impacts and ethical implications when engaging in government contracts, showcasing the influence of employee activism in shaping ethical decisions.

Lessons Learned:

- The need for transparent communication and ethical guidelines in engaging with government projects.
- The impact of employee activism in influencing ethical decision-making.

By understanding the ethical landscape, assessing societal impacts, balancing innovation with ethical considerations, managing data responsibly, studying case studies, incorporating ethics into decision-making frameworks, fostering a culture of ethical innovation, leveraging certifications, monitoring continuously, and engaging transparently with stakeholders, leaders can navigate the intricate ethical dimensions of technology leadership and ensure that their innovations contribute positively to society.

6.7 Continuous Learning, Professional Development, and Leading Through Challenges

The commitment to ongoing learning not only enhances your credibility within your field but also ensures that as a professional you remain at the forefront of industry advancements. Credibility is earned through a demonstrable dedication to staying informed, adapting to change, and mastering new skills. As professionals continually expand their knowledge base, they not only gain the trust of colleagues and stakeholders but also position themselves as authorities within their respective domains.

The impact of continuous learning extends beyond personal credibility to the very core of knowledge acquisition and application. Professionals who engage in ongoing development contribute significantly to the knowledge capital of their organisations. By staying abreast of the latest trends, methodologies, and technologies, they become invaluable assets to their teams. This continuous influx of new knowledge fuels innovation and business growth. It equips individuals to identify emerging opportunities, propose creative solutions to challenges, and contribute to the development of cutting-edge products or services. In essence, professional development serves as a catalyst for fostering a culture of innovation within organisations.

From a career development perspective, the importance of continuous learning cannot be overstated. In an era where adaptability is synonymous with career longevity, professionals who invest in their own growth and skill enhancement stand out as resilient and forward-thinking leaders. Continuous learning not only opens doors to new opportunities but also provides a competitive edge in a crowded job market. Professionals who consistently develop their expertise are better positioned for career advancement, leadership roles, and executive responsibilities. It becomes a self-reinforcing cycle where ongoing learning propels career development, and a successful career, in turn, encourages a commitment to further professional growth. In essence, professional development and continuous learning are not just activities; they are integral components of a thriving, dynamic, and enduring career journey.

Continuous Learning: A Lifelong Commitment

Embracing a mindset of continuous learning is paramount for a CTO. This involves staying abreast of emerging technologies, industry trends, and evolving best practices. An example of an effective continuous learning strategy is the integration of online learning platforms like ***Coursera, Udacity***, or ***LinkedIn Learning*** into the professional development plan. These platforms offer a plethora of courses ranging from technical skills to leadership development, empowering CTOs to continually expand their knowledge base.

Leadership in Personal and Team Development

Effective leadership extends beyond technical proficiency; it involves nurturing the growth of oneself and the team. A CTO should champion a culture of professional development within the organisation. For instance, establishing mentorship programs, both internally and externally, can provide valuable insights and guidance to both seasoned professionals and emerging talent. This approach not only fosters individual growth but also cultivates a collaborative and knowledge-sharing culture within the team.

Examples of respected Mentorship Programmes

1. **TechWomen:**

    1. *Overview:* TechWomen is an international mentorship program that connects women in STEM (Science, Technology, Engineering, and Mathematics) fields from different parts of the world. Sponsored by the U.S. Department of State, the program pairs emerging leaders with mentors from leading tech companies. TechWomen focuses on cross-cultural exchange and skill development, providing a platform for women to excel in the global tech industry.

2. **Techstars Mentorship Programs:**

    1. *Overview:* Techstars, a renowned startup accelerator, offers mentorship programs designed to support and guide early-stage entrepreneurs and tech professionals. With a network of experienced mentors, including successful entrepreneurs and industry leaders, Techstars provides valuable insights, feedback, and connections to help participants navigate the challenges of building and scaling innovative tech ventures.

Strategies for Navigating Technological Challenges and Crises

In the face of technological challenges and crises, a CTO must exemplify resilience, stability, and foresight. One notable tool for navigating challenges is scenario planning. By considering various potential scenarios and developing contingency plans, a CTO can prepare the organisation for unforeseen disruptions. Notable scenario planning software, such as ***Strategic Decision Group's Decisioneering*** or **IBM's Watson Studio**, provides a digital platform for CTOs to simulate and evaluate diverse scenarios, aiding in the development of comprehensive and adaptive contingency plans.

Furthermore, scenario planning enables CTOs to cultivate a culture of preparedness within their technology teams. By involving key stakeholders in the scenario planning process, a CTO fosters a collaborative environment where employees become adept at identifying potential threats and developing innovative solutions. The participatory nature of scenario planning software, such as ***Detecon's PlanB***, encourages team engagement in crafting contingency plans, ensuring a collective and well-rounded approach to addressing technological challenges. This approach not only enhances the organisation's overall resilience but also positions the CTO and the technology team as proactive leaders capable of navigating uncertainties with agility and strategic acumen.

In addition to utilising scenario planning software, CTOs can draw on industry-specific methodologies and frameworks to augment their crisis preparedness. Frameworks like the ***National Institute of Standards*** and ***Technology (NIST) Cybersecurity Framework*** or the ***Business Continuity Institute's Good Practice Guidelines*** offer structured approaches to identifying, managing, and mitigating technological risks. By integrating these methodologies with scenario planning software, CTOs can develop comprehensive strategies that not only navigate crises effectively but also drive long-term technological resilience and growth for the organisation.

Comparative Analysis of Executive CTO Programmes

The role of Chief Technology Officer (CTO) has evolved significantly over the years, becoming a crucial position in guiding organisations through the complexities of the digital age. Recognising the need for specialised training for CTOs, various executive education programs have emerged to equip these leaders with the skills and knowledge necessary to navigate the ever-changing technology landscape. In this chapter, we will delve into prominent executive CTO programmes.

Ranked Worldwide Programmes

1. Marlow Chief Technology Officer Programme
2. Cambridge Chief Technology Officer Programme
3. MIT Sloan CIO Leadership Program
4. Wharton Chief Technology Officer Programme
5. Stanford Advanced Project Management Program for Technology Executives

Comparative Analysis of Executive CTO Programmes

These reviews are made from a personal review process. Different courses will be more appropriate for different needs. For example the Cambridge programme has an academic approach which will suit those who learn well in an academic setting, whereas The Marlow programme recognises the existing experience of the CTO as a key part of its accreditation.

No.1. Marlow Chief Technology Officer Programme:

- **Focus:** Practical orientation with a curriculum designed and delivered by top industry CTO's rather than academics. Strong focus on Strategy, Leadership Innovation and Entrepreneurship. Focus on _acknowledging what the CTO already knows_ rather than expecting all learning to be new.

- **Strength:** Industry recognition as being on the pulse of accelerating change and emerging technologies and trends. Certificate most recognised internationally.

- **Practical Application:** Emphasises hands-on experience and real-world challenges. Strong peer to peer community for collaboration.

- **Business Collaboration:** Facilitates collaboration with industry experts, fostering real-world connections. Exemplary "buddy" programme.

- **Curriculum:** International focus. Tailored by leading CTO's, prides itself as being the most up to date, designed to address current challenges in technology leadership. Also covers start-up, accelerators, mergers & acquisitions and Private Equity

- **Value for Money:** Ranked No.1

- **Innovation and Entrepreneurship:** Focus on current practical skills that contribute to innovation within organisations.

No.2. Cambridge CTO Programme:

- **Focus:** Comprehensive exploration of technology strategy and organisational leadership.

- **Strength:** Organisational Leadership. *Academic approach*. Name is well known particularly in the UK. Lecturers have a good reputation.

- **Practical Application:** Emphasises case studies and collaborative projects.

- **Business Collaboration:** Strong connections with UK business and technology community. Has networking events, peer learning.

- **Curriculum:** Covers technology strategy, innovation management, and organisational leadership. Choice of electives

- **Innovation and Entrepreneurship:** UK Focus, fostering a culture of innovation.

- **Value for Money:** Ranked 3

No.3. MIT Sloan CIO Leadership Program:

- **Focus:** Leadership in the digital era, digital transformation, and data-driven decision-making.

- **Practical Application:** Extensive exploration of leadership and strategic concepts.

- **Business Collaboration:** Emphasises thought leadership, connecting participants with influential figures.

- **Curriculum:** Broad spectrum, covering technology, business strategy, and organisational leadership.

- **Innovation and Entrepreneurship:** Strong emphasis on fostering a culture of research. US Focus

- **Value for Money:** Ranked 2

No.4. Wharton CTO Programme:

- **Focus:** Integration of business and technology, strategic leadership, and innovation.

- **Practical Application:** Strong emphasis on hands-on projects and simulations.

- **Business Collaboration:** Aligns technology initiatives with business goals.

- **Curriculum:** Emphasis on strategic leadership, digital transformation, and innovation.

- **Innovation and Entrepreneurship:** Focus on driving innovation within organisational contexts.

- **Value for Money:** Ranked 5

No.5. Stanford Advanced Project Management Program for Technology Executives:

- **Focus:** Advanced project management skills for technology executives.

- **Practical Application:** Emphasis on real-world application through case studies.

- **Business Collaboration:** Encourages cross-disciplinary collaboration.

- **Curriculum:** Specialised in project management methodologies, risk assessment, and stakeholder communication.

- **Innovation and Entrepreneurship:** Integrates project management principles into technology project scenarios.

- **Value for Money:** Ranked 4

Comparative Overview:

- **Focus:** While Cambridge and Wharton offer an academic view, MIT Sloan emphasises leadership, and Stanford focuses on project management. The Marlow program leans towards practical application, recognising pre-existing experience.

- **Practical Application:** Wharton, MIT Sloan, and Marlow place a strong emphasis on hands-on experience. Cambridge and Stanford integrate practical application through case studies and real-world projects.

- **Business Collaboration:** All programs value collaboration, but Marlow and MIT Sloan enhance this by connecting participants with influential figures in business and academia.

- **Curriculum:** The Cambridge and Wharton programs provide a well-rounded curriculum. MIT Sloan offers a broad spectrum, while Stanford specialises in project management. The Marlow program tailors its curriculum to current and emerging industry challenges.

- **Innovation and Entrepreneurship:** All programs recognise the importance of fostering innovation. Wharton, and MIT Sloan, focus on driving innovation within organisations, while Marlow and Cambridge emphasise a global innovation perspective.

Conclusion:

The choice among these executive CTO programs depends on individual curriculum preferences, value for money, and recognition of certified status. Each program contributes uniquely to the development of CTOs, offering a diverse set of skills, perspectives, and practical experiences to navigate the complexities of the digital landscape.

* All views expressed are those of the author

6.8 Pioneering Thought Leadership in the Tech Industry

Establishing thought leadership is not just about individual recognition; it's about shaping the discourse of the entire industry. This section explores the significance of personal and organisational thought leadership. It provides insights into influencing industry trends and discourse. Practical guidance helps CTOs position themselves and their organisations as leaders in shaping the technological narrative.

The Essence of Thought Leadership

Individual Thought Leadership

Individual thought leadership in the tech industry involves professionals articulating and sharing their insights, experiences, and perspectives on relevant topics. A notable example is Elon Musk, whose visionary ideas and bold initiatives have not only propelled Tesla and SpaceX but have also influenced discussions on renewable energy and space exploration globally.

Organisational Thought Leadership

Organisational thought leadership extends beyond individual efforts, requiring companies to actively contribute to industry conversations. IBM's "Smarter Planet" initiative was a prime example. By advocating for the intelligent use of technology to address global challenges, IBM positioned itself as a leader in shaping discussions around the role of technology in creating a more sustainable and interconnected world.

Influencing Industry Trends and Discourse

Content Creation and Dissemination

- Thought leaders leverage various platforms to share their insights. Blog posts, whitepapers, and engaging social media content are powerful tools. Neil deGrasse Tyson, an astrophysicist and science communicator, effectively utilises platforms like X (Twitter) and podcasts to disseminate scientific knowledge and influence public discourse.

Participation in Industry Events

- CTOs can establish thought leadership by actively participating in and contributing to industry conferences, webinars, and panel discussions. Sundar Pichai's keynote addresses at Google I/O not only showcase Google's latest innovations but also position him as a key influencer in the tech space.

Collaborations and Partnerships

- Strategic collaborations amplify thought leadership. OpenAI's collaboration with leading research institutions and its commitment to advancing artificial general intelligence has positioned the organisation at the forefront of discussions on the future of AI.

Practical Guidance for CTOs

- **Define Your Vision and Message**
- Articulate a clear vision for the future of your industry or technology domain. Bill Gates, through his annual letters and philanthropic efforts, has consistently communicated his vision of a world with improved health and reduced poverty through innovation.

Build a Strong Online Presence

- Invest in a compelling online presence through a well-crafted personal website, active engagement on social media, and contributions to reputable industry publications. This helps in reaching a broader audience and establishes credibility.

Foster a Culture of Innovation

- Organisations that prioritise innovation naturally become thought leaders. Apple, under the leadership of Steve Jobs, consistently introduced groundbreaking products, setting trends and influencing the entire tech industry.

Resources and Data Sources

Industry Reports and Surveys

- Referencing reputable industry reports and surveys provides data-driven insights. For example, the ***Gartner Magic Quadrant*** reports are widely recognised in the technology industry and can be valuable sources for CTOs looking to substantiate their perspectives.

Academic Research

- Engaging with academic research helps in staying abreast of the latest advancements. Platforms like ***arXiv.org*** host preprints of academic papers, offering a wealth of knowledge that can be applied to real-world challenges.

Collaborative Platforms

- Participate in collaborative platforms like GitHub, fostering open-source contributions and collaboration. Linus Torvalds, the creator of Linux, exemplifies how collaborative development can drive innovation and industry discussions.

Case Study 1: Demis Hassabis - Advancing AI with DeepMind

Background: Demis Hassabis, Co-founder and CEO of DeepMind.

Thought Leadership Highlights:

1. **AI for Scientific Discovery:** Hassabis has expressed a vision for AI to contribute to scientific breakthroughs. DeepMind's AlphaFold, under his leadership, made headlines by predicting protein structures with remarkable accuracy, opening avenues for advancements in biology and medicine.

2. **Ethical AI Framework:** DeepMind, guided by Hassabis, has been at the forefront of promoting ethical AI. They have developed frameworks for responsible AI deployment and are actively engaged in the broader discourse on AI ethics, showcasing thought leadership in ensuring the responsible development of powerful technologies.

3. **Healthcare Innovation:** Demis Hassabis has led DeepMind in collaborations with healthcare institutions, aiming to apply AI to improve patient outcomes. By sharing insights and progress updates, he contributes to the narrative of how AI can positively impact healthcare delivery and patient care.

Impact: Demis Hassabis' thought leadership has positioned DeepMind as a global leader in AI research and application, influencing discussions on the responsible and impactful use of artificial intelligence.

Case Study 2: Werner Vogels - AWS and Cloud Computing Innovation

Background: Werner Vogels, Chief Technology Officer at Amazon Web Services (AWS), is a key figure in the evolution of cloud computing. His thought leadership has played a crucial role in establishing AWS as a leading cloud services provider.

Thought Leadership Highlights:

1. **Evangelising Cloud Computing:** Werner Vogels has been a vocal advocate for the adoption of cloud computing. Through blog posts, keynote addresses, and interviews, he has positioned AWS as a pioneer in cloud technology, influencing the broader tech industry's shift towards cloud-based solutions.

2. **Serverless Computing:** Vogels has been at the forefront of promoting serverless computing architectures. His thought leadership on this topic has shaped industry discussions and influenced the development of serverless technologies, contributing to a paradigm shift in how applications are built and deployed.

3. **Security in the Cloud:** Security is a paramount concern in cloud computing. Vogels has provided thought leadership on AWS's approach to cloud security, contributing to the development of best practices and standards. His influence has been instrumental in building trust among enterprises transitioning to the cloud.

Impact: Werner Vogels' thought leadership has been instrumental in establishing AWS as a leader in cloud computing, shaping industry practices, and influencing the trajectory of cloud technology adoption worldwide.

Case Study 3: Dr. Fei-Fei Li - Advancing Ethical AI at Stanford

Background: Dr. Fei-Fei Li, the former Chief Scientist of AI/ML at Google Cloud, is a prominent figure in the field of artificial intelligence. She is known for her contributions to computer vision and her efforts to advance ethical considerations in AI.

Thought Leadership Highlights:

1. **AI for All:** Dr. Li co-founded the AI4ALL initiative, aimed at increasing diversity and inclusion in the field of artificial intelligence. Through thought leadership talks, interviews, and op-eds, she has been a driving force behind the narrative of making AI accessible and beneficial for all.

2. **Ethical AI Education:** As the director of the Stanford Artificial Intelligence Lab, Dr. Li has been a vocal proponent of integrating ethics into AI education. Her

thought leadership emphasises the importance of responsible AI development and the need for ethical considerations from the early stages of technology creation.

3. **National AI Strategy:** Dr. Li has been actively involved in shaping national AI strategies, providing thought leadership on the need for collaborative efforts between academia, industry, and policymakers. Her influence has contributed to a more nuanced conversation about the societal implications of AI.

Impact: Dr. Fei-Fei Li's thought leadership has had a profound impact on discussions around diversity in AI, ethical considerations, and the role of academia in shaping the future of artificial intelligence.

Conclusion

Thought leadership in the tech industry is a powerful force that can shape the trajectory of innovation, influence public opinion, and position individuals and organisations as driving forces in the digital era. By defining a clear vision, leveraging various communication channels, and actively participating in industry conversations, CTOs can not only contribute to the discourse but lead the way in shaping the future of technology. The examples, resources, and practical guidance provided in this chapter offer a roadmap for CTOs aspiring to become pioneers in thought leadership within the tech industry.

6.9 Exploring Environmental Sustainability & ESG

There has never been a more exciting time for technology professionals. Advances in digital technologies are converging to offer what many believe will be the 4th Industrial Revolution. Advances in AI, Robotics, Gene Therapy, IoT, Drones, Virtual Reality and Clean Energy have the opportunity to create huge good for our world, however over the last 35 years we have lost over half the wild animals on the planet and we are destroying habitats and food stocks through over grazing and pollution. We created the internet but screwed the planet. Well done us!

Our world population is growing exponentially and, with longer anticipated lifespans, this means increased pressure on our resources. Not all will have the ability to invest in technologies that will drive or benefit from this 4th revolution and the gap in society could widen rather than diminish.

This section delves into the imperative of integrating sustainability into technological strategies and leading green technology initiatives. Practical examples showcase successful initiatives, emphasising the role of technology leaders in contributing to environmental sustainability and corporate social responsibility.

The Imperative of Sustainable Technology Leadership

In recent years, there has been a paradigm shift in the business landscape, with organisations recognising the need to embrace environmental sustainability and ESG principles. Technology leaders, including Chief Technology Officers (CTOs), find themselves at the forefront of this change. It's no longer sufficient for technology initiatives to be solely efficient and innovative; they must also align with sustainable practices and contribute positively to the environment.

Successful Initiatives: Examples of Sustainability Leadership

1. **Google's Commitment to Renewable Energy:**

   - *Initiative:* Google has committed to operating on 100% renewable energy. This initiative includes direct investment in renewable projects, such as wind and solar farms, to power their data centres.

   - *Leadership Impact:* Sundar Pichai, CEO of Google, has emphasised the importance of sustainability, demonstrating how a tech giant can lead the way in adopting eco-friendly practices.

2. **Microsoft's Carbon Negative Pledge:**

   - *Initiative:* Microsoft has pledged to become carbon negative by 2030, meaning they will remove more carbon from the atmosphere than they emit. This ambitious goal extends to their entire supply chain.

   - *Leadership Impact:* Satya Nadella, CEO of Microsoft, advocates for corporate responsibility, showcasing how technology leadership can drive significant environmental commitments.

3. **Tesla's Electric Vehicle Revolution:**

   - *Initiative:* Tesla, led by Elon Musk, has been a trailblazer in promoting electric vehicles. Their efforts go beyond cars, encompassing energy storage solutions and solar products.

   - *Leadership Impact:* Musk's visionary approach not only revolutionised the automotive industry but also highlighted the potential for sustainable technology to disrupt traditional sectors.

Resources and Data Sources

1. **World Economic Forum (WEF):**

   1. *Resource:* The WEF provides valuable insights into the intersection of technology and sustainability, offering reports and case studies on how businesses are incorporating sustainable practices.

2. **Global Reporting Initiative (GRI):**

   1. *Resource:* GRI offers comprehensive standards for reporting environmental, social, and governance impacts. This resource assists technology leaders in establishing transparent reporting mechanisms.

3. **Sustainability Accounting Standards Board (SASB):**

   1. *Resource:* SASB provides industry-specific standards for reporting on sustainability factors. This assists technology leaders in aligning their initiatives with sector-specific sustainability goals.

4. **Renewable Energy World:**

   1. *Resource:* For technology leaders looking to stay informed about the latest trends and advancements in renewable energy, Renewable Energy World provides news, articles, and insights.

5. **UN Sustainable Development Goals (SDGs):**

   1. *Resource:* The UN SDGs offer a framework for aligning technology initiatives with global sustainability objectives. Technology leaders can leverage these goals to guide their organisations toward positive social and environmental impacts.

Practical Initiatives for CTOs: Driving Sustainability in Technology

As technology leaders, Chief Technology Officers (CTOs) have the unique opportunity to champion sustainability initiatives within their organisations. Engaging in practical and impactful efforts not only aligns with broader ESG goals but also contributes to the development of a more sustainable future. Here, we explore specific initiatives that CTOs can get involved in.

1. Energy-Efficient Data Centres:

- **Initiative:** Implementing energy-efficient practices in data centers can significantly reduce environmental impact. This includes optimising cooling systems, adopting renewable energy sources, and leveraging advanced technologies for efficient server utilisation.

- **Involvement:** Collaborate with data centre teams to implement energy-efficient technologies and explore partnerships with renewable energy providers.

2. Renewable Energy Integration:

- **Initiative:** Actively pursue the integration of renewable energy sources to power organisational operations. This may involve investing in on-site solar or wind energy systems or procuring renewable energy from external sources.

- **Involvement:** Collaborate with energy procurement teams to explore renewable energy partnerships and technologies, ensuring a seamless integration into the organisation's infrastructure.

3. Sustainable Software Development Practices:

- **Initiative:** Promote sustainable software development practices, including optimising code for energy efficiency, adopting green hosting solutions, and incorporating eco-friendly design principles.

- **Involvement:** Work closely with development teams to integrate sustainability considerations into the software development life cycle. Encourage the adoption of frameworks that prioritise energy efficiency.

4. Green Supply Chain Management:

- **Initiative:** Extend sustainability efforts beyond internal operations by collaborating with suppliers and ensuring that the entire supply chain adheres to environmentally responsible practices.

- **Involvement:** Engage with procurement teams to assess and select suppliers committed to sustainable practices. Leverage technology for real-time tracking and monitoring of supply chain sustainability metrics.

5. Employee Engagement and Education:

- **Initiative:** Foster a culture of sustainability within the organisation by providing employees with the knowledge and tools to make environmentally conscious choices.

- **Involvement:** Support the development of internal educational programs, leveraging technology for training and awareness campaigns. Implement digital tools that enable employees to track and reduce their environmental footprint.

6. Collaboration with Energise Resources:

- **Initiative:** Partner with Energise Resources (www.energiseresources.org) to access valuable insights, resources, and networking opportunities in the sustainability space.

- **Involvement:** Explore Energise Resources for best practices, case studies, and industry collaborations. Engage with the platform to stay informed about the latest trends and innovations in sustainable technology.

7. IoT for Environmental Monitoring:

- **Initiative:** Utilise Internet of Things (IoT) technologies to monitor and optimise environmental conditions within the organisation. This can include smart building systems for energy efficiency and real-time monitoring of resource consumption.

- **Involvement:** Lead the implementation of IoT solutions, collaborating with relevant teams to deploy sensors and analytics platforms that enable data-driven decision-making for sustainability.

CTOs play a pivotal role in shaping the technological landscape of their organisations. By actively engaging in sustainability initiatives, they not only contribute to environmental conservation but also align their organisations with the values of corporate social responsibility. Energise Resources provides an additional avenue for CTOs to access valuable resources, network with like-minded professionals, and stay informed about the latest developments in sustainable technology. As technology leaders, embracing these practical initiatives can drive positive change and set the standard for a more sustainable future.

As technology continues to evolve, so does the responsibility of technology leaders. The integration of sustainability into technological strategies is not just an ethical choice but a strategic imperative. The examples of Google, Microsoft, and Tesla demonstrate the transformative impact that technology leaders can have on environmental sustainability. By leveraging the provided resources and staying informed, CTOs and technology executives can guide their organisations toward a more sustainable and socially responsible future.

6.10 Remote and Hybrid Team Management

Effective leadership for Chief Technology Officers (CTOs) extends beyond the traditional confines of the office. This chapter delves into the intricacies of leading distributed technology teams, offering strategies for effective leadership in remote work environments. Best practices, essential tools, and relevant case studies provide valuable insights into successfully managing remote and hybrid teams while maintaining optimal levels of productivity and collaboration.

The Evolution of Work Structures

Understanding the Remote and Hybrid Landscape

The workplace landscape has undergone a seismic shift with the rise of remote and hybrid work structures. CTOs are now tasked with leading teams scattered across geographical locations, requiring a nuanced approach to team management.

Strategies for Effective Leadership

1. **Clear Communication:**

   - Emphasise transparent and consistent communication.

   - Utilise tools like Slack, Microsoft Teams, or Zoom for real-time collaboration.

   - **Example:** GitLab's Handbook (https://about.gitlab.com/handbook/) serves as a reference for fostering transparent communication in a fully remote environment.

2. **Result-Oriented Approach:**

   - Focus on outcomes rather than micromanagement.

   - Implement project management tools such as Jira or Asana.

   - **Reference:** The book "Remote: Office Not Required" by Jason Fried and David Heinemeier Hansson explores the benefits of a results-oriented approach to remote work.

3. **Digital Collaboration Tools:**

   - Leverage collaboration tools like Miro or Trello for virtual whiteboarding and project planning.

   - Encourage the use of shared documents through platforms like Google Workspace or Microsoft 365.

- **Case Study:** Atlassian's remote work toolkit provides practical insights into using digital collaboration tools effectively (https://www.atlassian.com/team-playbook).

4. Regular Virtual Check-Ins:

- Schedule regular video meetings to maintain team cohesion.

- Use platforms like Zoom or Microsoft Teams for virtual check-ins.

- **Advice:** Consider a mix of one-on-one sessions and team meetings to cater to individual needs.

5. Cultural Reinforcement:

- Reinforce company culture through virtual events and team-building activities.

- Platforms like Donut or TeamBonding facilitate remote team-building.

- **Example:** Automattic, the company behind WordPress, maintains a strong remote culture, with regular meetups and a distributed work ethos.

6. Flexible Work Policies:

- Implement flexible work hours to accommodate different time zones.

- Encourage asynchronous communication where applicable.

- **Reference:** Buffer's State of Remote Work report (https://buffer.com/state-of-remote-work) provides insights into the advantages of flexible work policies.

Additional Data Sources

1. **Remote.co:** A valuable resource offering guides, case studies, and best practices for remote work (https://remote.co/).

2. **Gartner Research:** Gartner provides research and insights on remote work trends and technology adoption (https://www.gartner.com/en).

3. **Harvard Business Review (HBR):** HBR offers articles and research on effective leadership in remote and hybrid work environments (https://hbr.org/).

4. **Owl Labs' State of Remote Work Report:** An annual report providing in-depth insights into the state of remote work globally (https://www.owllabs.com/state-of-remote-work).

The power of bringing people together

Automattic (WordPress):

Approach: Automattic, the company behind WordPress.com, is known for its distributed workforce. Despite having remote employees scattered globally, Automattic brings its team members together for bi-annual company-wide meet-ups known as "Grand Meet-ups." During these events, employees engage in team-building activities, workshops, and socialising. The physical gatherings foster a sense of community, strengthen interpersonal relationships, and enhance communication. The face-to-face interactions contribute to a more cohesive and connected remote team, improving collaboration and teamwork.

GitLab:

Approach: GitLab, a web-based Git repository manager, follows an all-remote model with team members located across different continents. To maintain a strong sense of company culture and collaboration, GitLab organises an annual event called the "GitLab Contribute Summit." This summit brings together team members for workshops, discussions, and team-building activities. The physical gathering provides an opportunity for employees to connect beyond the virtual workspace, fostering a sense of belonging and shared purpose. The benefits include improved team dynamics, increased morale, and a deeper understanding of individual roles within the organisation.

Buffer:

Approach: Buffer, a social media management company with a distributed team, places a strong emphasis on in-person team building. The company organises an annual "Buffer Retreat" where employees from around the world come together. This retreat combines work-related sessions with recreational activities, allowing team members to build stronger relationships and recharge. The benefits include enhanced communication, increased trust among team members, and a renewed sense of motivation and purpose. Buffer's approach to periodic physical gatherings supports a positive remote work culture and contributes to the overall well-being of its distributed workforce.

Useful tools to aid remote or hybrid working

Communication Tools:

Slack: A popular messaging platform for team communication. Enables real-time chat, file sharing, and integrations with other tools.

Microsoft Teams: Integrated into Microsoft 365, Teams offers chat, video conferencing, file sharing, and collaborative workspace features.

Zoom: Video conferencing software that facilitates virtual meetings, webinars, and online collaboration.

Google Meet: Google's video conferencing tool, seamlessly integrated with Google Workspace for virtual meetings.

Project Management Tools:

Trello: A user-friendly project management tool that uses boards, lists, and cards to organise tasks and projects.

Asana: A versatile project management tool that helps teams plan, track, and manage work in a visual manner.

Jira: Particularly useful for software development teams, Jira is an agile project management tool by Atlassian.

Monday.com: A work operating system that powers teams to run projects and workflows with confidence.

Collaboration and Document Sharing:

Google Workspace (formerly G Suite): Offers a suite of cloud-based productivity tools including Google Docs, Sheets, Slides, and Drive for real-time collaboration.

Microsoft 365: Includes Microsoft Word, Excel, PowerPoint, and OneDrive for document creation, sharing, and collaboration.

Dropbox: A cloud-based file storage and collaboration platform for securely sharing files and documents.

Virtual Whiteboarding and Collaboration:

Miro: A digital whiteboard platform that enables teams to collaborate visually, brainstorm ideas, and organise information.

Conceptboard: Online visual collaboration tool for brainstorming ideas, mapping processes, and collaborating on projects.

Employee Engagement and Team Building:

Donut: Slack integration that randomly pairs team members for virtual coffee or lunch meetings to foster casual interactions.

TeamBonding: Offers virtual team-building activities and events to strengthen team connections.

Time Tracking and Productivity:

Toggl: A simple time tracking tool that helps teams monitor work hours and improve productivity.

RescueTime: Tracks time spent on digital devices, providing insights into individual and team productivity.

Employee Recognition and Feedback:

Bonusly: An employee recognition platform that allows team members to give and receive recognition for good work.

15Five: Facilitates continuous feedback and employee performance management through regular check-ins.

Wellness and Mental Health:

Headspace: A meditation and mindfulness app to support employee well-being and mental health.

Calm: Offers guided meditations, sleep stories, and relaxation tools to promote mental wellness.

Case Studies

Case Study 1: Automattic (WordPress) Remote Working

Overview:

- Company: Automattic, the company behind WordPress.com and other web-related products.

- Industry: Technology, Web Development. United Kingdom

- Remote Work Status: Fully remote since its inception.

Background: Automattic is renowned for its commitment to a distributed work model. The company was founded in 2005, and from the beginning, it embraced a remote-first culture. Its workforce spans the globe, with employees working from various locations.

Key Points:

1. Communication and Collaboration: Automattic heavily relies on tools like Slack, Zoom, and P2 (a WordPress-powered platform for team collaboration) to ensure effective communication and collaboration among its globally dispersed team.

2. Results-Oriented Work Environment: The company focuses on outcomes rather than traditional work hours. This approach empowers employees to manage their time effectively, fostering a culture of trust and accountability.

3. Regular Meetups: Automattic organises company-wide meetups where employees from around the world gather to collaborate, share ideas, and build strong personal connections. These meetups play a crucial role in maintaining team cohesion.

4. Wellness and Flexibility: Emphasising employee well-being, Automattic offers a high degree of flexibility in work schedules. The company understands the importance of accommodating different time zones and personal circumstances.

Outcomes:

- Automattic has been highly successful in maintaining a thriving remote culture, contributing to the company's growth and the widespread adoption of WordPress globally.

Case Study 2: Salesforce

Overview:

- Company: Salesforce, a global leader in customer relationship management (CRM) solutions.

- Industry: Cloud Computing, Software. USA

- Hybrid Work Status: Adopted a hybrid working model.

Background: Salesforce, headquartered in San Francisco, implemented a hybrid work model in response to the changing nature of work dynamics, aiming to provide flexibility and cater to employee preferences.

Key Points:

1. Flexibility and Employee Choice: Salesforce's hybrid model allows employees to choose from a variety of working arrangements, including remote, office-based, or a combination of both. This flexibility empowers employees to customise their work environments.

2. Technology Integration: The company invested in technology to facilitate seamless collaboration between remote and in-office workers. Tools like Salesforce Anywhere and Slack integration play a vital role in connecting teams regardless of their physical locations.

3. Office Redesign: Salesforce has been redesigning its physical offices to accommodate the needs of a hybrid workforce. The emphasis is on creating spaces that foster collaboration and engagement when employees choose to work from the office.

4. Well-Being Initiatives: Salesforce places a strong emphasis on employee well-being. The company provides resources and initiatives to support mental health, including virtual wellness sessions, counselling services, and flexible time-off policies.

Outcomes:

- Salesforce's hybrid working model aims to combine the best aspects of in-office collaboration and remote flexibility. The company is positioned to adapt to the evolving expectations of the workforce while maintaining a strong corporate culture.

-

These case studies highlight the diversity of approaches to remote and hybrid working, showcasing successful implementations in different geographic contexts and industries. They underscore the importance of effective communication, flexibility, and employee well-being in navigating the complexities of modern work structures.

Conclusion

Effectively managing remote and hybrid teams is an essential skill for today's CTOs. By embracing clear communication, leveraging digital collaboration tools, and reinforcing company culture, CTOs can foster an environment conducive to productivity and collaboration. The examples, references, and advice presented in this chapter serve as a guide for CTOs navigating the challenges and opportunities of remote and hybrid team management in the contemporary technological landscape.

6.11 Strategies for Diversity and Inclusion in Technology Leadership

Diversity and inclusion are not just moral imperatives but crucial elements for building robust and innovative technology teams. This section explores practical strategies for fostering inclusive leadership practices and leading diversity initiatives within the technology sector. Drawing from real-world examples, case studies, and success stories, we delve into the positive impact of diversity on innovation, team dynamics, and the overall health of the tech industry.

Understanding the Business Case for Diversity

Diversity and inclusion go beyond mere compliance; they are strategic imperatives that contribute to organisational success. Numerous studies, including *McKinsey's "Diversity Wins"* reports, have consistently shown a positive correlation between diverse leadership teams and financial performance. In the context of technology, diversity becomes a catalyst for innovation, driving creativity and problem-solving.

Practical Strategies for Inclusive Leadership

1. **Establishing Inclusive Hiring Practices:**

   - *Example:* IBM's commitment to inclusive hiring is evident through its P-TECH (Pathways in Technology Early College High School) program, providing diverse students with the education and skills needed for tech careers.

   - *Advice:* Implement blind recruitment processes, diverse interview panels, and partnerships with educational institutions to attract a broad talent pool.

2. **Cultivating Inclusive Work Environments:**

   - *Example:* Salesforce's focus on equality includes the implementation of the "Ohana" culture, fostering a sense of family and belonging for employees of all backgrounds.

   - *Advice:* Establish employee resource groups, mentorship programs, and inclusive policies to create an environment where every team member feels valued and supported.

3. **Promoting Diverse Leadership Development:**

- *Example:* Microsoft's LEAP program (Leading and Executing for Allyship and Progress) aims to empower diverse talent within the company for leadership roles.

- *Advice:* Implement leadership development programs that identify and nurture diverse talents, providing mentorship and sponsorship opportunities.

4. **Embracing Remote Work Inclusivity:**

- *Example:* GitHub's commitment to building a more inclusive remote work culture includes initiatives like "Distributed, Together," emphasising the importance of connection in virtual environments.

- *Advice:* Ensure remote work policies consider inclusivity, provide necessary resources, and foster virtual collaboration to bridge geographical and cultural gaps.

Useful Data Sources and References

McKinsey & Company - "Diversity Wins" Report:

Reference: Explore the insights provided in McKinsey's report on the business case for diversity and its positive impact on financial performance.

Tech Nation - Diversity & Inclusion Toolkit:

Resource: The Diversity & Inclusion Toolkit by Tech Nation provides practical guidance and resources for tech leaders looking to foster diversity in their organisations.

CIPD - Diversity and Inclusion:

Reference: The Chartered Institute of Personnel and Development (CIPD) offers research and insights on diversity and inclusion in the workplace.

Inclusive Tech Alliance:

Resource: The Inclusive Tech Alliance provides tools, resources, and events to support diversity and inclusion efforts in the tech industry.

Guidance on Measuring Diversity and Inclusion

Measuring diversity and inclusion (D&I) is a critical aspect of creating an equitable workplace. It involves tracking various demographic factors, assessing workplace policies, and evaluating the overall inclusivity of an organisation. Here's a guide on how to measure D&I:

1. Demographic Data Collection:

- **Guidance:** Begin by collecting data on key demographic factors, including gender, race, ethnicity, age, sexual orientation, disability status, and more.

- **Best Practices:** Use voluntary self-disclosure, ensure data privacy, and employ inclusive categorisation to respect individual identities.

2. Representation Metrics:

- **Guidance:** Analyse the representation of diverse groups across different levels of the organisation, from entry-level positions to leadership roles.

- **Best Practices:** Set benchmarks for representation, identify areas for improvement, and track progress over time.

3. Employee Surveys:

- **Guidance:** Conduct regular surveys to gather insights into the workplace experience, perceptions of inclusivity, and satisfaction among diverse groups.

- **Best Practices:** Use anonymised surveys, include open-ended questions, and act on the feedback received.

4. Pay Equity Analysis:

- **Guidance:** Regularly assess pay equity to ensure that employees are compensated fairly regardless of demographic factors.

- **Best Practices:** Use standardised job evaluations, conduct regular pay audits, and address any identified disparities promptly.

- **Examples:** *UK - https://www.reed.com/tools/uk-salary-guide-2023 USA – https://www.splunk.com/en_us/blog/learn/it-salaries.html*

5. Promotion and Advancement Tracking:

- **Guidance:** Monitor the progression of employees through the organisation to identify any disparities in promotion rates among different demographic groups.

- **Best Practices:** Examine promotion rates at each level, provide mentorship opportunities, and implement transparent promotion processes.

6. Inclusive Policies and Practices:

- **Guidance:** Evaluate the inclusivity of workplace policies, including flexible work arrangements, parental leave, and accommodation policies.

- **Best Practices:** Regularly review and update policies, communicate inclusivity efforts, and ensure accessibility for all employees.

Government Practices in the UK and US

United Kingdom:

1. **Equality Act 2010:**

    1. **Data Collection:** The UK Equality Act mandates employers to collect and report gender pay gap data annually for organisations with 250 or more employees. While not explicitly requiring other demographic data, companies are encouraged to voluntarily disclose additional diversity metrics.

2. **Corporate Governance Code:**

    1. **Reporting:** The UK Corporate Governance Code recommends that companies disclose the gender balance within their organisation, both in the boardroom and across management levels. Companies are encouraged to explain their diversity policies and practices.

United States:

1. **Equal Employment Opportunity (EEO) Reports:**

    1. **Data Collection:** The U.S. government mandates certain employers to submit EEO reports, providing data on the gender, race, and ethnicity of their workforce. This data helps monitor compliance with equal employment opportunity laws.

2. **Diversity and Inclusion Efforts:**

    1. **Reporting:** The U.S. government encourages federal contractors to report on their diversity and inclusion efforts. This includes initiatives to promote equal opportunities and eliminate barriers for underrepresented groups.

Conclusion

Measuring diversity and inclusion requires a combination of quantitative and qualitative approaches. By collecting demographic data, tracking representation metrics, conducting employee surveys, and regularly assessing policies, organisations can gain insights into their D&I efforts. UK, EU and US governments emphasise the importance of transparency and reporting to drive diversity and inclusion in workplaces, providing guidelines and frameworks for organisations to follow. Continuous evaluation and improvement based on these metrics are essential for creating workplaces that are not only diverse but truly inclusive.

By implementing inclusive leadership practices, embracing diversity initiatives, and learning from successful case studies, CTOs can foster an environment where every team member can thrive. Embracing diversity not only contributes to a more innovative and creative tech industry but also aligns with the ethical responsibility of technology leaders to create a workplace that is equitable and inclusive.

6.12 Leadership in Mergers and Acquisitions

In this section we delve into strategies for leading technology integration during M&A and offer practical insights, case studies, and key considerations. These resources aim to equip CTOs to play a pivotal role in ensuring the success of technological transitions amid organisational mergers and acquisitions.

Understanding the Landscape of M&A

1. The Strategic Imperative:

- **Example:** SoftBank's acquisition of ARM Holdings strategically positioned the company in the semiconductor industry. M&A can drive long-term technological advantages.

2. Navigating Cultural Integration:

- **Example:** When IBM acquired Red Hat, the successful integration was not just technological but also cultural. Red Hat's open-source ethos was maintained, illustrating the importance of aligning organisational cultures.

Strategies for CTOs in M&A

Early Due Diligence: Conducting thorough due diligence on the technology landscape of the target company early in the M&A process is pivotal. Early identification of technological synergies and challenges sets the stage for effective integration. Understanding the technology infrastructure, software applications, and potential compatibility issues enables the acquirer to make informed decisions about the feasibility and risks associated with the merger. To achieve this, engaging technology experts, conducting comprehensive audits, and leveraging due diligence tools contribute to a detailed understanding of the target's technology assets and liabilities.

Establish a Cross-Functional Team: Building a cross-functional team is essential for a holistic approach to integration. An exemplary case is Cisco's USD 28 Billion acquisition of Splunk in 2023. Cisco hasn't always had the best reputation for acquisitions that live up to expectations but on this occasion they put together a diverse team comprising engineers, sales, and support staff who collaborated seamlessly. This approach ensured that challenges from various perspectives were addressed, and the integration plan aligned with the goals of different departments. Creating a cross-functional team involves identifying key stakeholders, involving representatives from different departments early in the process, and fostering effective communication among team members. This diversity of skills and expertise helps in navigating the complexities of technology integration.

Develop a Business Integration Technology Roadmap: Creating a clear and detailed technology roadmap is crucial for a successful M&A integration, as highlighted in "***M&A Technology Integration: A Practical Guide" by Marc J. Schiller***. The roadmap outlines the integration process and aligns it with the overall business strategy. Although we call it a "Technical" roadmap, it should consider both cultural and technical components to ensure a comprehensive approach. Developing a roadmap involves defining key milestones, establishing timelines, and incorporating feedback from the cross-functional team. This planning minimises disruptions, manages expectations, and guides the entire organisation through the integration process.

Communication is Key: Open and transparent communication is essential throughout the M&A process. Microsoft's acquisition of LinkedIn serves as a case study where effective communication provided a seamless experience for users and shareholders. Key considerations involve keeping internal and external stakeholders informed about the progress, challenges, and benefits of the integration. This transparency fosters trust, alleviates concerns, and ensures alignment with the overall strategic objectives. Implementing a communication strategy that includes regular updates, Q&A sessions, and addressing concerns promptly contributes to a smoother transition. If this topic is of particular interest ***the Ultimate Guide to M&A : Microsoft + LinkedIn Case Study by the Wall Street Prep*** is an excellent read. Available by search on the internet.

Agility in Technological Transitions: Adopting an agile approach to technological transitions allows for flexibility and quick adjustments in response to unforeseen challenges. Referenced in *"Agile M&A: Proven Techniques to Close Deals Faster and Maximise Value" by J. Stewart Black and Donald E. Hess*, this insight encourages a mindset that embraces change. Implementing agile practices involves iterative testing, continuous feedback loops, and adaptive project management methodologies. This agility ensures that the integration process remains responsive to evolving circumstances, technological advancements, and the dynamic nature of the business environment.

Legacy System Integration: In the Vodafone-Idea merger, addressing the integration of legacy systems emerged as a critical technological challenge. Developing a comprehensive plan for integrating and modernising legacy systems is a key consideration. This involves assessing the compatibility of existing systems, prioritising modernisation efforts based on strategic goals, and planning for a phased transition. The integration plan should include measures to mitigate risks associated with legacy systems, ensuring a smooth transition while maintaining operational efficiency and data integrity.

Additional Useful Data Sources

1. **Harvard Business Review (HBR):**

    1. *Resource:* HBR offers a wealth of articles on M&A strategies, technological transitions, and leadership during organisational change.

2. **Mergers & Acquisitions Journal:**

    1. *Resource:* Stay informed about the latest trends, case studies, and best practices in M&A through reputable journals like the Mergers & Acquisitions Journal.

3. **Deloitte's M&A Technology Integration Guide:**

    1. *Resource:* Deloitte provides a comprehensive guide on technology integration during M&A, offering practical frameworks and insights.

4. **TechCrunch and Financial Times:**

    1. *Resource:* Regularly follow technology and business news from reputable sources like TechCrunch and Financial Times for insights into recent M&A activities and their technological implications.

Keeping Staff after the Acquisition

Retaining key staff post-acquisition is a pivotal aspect in ensuring the success and continuity of the integrated entity. Communication is paramount during this period, as uncertainty can lead to anxiety among employees. A transparent and comprehensive communication strategy, outlining the vision and goals of the merged organisation, can help allay concerns and foster a sense of belonging. Providing clarity on the future trajectory, job roles, and the value each employee brings to the newly formed entity is instrumental in retaining key personnel who might be uncertain about their roles post-acquisition.

In particular for technology talent post-acquisition, where skilled professionals are in high demand, organisations must recognise the intrinsic value of their technology teams. A strategy that articulates the exciting technical prospects that sits alongside the business vision and goals, and the role each individual plays in shaping the future technical journey can serve as a powerful incentive. By providing a clear vision of upcoming projects, technological advancements, career enhancement through new technologies, innovation opportunities, and the collective impact of the integrated team, organisations can instil a sense of purpose and excitement among technology professionals, reinforcing their commitment to the newly formed entity.

Organisations can enhance the allure of the future technical journey by aligning it with attractive financial incentives. Offering technology-focused retention bonuses, career development opportunities, and participation in cutting-edge projects can be instrumental. Technology professionals stand out from other professions as they often seek roles that not only challenge their skills but also offer a pathway for career growth and continuous learning. Ensuring that the post-acquisition period brings forth stimulating technical challenges, professional development initiatives, and a supportive environment can be a very compelling strategy to retain key technology talent, making them feel valued and invested in the exciting technological future of the integrated organisation.

In addition to effective communication, offering competitive and attractive financial incentives can be a powerful tool for retaining key staff. Employee retention bonuses, performance-based incentives, and stock options are commonly employed financial options. Retention bonuses provide a one-time payout to employees who stay with the company for a specified period post-acquisition. Performance-based incentives tie financial rewards to individual or team achievements, aligning employee goals with the overall success of the integrated organisation. Stock options, especially in the form of restricted stock units (RSUs) or employee stock ownership plans (ESOPs), not only provide financial incentives but also create a sense of ownership and alignment with the company's long-term success.

Additionally, Long-Term Incentive Plans (LTIPs) can be instrumental in retaining key talent. LTIPs typically involve granting employees a stake in the company's equity over an extended period, contingent on meeting specific performance benchmarks. This approach not only provides financial rewards but also fosters a sense of loyalty and dedication among staff. By implementing LTIPs alongside other financial retention strategies, organisations can create a comprehensive and attractive package that addresses both short-term and long-term employee motivations, thereby enhancing the likelihood of retaining valuable talent post-acquisition.

Furthermore, organisations can consider implementing retention agreements that outline the terms and conditions under which employees receive additional benefits for staying with the company during the post-acquisition period. These agreements may include specific milestones or timelines and can be customised based on the unique needs of the organisation. By tailoring financial options to recognise and reward the contributions of key staff, organisations can significantly increase the likelihood of retaining talent critical to the ongoing success of the integrated entity.

Do you understand the SaaS components in an acquisition?

The prevalence of Software as a Service (SaaS) in contemporary business operations underscores its indispensable role, establishing it as a necessary cost of conducting business. While determining a precise value for SaaS, in general, can be challenging and varies based on specific software solutions, the fact remains that understanding SaaS has to be higher on the agenda during M&A. In the current landscape, businesses allocate a substantial portion of expenditure, typically ranging between 5% to 15%, towards SaaS solutions. This significant share of the financial pie emphasises the critical need to understand if organisations manage and optimise their SaaS investments effectively.

However, managing SaaS costs is not a straightforward task, especially considering that approximately 70% of vendors include clauses in their contracts allowing them to alter service prices whenever they wish. This lack of price predictability leaves organisations uncertain about the future cost implications of their SaaS software. A study by Vertice sheds light on the negotiation landscape in the SaaS realm, revealing that all SaaS providers analysed reported a willingness to settle for more budget-friendly rates when clients engaged in negotiation, yet few did. This negotiation process yielded substantial savings, with reported reductions ranging from 15% to 34% per user, underscoring the potential for organisations to optimise their SaaS expenditure through effective

307

negotiation strategies.

In essence, the financial commitment to SaaS is a critical consideration for businesses, and the dynamic pricing environment necessitates strategic approaches to cost management. The ability to negotiate with SaaS providers not only empowers organisations to gain better control over their expenditure but also highlights the flexibility within the SaaS market. As businesses continue to rely on SaaS for essential functionalities, understanding and actively managing the costs associated with these services become paramount for financial stewardship and operational efficiency.

Due Diligence Approaches for Mergers and Acquisitions

As technology evolves, mergers and acquisitions become strategic pathways for growth and innovation. For Chief Technology Officers (CTOs), overseeing due diligence in such transactions is paramount to ensure seamless integration and mitigate risks.

Why Due Diligence Matters

Due diligence is the process of thorough investigation and analysis to evaluate the financial, operational, and technological aspects of a target company. For technology companies, successful due diligence ensures compatibility, risk mitigation, and the realisation of synergies. Let's delve into the key considerations for CTOs:

Approaches to Technical Due Diligence

1. Infrastructure Assessment:

Conducting an Infrastructure Assessment in the acquisition of another organisation aims to thoroughly evaluate the strength and scalability of the target company's underlying technological framework. This assessment is a critical phase in the pre-acquisition due diligence process, seeking to provide a comprehensive understanding of the target's IT infrastructure. Robustness, in this context, refers to the resilience and reliability of the systems in place, ensuring they can effectively support ongoing operations without frequent disruptions or vulnerabilities. Scalability, on the other hand, assesses the capacity of the infrastructure to adapt and expand in response to growing business needs, ensuring that it can accommodate increased data loads, user traffic, and other demands as the organisation evolves.

During the Infrastructure Assessment, various elements come under scrutiny, including hardware, software, networking, and security protocols. The goal is to identify any weaknesses, potential points of failure, or limitations that could impact the target company's operational efficiency or hinder seamless integration with the acquiring organisation. This evaluation provides invaluable insights into the compatibility of the IT environments, aids in the formulation of integration strategies, and ultimately informs decision-making processes surrounding the acquisition. By ensuring that the target company's infrastructure aligns with the acquiring organisation's standards and future growth plans, the Infrastructure Assessment becomes a strategic tool for mitigating risks and enhancing the overall success of the acquisition.

2. **Codebase Analysis:**

Codebase analysis is a critical phase in the due diligence process during an acquisition, with its primary objective being to comprehensively assess the quality, security, and maintainability of the target company's software codebase. This thorough examination aims to uncover the intrinsic strengths and weaknesses inherent in the code, providing valuable insights into the overall health of the software infrastructure. An exemplary method within this process is conducting a meticulous code review, whereby experienced developers delve into the intricacies of the target's source code. This examination is geared towards identifying potential vulnerabilities that could pose security risks and assessing the extent of technical debt – the accumulated cost of delayed or postponed software maintenance. By scrutinising the codebase, the acquiring entity gains a comprehensive understanding of the software's robustness, potential areas for improvement, and the long-term viability of the technology stack.

A nuanced codebase analysis not only evaluates the immediate condition of the software but also aids in predicting future challenges and opportunities. Through this examination, the acquiring organisation can proactively address issues related to code quality and security, mitigating potential risks before integration. Additionally, insights gained from codebase analysis inform strategic decisions regarding the scalability and adaptability of the software in alignment with the acquirer's broader technological goals. Thus, the codebase analysis serves as a pivotal due diligence step, ensuring a well-informed approach to the acquisition and setting the foundation for a successful integration of technology assets.

3. **Technology Stack Alignment:**

Technology Stack Alignment is a critical phase in the merger and acquisition process, aiming to harmonise the technological infrastructures of the acquiring and target companies. The primary objective is to guarantee seamless compatibility and integration feasibility between the distinct technology stacks employed by the entities involved. This entails a meticulous evaluation of the software, hardware, programming languages, frameworks, and databases in use. By conducting a comprehensive analysis, organisations can identify areas of synergy and divergence, allowing them to make informed decisions about how to align these stacks for optimal operational efficiency and interoperability. Successful alignment ensures that the technological foundation of the integrated entity is robust, minimising potential disruptions and enabling a smooth transition as both companies converge into a unified technological ecosystem.

In practical terms, Technology Stack Alignment involves assessing the current state of technology assets in both the acquiring and target companies. This includes an evaluation of the compatibility of existing systems, identification of redundant technologies, and determining potential areas of enhancement. The alignment process is not only about achieving technical cohesion but also ensuring that the integrated technology stack aligns with the overarching business objectives and future strategic goals of the combined entity. By achieving a seamless integration of technology stacks, organisations set the stage for a more efficient and collaborative post-acquisition environment, unlocking the full potential of the consolidated technological resources.

4. IP and Software Asset Examination:

Conducting an IP and Software Asset Examination is a critical component of due diligence in business transactions, ensuring that an organisation's intellectual property (IP) and software assets are legally owned and compliant. The primary objective of this examination is to verify the authenticity and legality of these assets, providing a comprehensive understanding of their ownership status. An example of how this objective is achieved involves conducting an IP audit, a systematic review process that confirms the rightful ownership of proprietary technologies. This audit encompasses an in-depth assessment of patents, trademarks, copyrights, and any other IP elements to ascertain their legal standing and identify potential risks or discrepancies. By conducting such an examination, businesses seeking to engage in mergers, acquisitions, or partnerships can mitigate legal risks, protect their interests, and ensure that the transaction is founded on a secure and legal foundation.

In the context of business transactions, an IP and Software Asset Examination serves as a safeguard against potential legal complications and enhances the overall risk management strategy. Verifying the ownership and legality of intellectual property and software assets is crucial not only for compliance with legal frameworks but also for protecting the inherent value and competitive edge that these assets bring to an organisation. This diligence ensures that the acquiring party has a comprehensive understanding of the IP landscape, reducing the likelihood of future disputes and contributing to a more informed decision-making process throughout the business transaction.

5. **Cybersecurity Evaluation:**

Conducting a comprehensive cybersecurity evaluation is a pivotal step in scrutinising the digital resilience of a target company during an acquisition. The primary objective is to assess the effectiveness of the target's cybersecurity measures and identify potential vulnerabilities or risks that could pose threats to data integrity and confidentiality. A robust cybersecurity evaluation involves various methodologies, such as penetration testing, which simulates cyber-attacks to uncover weaknesses in the target's security infrastructure. This method helps in identifying potential entry points for malicious actors and gauging the effectiveness of existing defensive mechanisms. Additionally, the evaluation entails a meticulous review of encryption protocols employed by the target, ensuring that data transmission and storage adhere to industry best practices for safeguarding sensitive information. Lastly, scrutinising the incident response plans of the target company is essential to ascertain their preparedness and effectiveness in mitigating and recovering from cybersecurity incidents promptly.

For example, the cybersecurity evaluation might employ penetration testing tools and ethical hacking techniques to simulate real-world cyber threats. By systematically assessing the target's network, applications, and systems, the evaluation can reveal vulnerabilities that need immediate attention. The thorough examination of encryption protocols ensures that sensitive data, whether in transit or at rest, remains adequately protected against unauthorised access. Additionally, reviewing incident response plans helps in determining the target's capability to detect, respond, and recover from security incidents. This multifaceted approach to cybersecurity evaluation not only provides insights into the target company's current security posture but also informs the acquirer about potential areas for improvement and strategic investments to fortify the overall cybersecurity resilience of the integrated entity.

6. **Regulatory Compliance Check:**

Conducting a Regulatory Compliance Check is a vital step in the due diligence process during mergers and acquisitions. The primary objective is to verify that the target company strictly adheres to pertinent industry regulations and standards. This encompasses a thorough examination of the target's compliance with data protection laws, industry-specific regulations, and international standards. For instance, confirming the adherence to data protection laws ensures that the target handles sensitive information ethically and legally. Industry-specific regulations may include guidelines that govern manufacturing processes, quality control, or service delivery, ensuring that the target operates within the defined parameters of its sector. Additionally, aligning with international standards enhances the target company's global credibility and minimises risks associated with regulatory non-compliance, making the Regulatory

Compliance Check a critical aspect of the overall due diligence process.

As an illustrative example, during the Regulatory Compliance Check, the acquiring company may investigate whether the target has implemented robust measures to comply with data protection laws such as GDPR or HIPAA. Additionally, if the target operates in a regulated industry like finance or healthcare, the check would extend to confirming adherence to specific regulations governing those sectors. International standards such as ISO certifications may also be scrutinised, indicating the target's commitment to maintaining high-quality standards in its operations. By thoroughly examining the target's regulatory compliance, the acquiring company can make informed decisions, mitigate potential legal risks, and ensure a smoother integration *process, ultimately contributing to the overall success of the merger or acquisition.*

References and Resources

1. **NIST Cybersecurity Framework:**

   1. *Resource:* The National Institute of Standards and Technology (NIST) provides a framework for improving cybersecurity that can guide the evaluation of cybersecurity measures.

2. **ISO/IEC 27001:**

   1. *Resource:* The international standard for information security management systems offers a comprehensive framework for evaluating security practices.

3. **The Open Web Application Security Project (OWASP):**

   1. *Resource:* OWASP provides resources and guidelines for assessing and improving the security of software applications.

Technology Considerations in Due Diligence: Red Flags

During the due diligence process, Chief Technology Officers (CTOs) must be vigilant in uncovering potential strategies that companies may employ to present themselves as more profitable than their actual financial standing. In the lead-up to a sale, there is a risk that the investment in technology may be manipulated or diminished to create a deceptive appearance of higher profitability. CTOs should pay keen attention to the following aspects during due diligence:

1. Technology Infrastructure Investments:

- *Red Flags:*

    - Abrupt Reductions: Sudden and substantial cuts in technology infrastructure investments without apparent strategic reasons.

    - Deferred Upgrades: Postponing essential infrastructure upgrades or necessary technology refresh cycles.

- *Implications:* Such tactics may lead to a short-term boost in profitability by reducing immediate costs but can result in long-term technical debt and operational inefficiencies.

2. Research and Development (R&D) Spending:

- *Red Flags:*

    - Slashed R&D Budgets: Drastic reductions in R&D budgets or a shift towards projects with quicker returns on investment.

    - Deferred Innovation: Delaying innovative projects or deprioritising long-term research initiatives.

- *Implications:* Diminished R&D spending may create an illusion of increased short-term profits, but it raises concerns about the company's ability to stay competitive in the long run.

3. Cutbacks in Cybersecurity Measures:

- *Red Flags:*

    - Reduced Security Personnel: Downsizing or restructuring cybersecurity teams.

    - Postponed Security Investments: Delaying crucial cybersecurity investments, such as advanced threat detection systems.

- *Implications:* Scaling back on cybersecurity measures may create short-term savings but exposes the company to increased cybersecurity risks, potentially impacting both data integrity and regulatory compliance.

4. Short-Term Cost-Cutting Measures:

- *Red Flags:*

    - Scaling Down IT Staff: Abrupt layoffs or reductions in IT personnel.

- Outsourcing Core Functions: Outsourcing critical technology functions to cut immediate costs.

- *Implications:* While these measures may show immediate savings, they pose risks to operational stability and may compromise the quality of technology services.

5. Asset Depreciation and Amortisation Tactics:

- *Red Flags:*

    - Aggressive Depreciation Schedules: Accelerated depreciation of assets to artificially boost profitability.

    - Manipulative Amortisation: Manipulating the amortisation of technology-related expenses.

- *Implications:* While these tactics can enhance short-term financials, they may misrepresent the true financial health of the company and impact its valuation.

6. Deferred Maintenance and Software Updates:

- *Red Flags:*

    - Delayed Maintenance: Postponing critical maintenance tasks for technology infrastructure.

    - Skipped Software Updates: Avoiding necessary software updates to cut costs.

- *Implications:* These actions may temporarily reduce expenses but can lead to increased risks of system failures, security vulnerabilities, and operational disruptions.

CTOs engaged in due diligence must remain vigilant and critically evaluate the technological aspects of a company's operations. While short-term cost-cutting measures may create the appearance of increased profitability, they often come at the expense of long-term sustainability and innovation. By carefully examining technology infrastructure investments, R&D spending, cybersecurity measures, staffing decisions, and financial tactics, CTOs can uncover potential facades and ensure that the true value of the technology assets aligns with the presented financial picture. This diligence is essential for making informed decisions and mitigating risks associated with deceptive financial strategies.

Technical due diligence is a complex process that demands a thorough examination of technological aspects during mergers and acquisitions. CTOs must consider infrastructure, codebase, technology compatibility, intellectual property, compliance, and cybersecurity. By utilising the provided checklist and referring to reputable resources and standards, CTOs can navigate the technical due diligence process with confidence, ensuring a seamless integration of technology and maximising the success of M&A activities.

There is a due diligence checklist to be found in Appendix B. This can also be found online at: https://www.marlowbusinessschool/resources

Technical Due Diligence Checklist

No1 Priority: The first thing to consider when doing due diligence on a company is what they may have done to make the company appear more profitable than they really are. In the months approaching a sale the investment in technology may have been reduced to an inappropriate or unsustainable level.

Infrastructure and Architecture:

- Server architecture and scalability assessment.

- Cloud service providers and configurations.

- Disaster recovery and business continuity plans.

Codebase and Development:

- Code review for quality, security, and maintainability.

- Identification of technical debt and potential refactoring needs.

- Assessment of development and release processes.

Technology Compatibility:

- Evaluation of programming languages, frameworks, and databases.

- Compatibility of API endpoints and data interchange formats.

- Assessment of middleware and integration points.

Intellectual Property and Compliance:

- IP audit for software and technology ownership.

- Evaluation of licenses, patents, and trademarks.

- Confirmation of compliance with data protection laws and industry regulations.

Cybersecurity and Data Security:

- Penetration testing and vulnerability assessments.

- Review of encryption protocols and data storage security.

- Assessment of incident response plans and security policies.

Chapter 7.

Digital Transformation and Re-Engineering

In an era defined by rapid technological advancements, businesses are compelled to embark on a journey of digital transformation and re-engineering to remain competitive and relevant. The convergence of emerging technologies, such as Artificial Intelligence (AI), the Internet of Things (IoT), and cloud computing, has catalysed a paradigm shift in how organisations operate and deliver value. This chapter explores the landscape of digital transformation, delving into its intricacies and showcasing compelling examples of businesses that have successfully navigated this transformative journey.

1. **"Laying the Groundwork: Preparing for a Digital Transformation Journey"**

Overview of the initial planning and assessment phases.

2. **"Selecting the Right Technologies: Navigating the Digital Landscape"**

Guidance on choosing technologies that align with the organisation's needs and goals.

3. **"Agile Methodologies in Action: Iterative Approaches to Transformation"**

Utilising agile methodologies for flexibility and adaptability in project execution.

4. **"Integration Challenges and Solutions: Ensuring Seamless Connectivity"**

Addressing the complexities of integrating new systems with existing infrastructure.

7.1 Embracing Digital Transformation and Re-engineering

Digital Transformation: A Holistic Business Shift

Digital transformation is a holistic and strategic overhaul that transcends mere technological upgrades; it entails a profound rethinking of entire business models, operational processes, and customer interactions. At its core, this shift signifies a departure from traditional approaches and embraces a comprehensive integration of digital technologies into every facet of an organisation. It's not just about adopting new tools; it's about leveraging technology to drive innovation, enhance agility, and ultimately reshape the way a business delivers value.

Amazon's Retail Revolution: A Case in Point

An exemplary case illustrating the transformative power of digital innovation is the journey of e-commerce giant Amazon. By pioneering seamless online experiences, harnessing the potential of predictive analytics, and optimising logistics through digital technologies, Amazon fundamentally redefined the retail landscape. The company's ability to anticipate customer needs, streamline supply chains, and personalise interactions showcases the transformative impact of digital integration. Amazon's success serves as a compelling testament to the potential rewards of embracing digital transformation at a comprehensive level.

Customer Expectations Redefined

One of the key outcomes of the digital imperative is the redefinition of customer expectations. Amazon's success was not solely due to the efficiency of its logistics or the sophistication of its algorithms; it was rooted in a fundamental shift in how customers perceive and engage with retail. The seamless, personalised experiences offered by Amazon set a new standard, shaping consumer expectations across industries. Digital transformation becomes a means not only to meet but to exceed evolving customer demands, making customer-centricity a cornerstone of successful business strategies in the digital era.

Navigating the Digital Landscape: Insights from McKinsey & Company

Understanding the intricacies of digital transformation is crucial for organisations navigating this paradigm shift. McKinsey & Company's insightful resource, "Digital strategy: The four fights you have to win," offers a comprehensive guide to the challenges and strategies involved in successful digital transformation. The resource delves into key aspects such as navigating the digital landscape, addressing organisational resistance, and aligning technology initiatives with overarching business goals. By examining the four strategic 'fights' outlined in the resource, businesses gain valuable insights to effectively strategise and execute their digital transformation initiatives.

Resource:

- McKinsey & Company. (2018). "Digital strategy: The four fights you have to win."

Re-engineering for Agility and Efficiency

Digital re-engineering is a strategic imperative for organisations seeking to harness the full potential of digital technologies, driving efficiency and agility in their operations. At its core, this transformative process involves a meticulous examination and restructuring of existing workflows and processes to align with the capabilities offered by emerging technologies.

Organisations, such as the multinational conglomerate Siemens, have successfully embraced digital re-engineering to elevate their operational efficiency. Siemens, a global leader in industrial manufacturing, employed a comprehensive digitalisation strategy to streamline operations across diverse business units. This involved deploying advanced technologies to automate repetitive tasks, enabling employees to redirect their focus towards strategic initiatives and creative endeavours. Siemens' commitment to digitalisation stands as a testament to the transformative power of re-engineering in large-scale enterprises.

Resource:

- Siemens. (2020). "Digitalisation at Siemens: Driving Innovation and Productivity."

Cultural Shift towards Innovation

Digital re-engineering extends beyond technology implementation; it necessitates a cultural shift within an organisation. Siemens' embrace of digitalisation reflects a commitment to fostering an innovative culture. By empowering employees to embrace change and adapt to new technologies, Siemens has created an environment conducive to continuous improvement and innovation. This cultural transformation has been instrumental in navigating the complexities of digital re-engineering and ensuring its sustained success.

Scalable Solutions for Diverse Business Units

Siemens' digitalisation journey showcases the adaptability of re-engineering initiatives across diverse business units. Whether in manufacturing, energy, or healthcare, the application of digital technologies has been customised to suit the unique needs of each sector. This scalability underscores the versatility of digital re-engineering, demonstrating that its principles can be tailored to fit the specific requirements and challenges faced by different industries within a large enterprise.

Customer-Centric Transformation

At the heart of digital transformation lies the imperative to create customer-centric experiences. The financial sector, with its adoption of mobile banking and digital wallets, exemplifies this shift towards customer-focused solutions. Fintech disruptors like Revolut have harnessed the power of digital transformation to provide seamless, user-friendly financial services, challenging traditional banking models and emphasising the importance of customer-centricity in the digital age.

Resource:

- Revolut. (2020). "Revolut - A better way to handle your money." [https://www.revolut.com/]

Navigating Challenges and Ensuring Cybersecurity

Navigating the intricate landscape of cybersecurity challenges within the realm of digital transformation demands a nuanced understanding and proactive strategies. As organisations embrace the benefits of digitisation, they become increasingly vulnerable to a spectrum of cyber threats. Beyond the immediate concerns of data breaches and unauthorised access, there is a pressing need for robust data governance to ensure the ethical and lawful use of information. Cybersecurity, therefore, emerges as a critical component of any comprehensive digital transformation strategy.

The World Economic Forum's seminal report, titled "Cyber Resilience: Playbook for Public-Private Collaboration," serves as an invaluable resource for organisations navigating the complexities of cybersecurity in the digital era. This comprehensive playbook not only acknowledges the evolving nature of cyber threats but also outlines a strategic framework for enhancing cyber resilience. By emphasising collaboration between public and private sectors, the report recognises that addressing cyber threats requires a collective effort, drawing on the expertise and resources of various stakeholders.

The playbook delves into the multifaceted aspects of cyber resilience, offering insights into risk assessment, incident response, and the integration of cybersecurity measures into organisational culture. It underscores the importance of a proactive and adaptive approach, recognising that cybersecurity is not a one-time initiative but an ongoing process that must evolve alongside the rapidly changing threat landscape. The report's emphasis on public-private collaboration aligns with the interconnected nature of cyber threats, acknowledging that effective cybersecurity strategies transcend organisational boundaries.

In practical terms, the playbook provides a roadmap for organisations to fortify their digital assets and maintain trust in an interconnected world. From establishing clear governance structures to fostering information sharing and collaboration, the recommendations outlined in the report offer tangible steps towards building a resilient cybersecurity posture. Organisations can leverage this resource to develop customised cybersecurity frameworks that align with their unique operational landscapes, industry-specific challenges, and risk appetites.

As the digital ecosystem continues to evolve, the insights from the World Economic Forum's report become increasingly relevant. Cybersecurity is not merely a technical challenge but a strategic imperative, woven into the fabric of successful digital transformation. By integrating the principles of the "Cyber Resilience: Playbook for Public-Private Collaboration," organisations can proactively mitigate cyber risks, enhance their overall resilience, and contribute to the collective effort of safeguarding the digital landscape.

Resource:

- World Economic Forum. "Cyber Resilience: Playbook for Public-Private Collaboration." [https://www.weforum.org/whitepapers/cyber-resilience-playbook-for-public-private-collaboration]

As we embark on this exploration of digital transformation and re-engineering, it becomes evident that success in the digital era demands a holistic approach, encompassing technology, culture, and strategy. The examples and resources highlighted in this chapter serve as guideposts for businesses seeking to not only survive but thrive in the dynamic landscape of digital innovation.

7.2 Selecting the Right Technologies: Navigating the Digital Landscape

Digital evolution, extends beyond mere technology adoption; it resides in the strategic selection of technologies that seamlessly align with an organisation's needs and overarching goals. This chapter serves as a comprehensive guide, offering nuanced insights into the intricate process of selecting technologies that not only keep pace with industry trends but strategically propel businesses forward.

Understanding the Dynamic Digital Ecosystem

The digital landscape is a multifaceted ecosystem, constantly reshaped by emerging technologies. From Artificial Intelligence (AI) and Cloud Computing to immersive technologies like Augmented Reality (AR), organisations face an array of choices. Decision-makers must possess a profound understanding of the strategic relevance of each technology within their industry and business model.

Conduct a comprehensive analysis of the technology ecosystem to identify the most suitable solutions. Consider the scalability, interoperability, and compatibility of potential technologies. Evaluate emerging trends and industry standards to future-proof the chosen technologies. This stage involves researching and benchmarking various solutions, understanding their strengths and weaknesses, and assessing their track record in similar industries or use cases. Resources such as industry reports, peer-reviewed journals, and case studies can provide valuable insights during this phase.

Understanding the Trends: Staying abreast of technology trends is paramount. Reports from reputable sources, such as Gartner's Technology Trends, offer invaluable insights into the emerging technologies shaping the digital landscape. (Reference: Gartner. (2024). Gartner Technology Trends)

Aligning Technologies with Organisational Goals

The success of technology adoption hinges on its alignment with organisational goals. A technology that propels one company to new heights may not necessarily yield the same results for another. To ensure synergy between technology investments and business objectives, organisations must conduct a thorough analysis of their current state, future aspirations, and industry dynamics.

Begin the selection process by conducting a thorough assessment of the organisation's business goals and needs. Understanding the specific challenges and objectives of the digital transformation is crucial. This involves engaging with key stakeholders, including executives, department heads, and IT professionals, to gain insights into the current state of operations and the desired future state. Identifying areas for improvement, operational bottlenecks, and the overall strategic vision will inform the selection of technologies that align with the organisation's objectives.

Example: Rolls-Royce's Technological Excellence Rolls-Royce, a stalwart in the aerospace and marine engineering industries, showcases how a focus on technological excellence contributes to a future-ready strategy. Their innovations in aircraft engines, marine propulsion systems, and digital twin technology illustrate a dedication to staying at the forefront of engineering advancements. (Reference: Rolls-Royce. (n.d.). Our Story)

Balancing Innovation with Practicality

While innovation is pivotal, the practicality of technology adoption cannot be overstated. Striking a balance between embracing cutting-edge technologies and ensuring seamless integration with existing systems is essential. This requires a nuanced approach, where organisations evaluate the scalability, interoperability, and potential risks associated with each technology choice.

Example: NHS Digital Transformation The National Health Service (NHS) in the UK provides a noteworthy example of balancing innovation with practicality. The NHS Digital Transformation initiative focuses on leveraging technology to improve patient care and operational efficiency while carefully addressing the unique challenges of a national healthcare system. (Reference: NHS Digital. (2022). Digital Transformation)

Building a Future-Ready Tech Stack

Selecting the right technologies is not merely about meeting current needs; it involves constructing a future-ready tech stack that can adapt to evolving market dynamics. Forward-thinking organisations strategically invest in technologies that not only address immediate challenges but also position them as leaders in an ever-changing digital landscape.

Example: Samsung's commitment to innovation across various sectors exemplifies the construction of a future-ready tech stack. From groundbreaking advancements in semiconductor technology to leading the market in consumer electronics, Samsung continually demonstrates its ability to anticipate and adapt to technological shifts, maintaining a prominent position in the global tech arena. (Reference: Samsung. (n.d.). Our Company)

Navigating the digital landscape demands a judicious approach to technology selection.

By comprehending industry trends, aligning technologies with organisational goals, balancing innovation with practicality, and building a future-ready tech stack, organisations can harness the power of technology to drive sustainable growth and success. The examples provided serve as detailed illustrations, highlighting diverse pathways organisations can take in their digital journey.

Right Technologies, Right Tech-Stack, Right Architecture

Selecting the right technologies, tech stack, and architecture for a digital transformation is a critical and process that requires careful consideration of various factors. Here, we delve into a comprehensive approach to guide decision-makers in making informed choices.

Strategic Development of an Agile Tech Stack

Building an agile tech stack is a foundational aspect of any successful digital transformation. This involves a strategic approach to developing a technology infrastructure that not only meets current needs but also adapts seamlessly to future requirements. It begins with a meticulous analysis of the organisation's existing technology landscape, understanding its strengths, weaknesses, and potential areas for enhancement. By engaging key stakeholders, including IT professionals and department heads, a comprehensive understanding of the technology requirements and the organisation's strategic goals is gained.

Seamless Integration for Synergistic Operations

The successful implementation of an agile tech stack hinges on the seamless integration of chosen technologies. It's imperative that these technologies work cohesively with each other and integrate smoothly with existing systems. This requires a deep understanding of the APIs, protocols, and data formats used by different systems. Compatibility assessments and pilot integrations can help identify potential challenges early in the process, ensuring a smoother transition. Additionally, a well-integrated tech stack promotes synergistic operations across departments, fostering collaboration and enhancing overall organisational efficiency.

Prioritising Interoperability for Future-Proofing

To future-proof the technology infrastructure, prioritising interoperability is paramount. This involves selecting technologies that can adapt and interoperate with emerging tools and industry standards. An agile tech stack should be designed with the foresight to accommodate evolving business needs and technological advancements. This not only mitigates the risk of technological obsolescence but also positions the organisation to harness the benefits of new innovations seamlessly. Staying abreast of industry trends and engaging in a continuous feedback loop with technology vendors are essential strategies for ensuring the long-term viability of the chosen tech stack.

Embracing Modularity and/or Microservices for Flexibility

Microservice architecture was heralded as a breakthrough a few years back and there is technical debate about their future and whether innovation has moved on from there. However they remain a prevalent architectural style for crafting software systems, particularly in managing large and intricate structures. Their enduring popularity is rooted in the emphasis on agility, scalability, and robust fault tolerance. This architectural approach involves constructing applications as a collection of small, independently deployable services, organised around business capabilities and communicating through APIs. Microservices architecture, tailored for enterprise systems, offers several key advantages that continue to make it a compelling choice.

The scalability feature of microservices is a pivotal advantage, allowing for the independent and scalable deployment of services. This flexibility enables organisations to adeptly handle fluctuations in traffic or demand without the need to scale the entire application. Furthermore, the architecture's inherent flexibility comes from decomposing applications into smaller services, facilitating independent work on individual services and allowing for agile development practices. Additionally, the high reliability of microservices is derived from the design's independence, ensuring that a failure in one service does not have a cascading effect on the entire application, thereby building more resilient applications. The fault isolation provided by modular services can enhance application maintainability, allowing developers to pinpoint and address issues without causing disruptions to the overall system.

The reusability aspect of microservices contributes to efficiency, enabling organisations to reuse individual services across multiple applications and use cases. This results in significant time and resource savings, fostering faster time-to-market and consistent functionality across various applications. Despite these advantages, some critiques of microservices architecture emphasise its complexity, potential increased latency due to network communication, and integration challenges, particularly with legacy systems. As organisations deliberate their software architecture, it becomes essential to carefully weigh these factors against the benefits offered by microservices.

The current technological landscape introduces alternatives such as serverless computing and Kubernetes, prompting a reconsideration of architectural choices. Serverless computing offers a paradigm shift by abstracting infrastructure management, allowing developers to focus solely on code without concerns about underlying servers. This can streamline development processes and reduce operational overhead. On the other hand, Kubernetes, an open-source container orchestration platform, provides a robust and scalable framework for deploying, managing, and scaling containerised applications. It simplifies container orchestration, ensuring consistent deployment and efficient resource utilisation. While these alternatives present viable options, organisations must carefully evaluate their specific needs, considering factors such as development speed, operational efficiency, and adaptability to determine the most suitable architectural approach for their unique requirements.

Building an agile tech stack is a complicated process where expertise is required. CEO's please don't think you can delegate that to a COO. This process involves strategic development, seamless integration, prioritising interoperability, and embracing modularity. A well-crafted tech stack is not merely a collection of tools but a dynamic framework that propels an organisation towards digital excellence. As organisations continue to navigate the evolving digital landscape, the agility and adaptability of their tech stack will play a pivotal role in determining their success in a rapidly changing business environment.

Consideration of Security and Compliance

Security is a cornerstone of any digital transformation initiative, and a comprehensive evaluation of potential technologies is essential. Begin by conducting a holistic assessment of the security features embedded in each technology under consideration. Scrutinise encryption capabilities, authentication mechanisms, and overall vulnerability management. A meticulous analysis will help in identifying how well these technologies can withstand evolving cyber threats. It's imperative to prioritise solutions with robust security features that not only meet current needs but also have the capacity to adapt to emerging cybersecurity challenges.

Adherence to Regulatory Frameworks

Adherence to industry regulations is non-negotiable. During the selection process, ensure that the chosen technologies align with the prevailing regulatory frameworks. This is particularly crucial in sectors subject to stringent regulations, such as finance and healthcare. Verify whether the technologies comply with data protection laws, like the General Data Protection Regulation (GDPR) in Europe or the Health Insurance Portability and Accountability Act (HIPAA) in the United States. Organisations must be proactive in understanding the legal landscape and consider the extraterritorial implications of regulations that may impact their operations.

Embedding Robust Cybersecurity Measures

The selected tech stack and architecture should not only comply with regulations but also embed robust cybersecurity measures into the fabric of the digital infrastructure. Prioritise technologies that offer features like multi-factor authentication, intrusion detection and prevention systems, and regular security updates. Implementing a defence-in-depth strategy, where multiple layers of security are employed, enhances the resilience of the digital ecosystem. Collaborate with cybersecurity experts to conduct penetration testing and vulnerability assessments to identify and address potential weaknesses in the chosen technologies, ensuring a proactive and adaptive security posture.

Collaboration with Legal and Compliance Experts

Navigating the legal and regulatory landscape requires close collaboration with legal and compliance experts. Engage with professionals who have up to date experience, not those within the organisation who are vocal about the law but actually may not have a clue. It is one area where vocal senior managers can lead you astray. I have been there, and it wastes a great deal of time and can impact solution design. Experts have a deep understanding of data protection laws and industry-specific regulations. Establish a robust communication channel between the IT team and legal/compliance departments to facilitate a seamless integration of technological solutions within the regulatory framework. This collaboration ensures that the organisation not only meets its legal obligations but also mitigates the risk of legal challenges and penalties. Regular updates and training sessions for staff on the evolving legal landscape can further enhance the organisation's ability to navigate compliance requirements effectively.

The consideration of security and compliance in the digital transformation journey goes beyond a checklist; it demands a proactive, dynamic, and collaborative approach. By holistically evaluating security features, ensuring adherence to regulatory frameworks, embedding robust cybersecurity measures, and fostering collaboration with legal and compliance experts, organisations can fortify their digital infrastructure against threats and regulatory pitfalls. This diligent approach not only safeguards sensitive information but also builds a resilient foundation for sustained digital success.

Selecting the right technologies, tech stack, and architecture for a digital transformation requires a holistic and strategic approach. By aligning choices with business goals, conducting a thorough analysis of the technology ecosystem, building an agile tech stack, and prioritising security and compliance, organisations can embark on a successful digital transformation journey. Regular reassessment and adaptation to evolving technologies and business needs will be key to maintaining the effectiveness of the chosen digital infrastructure over time.

Examples of Tech Stacks

Tech stacks can vary widely based on the specific needs and goals of an organisation. Here are four common tech stacks, each with its technological components:

These are just a few examples, *and before anyone shouts*, as we know - all of us techies think we are the one with correct answer! so these are not the only ones being recommended! The choice of a tech stack should align with the specific requirements and goals of a project or organisation. The technology landscape is diverse, and there are many other stacks and variations tailored to different needs and preferences.

1. LAMP Stack (Linux, Apache, MySQL, PHP/Python/Perl)

Components:

- **Linux (Operating System):** The foundation of the stack, providing a stable and open-source operating system.

- **Apache (Web Server):** A widely used web server that handles HTTP requests and serves web pages.

- **MySQL (Database):** An open-source relational database management system (RDBMS) for storing and retrieving data.

- **PHP/Python/Perl (Server-Side Scripting):** Server-side scripting languages used for dynamic web page generation and interacting with databases.

Use Case: LAMP is a classic stack for building dynamic websites and web applications. It is versatile and widely adopted for its open-source nature.

2. MEAN Stack (MongoDB, Express.js, Angular, Node.js)

Components:

- **MongoDB (Database):** A NoSQL database that stores data in a flexible, JSON-like format.

- **Express.js (Backend Framework):** A minimal and flexible Node.js web application framework that simplifies backend development.

- **Angular (Frontend Framework):** A JavaScript framework for building dynamic single-page web applications.

- **Node.js (Runtime Environment):** An open-source, cross-platform JavaScript runtime for executing JavaScript code server-side.

Use Case: MEAN is well-suited for building scalable and real-time applications, particularly those requiring a dynamic frontend and a flexible backend.

3. MERN Stack (MongoDB, Express.js, React, Node.js)

Components:

- **MongoDB (Database):** A NoSQL database for flexible and scalable data storage.

- **Express.js (Backend Framework):** Simplifies the development of backend applications using Node.js.

- **React (Frontend Library):** A JavaScript library for building user interfaces, particularly for single-page applications.

- **Node.js (Runtime Environment):** Executes JavaScript code on the server side.

Use Case: MERN is a popular choice for building modern web applications with a focus on a responsive user interface and real-time updates.

4. .NET Stack (Microsoft .NET, C#, ASP.NET, SQL Server)

Components:

- **Microsoft .NET (Framework):** A framework for building Windows applications, web applications, and services.

- **C# (Programming Language):** A versatile and modern programming language used with the .NET framework.

- **ASP.NET (Web Framework):** A framework for building dynamic web applications and services.

- **SQL Server (Database):** Microsoft's relational database management system.

Use Case: The .NET stack is often used in enterprise-level applications and businesses heavily invested in Microsoft technologies. It provides a comprehensive framework for various types of applications.

Integrating GraphQL with Hasura into your tech stack offers a compelling solution for modernising and optimising your application development. GraphQL provides a flexible and efficient way to query and manipulate data, allowing clients to request precisely the information they need. Paired with Hasura, a Backend as a Service (BaaS) platform, the integration becomes seamless, automatically generating a GraphQL API on top of your PostgreSQL database. Hasura simplifies the development process by handling real-time updates, authentication, and data relationships effortlessly. This combination significantly accelerates the development cycle, reducing boilerplate code and offering a unified, consistent API for your frontend. With the ability to seamlessly connect to existing databases and rapidly create scalable, real-time applications, GraphQL with Hasura becomes an attractive choice for developers aiming to enhance productivity, streamline workflows, and deliver feature-rich, performant applications with ease.

So what is Front-End, Back-End or Client and Server-Side Technologies?

Some useful descriptions to use with the organisation Senior Management Team

Front-End Components:

The front end of a system, also known as the client-side, encompasses all the elements and technologies that users interact with directly. This includes the user interface (UI), design, and functionalities that make the application visually engaging and user-friendly. Key front-end technologies include HTML (Hypertext Markup Language), CSS (Cascading Style Sheets), and JavaScript. HTML provides the structure of web pages, CSS defines their visual styling, and JavaScript adds interactivity and dynamic behaviour. Front-end developers focus on creating a seamless and intuitive user experience, ensuring that the application or website looks and feels appealing to users.

Back-End Components:

Contrastingly, the back end, or server-side, consists of all the components that operate behind the scenes to enable the functionality and data management of the application. This includes the server, database, and application logic. Server-side technologies such as Node.js, Ruby on Rails, or Django handle the business logic and communicate with the database to retrieve or store data. Databases like MySQL, MongoDB, or PostgreSQL are used to manage and organise the application's data. Back-end developers work on server-side scripting, server management, and database operations to ensure the application's functionality, security, and performance.

Differences Between Front-End and Back-End:

The primary distinction lies in their respective focuses. Front-end development is concerned with what users experience directly - visual elements, layouts, and interactions. It involves creating a responsive and visually appealing interface. In contrast, back-end development focuses on the server-side operations, managing databases, and implementing the business logic that ensures the application's functionality. While front-end development deals with the presentation layer, back-end development handles the application's underlying infrastructure.

Integration Between Front-End and Back-End:

Effective integration between the front end and back end is crucial for a seamless user experience. This is achieved through APIs (Application Programming Interfaces) that enable communication and data exchange between the client-side and server-side components. Front-end developers make HTTP requests to the back end, triggering server-side operations, and receive responses to update the user interface accordingly. The back end, in turn, processes these requests, interacts with the database, and sends back the necessary data. Technologies like RESTful APIs or GraphQL facilitate this communication, ensuring a cohesive and efficient interaction between the user interface and the underlying system logic.

The front end and back end of a system represent distinct aspects of application development. While the front end focuses on the user interface and client-side interactions, the back end manages server-side operations, databases, and business logic. The integration of these components is achieved through well-defined APIs, ensuring a harmonious collaboration between the user interface and the underlying infrastructure for a complete and functional application.

Why include a Full Stack Developer in your team? Full-stack developers play a versatile role in the development process, possessing the skills to work on both the frontend and backend aspects of an application. In the realm of web development, full-stack developers have expertise in handling client-side technologies, like HTML, CSS, and JavaScript, for building user interfaces, as well as server-side technologies, such as databases, server scripting, and application architecture. This dual proficiency allows them to bridge the gap between frontend and backend development, enabling them to understand the entire development stack and contribute to both sides of the application seamlessly.

In contrast to frontend developers who mainly focus on creating visually appealing and interactive user interfaces and backend developers who manage server-side logic and databases, full-stack developers are adept at navigating the entire development spectrum. Their comprehensive skill set empowers them to handle end-to-end development tasks, from designing user interfaces to implementing server-side functionalities and managing databases. The advantages of having full-stack developers in a team include enhanced communication and collaboration, as they can understand and contribute to both frontend and backend discussions. This versatility makes them valuable assets for projects with resource constraints, fostering efficient development processes, and enabling them to take on a variety of roles within a team.

7.3 Agile Methodologies in Action: Iterative Approaches to Transformation

Agile methodologies represent a transformative paradigm in project management, promoting flexibility and adaptability in the execution of initiatives. In this context, "Agile" refers to a set of principles and practices that prioritise iterative development, collaboration, and responsiveness to change. Unlike traditional project management methodologies that follow a linear and rigid approach, Agile embraces an iterative process, allowing teams to respond to evolving requirements and uncertainties inherent in complex projects.

Understanding Agile Principles

At the heart of Agile methodologies are the twelve principles outlined in the Agile Manifesto. These principles emphasise customer collaboration, responding to change over following a plan, and the importance of individuals and interactions. Agile places a premium on delivering functional increments of a project in short, iterative cycles known as sprints. This iterative process ensures that the project remains adaptable, providing opportunities for continuous feedback and refinement.

1. **Prioritising People:** The first principle underscores the significance of individuals and their interactions over rigid plans and processes. It advocates that people are the most crucial element of the team, and the focus should be on working together effectively.

2. **Embracing Change:** Agile welcomes changes in requirements, even late in the development process. Instead of strictly adhering to a plan, Agile teams adapt to new information and make necessary adjustments. It's akin to being prepared to alter your route if you discover a faster way to reach your destination.

3. **Regularly Delivering Value**: Agile teams aim to create useful and valuable software frequently. This involves delivering parts of a project in small, regular doses, rather than waiting a long time to showcase progress. It's like receiving updates on a book you're reading, keeping you engaged as the story unfolds.

4. **Collaboration is Essential**: Agile encourages constant collaboration between team members and stakeholders. By working together and sharing ideas, the team can create a better product. It's like everyone contributing their puzzle pieces to build the complete picture.

5. **Building Projects around Motivated Individuals:** The fifth principle highlights the importance of having motivated and trusted team members. When people are enthusiastic about their work and trusted to make decisions, projects tend to be more successful. It's like having a team of superheroes, each with their unique powers, working together.

6. **Face-to-Face Communication Matters:** Direct communication is often more effective than written messages. Agile teams prioritise face-to-face conversations to ensure everyone is on the same page. It's like having a chat with a friend to make sure you understand each other well.

7. **Progress is Measured by Working Software:** Agile values working software over detailed documentation. Instead of focusing too much on paperwork, the team's progress is measured by the actual working product. It's like judging a chef by the delicious meals they serve, not just by the recipes they write.

8. **Maintaining a Sustainable Pace:** Teams need to work at a steady pace to avoid burnout. Agile recognises that a sustainable pace leads to better results over the long run. It's like running a marathon – you want to keep a steady pace to reach the finish line without exhausting yourself.

9. **Paying Attention to Technical Excellence:** Creating high-quality work is crucial. Agile teams focus on maintaining technical excellence to ensure the product is reliable and efficient. It's like building a sturdy bridge – you want it to be strong and reliable.

10. **Keeping It Simple:** Agile encourages simplicity in both the product and the process. It's like making a straightforward recipe with just the right ingredients – not too complicated but still delicious.

11. **Encouraging Self-Organising Teams:** Teams work best when they organise themselves. Agile trusts teams to make their own decisions, empowering them to achieve their goals. It's like playing in a sports team where each player knows their role and works together to win.

12. **Reflecting and Adjusting:** The final principle emphasises the importance of regular reflection and adjustment. Agile teams continuously look at what's working well and what can be improved. It's like regularly checking the map on a road trip to make sure you're on the right track.

Resource: Agile Manifesto: The Agile Manifesto serves as the foundational document outlining the principles that underpin Agile methodologies. Teams looking to understand and implement Agile principles can refer to the manifesto as a guiding resource. (Reference: Agile Manifesto. (n.d.). Manifesto for Agile Software Development)

Flexibility and Adaptability in Project Execution

Agile methodologies shine in situations where change is inevitable, and requirements evolve. By breaking down a project into smaller, manageable iterations, Agile allows teams to address changing priorities and incorporate feedback swiftly. This flexibility not only accommodates uncertainties but also ensures that the end product aligns more closely with stakeholder expectations.

Case Study: Spotify's Agile Transformation Spotify's journey towards an Agile framework provides a compelling case study. As the music streaming giant expanded, it faced the need for greater flexibility in development. Adopting the Agile model allowed Spotify to iterate quickly, respond to user feedback promptly, and continuously enhance its platform. This approach has become a benchmark for Agile success in large-scale organisations. (Reference: Agile Alliance. Agile in the Real World: Overcoming challenges on the Spotify journey)

Collaborative and Cross-Functional Teams

Agile methodologies emphasise the importance of collaborative and cross-functional teams. Unlike traditional hierarchical structures, Agile teams are self-organising and empowered to make decisions. This collaborative environment fosters open communication, knowledge sharing, and collective ownership of project outcomes. Cross-functional teams bring together individuals with diverse skill sets, ensuring a holistic approach to problem-solving.

Resource: Scrum Framework Scrum is a popular Agile framework that provides a structured approach to project management. Scrum emphasises collaboration, accountability, and iterative progress. Teams can leverage the Scrum framework to implement Agile methodologies effectively. (Reference: Scrum.org. (n.d.). What is Scrum?)

Continuous Innovation through Retrospectives

Agile methodologies promote a culture of continuous improvement through regular retrospectives. At the end of each iteration, teams reflect on what worked well, what could be improved, and how to enhance their processes. This introspective approach fosters a learning mindset, driving teams to adapt and refine their practices continuously.

UK Example: British Airways' Agile Adoption British Airways' adoption of Agile methodologies in its IT department showcases a commitment to continuous improvement. Facing the challenges of a rapidly changing aviation landscape, British Airways implemented Agile practices to enhance collaboration, deliver projects more efficiently, and respond swiftly to changing market demands. (Reference: Agile Business Consortium. (n.d.). Agile Adoption at British Airways)

Agile Ceremonies Unveiled

Agile methodology is not just a set of principles; it's a well-orchestrated set of ceremonies designed to facilitate collaboration, communication, and adaptability. In this chapter, we explore the key ceremonies that form the backbone of Agile development.

Refinement: Crafting the User Stories

Refinement, although not a formal ceremony in itself, is a crucial function within Agile that deserves attention. Also known as "Backlog Grooming" or "Story Refinement," this ongoing process involves reviewing and clarifying user stories and tasks before they make it into a sprint. Teams collaborate to add details, break down large stories into manageable tasks, and ensure that each item in the backlog is well-defined and understood. This iterative process contributes to smoother sprint planning sessions and ensures that the team is always equipped with a refined backlog ready for action. It's like carefully preparing the ingredients before cooking a meal – ensuring everything is in order and well-understood before the team embarks on their sprint journey. Refinement promotes clarity and efficiency, allowing the team to focus on delivering value during each sprint.

Sprint Planning: Setting the Stage

At the beginning of each sprint – a short, fixed time frame in which a specific set of work must be completed – Agile teams gather for the Sprint Planning ceremony. During this collaborative session, the team discusses and agrees upon the user stories or tasks to be undertaken in the upcoming sprint. The product owner brings insights into the priority of features, and the team collectively decides how much work they can commit to completing. It's akin to a team huddle before a match, strategising on how to achieve their goals in the upcoming sprint.

Daily Stand-ups: Keeping in Sync

A cornerstone of Agile communication is the Daily Stand-up or Daily Scrum. Held at the same time every day, team members stand (to keep it short and focused) and answer three key questions: What did I accomplish yesterday? What am I planning to do today? Are there any obstacles in my way? This brief but regular check-in ensures everyone is on the same page, promotes transparency, and quickly addresses any roadblocks. It's like a quick team huddle on the field, ensuring everyone knows their role and any challenges are swiftly addressed.

Sprint Review: Showcasing Progress

At the end of each sprint, the team holds a Sprint Review to showcase the work completed. Stakeholders, including customers and product owners, attend this ceremony to see a demonstration of the features or user stories developed during the sprint. Feedback is collected, and adjustments are made based on the insights gained. It's like an artist unveiling their latest masterpiece, with the audience providing valuable input for refinement.

Retrospective: Learning and Improving

The Retrospective is the reflective ceremony that takes place after each sprint. Team members gather to discuss what went well, what could be improved, and actions to take in the next sprint. This open and honest conversation fosters a culture of continuous improvement, allowing the team to adapt and grow. It's like a sports team reviewing game footage after a match, learning from both successes and mistakes to enhance future performance.

In summary, Agile ceremonies are the carefully orchestrated events that keep the Agile methodology in motion. From planning and daily check-ins to showcasing progress and continuous improvement, these ceremonies embody the collaborative spirit and adaptive nature of Agile development, ensuring that teams deliver value consistently and adapt to changing requirements with ease.

Agile Development – Estimation and Story Points

In Agile development, effectively scoping the size of development tasks is crucial for planning and managing work within sprints. Here are some techniques and tools that developers can employ for this purpose:

Story Points Estimation:

Technique: Agile teams often use story points, a relative unit of measure, to estimate the size and effort required for tasks. During sprint planning, developers discuss each user story or task and assign story points based on the perceived complexity, effort, and risk.

Tool: Planning Poker: This is a collaborative estimation technique where team members use cards with numbers representing story points to cast their votes anonymously. The team discusses the differences in their estimates and converges towards a consensus. Digital tools like "***Planning Poker***" apps or integrations within Agile project management tools facilitate virtual estimation sessions for distributed teams.

T-shirt Sizing:

Technique: This approach involves assigning sizes like Small, Medium, Large, or Extra Large to tasks based on their complexity. It provides a quick and straightforward way to estimate without getting into detailed analysis.

Tool: Dot Voting: Developers can use dot voting to collectively agree on the size of tasks. Each team member is given a set number of dots, and they distribute these dots among the task sizes they believe are appropriate. This visual representation helps identify the majority view on the task size.

Relative Sizing:

Technique: Tasks can be sized relative to each other, indicating their complexity in comparison rather than an absolute measure. For example, a task might be considered twice as complex as another.

Tool: Fibonacci Sequence: Developers can use the Fibonacci sequence (1, 2, 3, 5, 8, 13, etc.) to assign relative sizes. This sequence acknowledges that estimating larger tasks becomes more uncertain and helps prevent unnecessary precision in sizing.

Three Amigos Session:

Technique: This technique involves a collaborative discussion between developers, testers, and product owners. By bringing together different perspectives, the team gains a more comprehensive understanding of the task's requirements and potential challenges.

Tool: Online Collaboration Tools: Virtual whiteboards, shared documents, or collaboration platforms enable remote teams to conduct Three Amigos sessions. These tools facilitate real-time discussions, ensuring that everyone is on the same page when sizing tasks.

Cycle Time Analysis:

Technique: Teams can analyse past performance and completion times of similar tasks to inform their estimates. This historical data provides insights into how long similar tasks have taken to complete in previous sprints.

Tool: Agile Project Management Tools: Tools like Jira, Trello, or Asana often provide features for tracking and analysing cycle times. Developers can leverage these tools to understand historical performance and use the insights to inform their estimations.

By incorporating these techniques and tools, developers in Agile teams can enhance their ability to scope the size of development tasks effectively. Regularly refining estimation practices based on feedback and experience contributes to improved accuracy in sizing, fostering a more predictable and efficient development process.

In summary, Agile methodologies represent a transformative approach to project management, focusing on flexibility, adaptability, collaboration, and continuous improvement. By embracing the iterative nature of Agile, organisations can navigate complex projects with a responsive mindset, ensuring successful outcomes in dynamic and ever-changing environments. The case studies and resources highlighted serve as valuable references for teams embarking on their Agile journey.

7.4 Integration Challenges and Solutions: Ensuring Seamless Connectivity

Addressing the complexities of integrating new systems with existing infrastructure.

Navigating the integration landscape poses a significant challenge in the dynamic terrain of business transformation. One primary concern lies in seamlessly integrating new systems with existing infrastructure. This integration challenge often arises due to the heterogeneity of legacy systems, varying data formats, and the need to maintain operational continuity during the transformation process.

To begin, legacy systems may employ outdated technologies that lack compatibility with modern solutions. These disparities can hinder the smooth integration of new systems, causing data silos and inefficiencies. Overcoming this challenge requires a meticulous assessment of existing technologies and strategic planning to bridge the gap between legacy and contemporary systems.

Moreover, the diversity of data formats across different systems exacerbates integration complexities. New systems might utilise different data structures, making it arduous to establish interoperability. Adopting data standardisation practices and employing middleware solutions, such as RabbitMQ, Apache Kafka or MuleSoft, facilitates the translation and seamless flow of data between disparate systems.

Another significant hurdle is the potential disruption to ongoing business operations during the integration process. The need for continuous service delivery demands careful orchestration of integration activities. Implementing phased integration, where components are introduced incrementally, allows organisations to maintain operational stability while gradually transitioning to the new system.

Additionally, security concerns loom large during integration. Merging systems can expose vulnerabilities, making data susceptible to breaches. Employing robust cybersecurity measures and conducting thorough risk assessments are essential. Compliance with data protection regulations, such as GDPR, ensures that data integrity and privacy are prioritised throughout the integration journey.

Addressing these challenges requires a multifaceted approach. Organisations can benefit from establishing a dedicated integration team responsible for planning, executing, and monitoring the integration process. This team can leverage application programming interfaces (APIs) to facilitate communication between systems, streamlining data exchange and ensuring a more cohesive integration.

Integration challenges in business transformation demand strategic planning, technological innovation, and a commitment to maintaining operational continuity. By carefully navigating the complexities of legacy systems, data formats, operational disruptions, and security concerns, organisations can foster a seamless integration process that positions them for success in the ever-evolving business landscape.

Integration Tools: Making life simple – Avoiding disruption

In the realm of business transformation, the integration of new systems with existing legacy infrastructure requires the judicious use of a variety of tools and technologies. This overview explores key tools that facilitate the seamless merging of diverse systems, ensuring operational continuity and efficiency during the transformation process.

1. Middleware Solutions:

Middleware plays a crucial role in bridging the gap between new and legacy systems by providing a communication layer that facilitates data exchange. Tools such as Apache Kafka, MuleSoft, Amazon Kinesis, and RabbitMQ act as middleware platforms, enabling smooth integration through data translation, routing, and orchestration. These solutions streamline communication between disparate systems, ensuring compatibility and interoperability.

2. API Management Platforms:

Application Programming Interfaces (APIs) serve as a fundamental component for connecting different systems. API management platforms like Apigee, AWS API Gateway, and Microsoft Azure API Management offer tools for creating, managing, and securing APIs. By standardising communication protocols, APIs facilitate seamless integration, allowing new systems to interact with legacy applications in a controlled and secure manner.

3. Enterprise Service Bus (ESB):

ESB solutions act as a central hub for integrating various systems within an enterprise architecture. Products like Peregrine Connect, Apache ServiceMix and Red Hat Fuse provide ESB capabilities, offering a unified platform for message routing, transformation, and mediation. ESBs simplify the integration process by providing a consistent framework for connecting diverse applications and services.

4. Data Integration Platforms:

Data integration is often a significant challenge in merging new and legacy systems with disparate data formats. Tools such as Informatica PowerCenter, Talend, Fivetran, and Microsoft SSIS (SQL Server Integration Services) facilitate the extraction, transformation, and loading (ETL) of data. These platforms enable organisations to harmonise data from different sources, ensuring consistency and accuracy.

5. Containerisation and Orchestration:

Containerisation tools like Docker and container orchestration platforms like Kubernetes offer a standardised and portable environment for deploying and managing applications. These tools enhance the flexibility of system integration by encapsulating applications and their dependencies, enabling seamless deployment across various environments.

6. Integration Platform as a Service (iPaaS):

iPaaS solutions, such as Dell Boomi, Jitterbit, and SnapLogic, provide cloud-based integration capabilities. These platforms offer pre-built connectors, allowing organisations to rapidly integrate cloud and on-premises applications. iPaaS solutions streamline the integration process, particularly in hybrid environments where both new and legacy systems coexist.

7. Legacy Modernisation Tools:

Tools specifically designed for legacy system modernisation help organisations update and enhance their existing applications. Solutions like Micro Focus Visual COBOL and TmaxSoft OpenFrame aid in transforming legacy codebases to modern programming languages and architectures, facilitating integration with contemporary systems.

A successful integration of new and legacy systems as part of business transformation relies on the strategic adoption of these tools. Middleware solutions, API management platforms, ESBs, data integration platforms, containerisation tools, iPaaS, and legacy modernisation tools collectively empower organisations to navigate the complexities of integration, ensuring a smooth transition towards a more cohesive and technologically advanced business environment.

Chapter 8:

Design Thinking & User-Centric Design: Enhancing Experiences in the Digital Age

8.1 Design Thinking

Design Thinking is a human-centric problem-solving methodology that places empathy and creativity at its core. Originating from the world of product design, this approach has evolved into a versatile framework applicable to various fields. The process typically unfolds in a series of iterative stages, each emphasising a deep understanding of user needs and experiences.

Stage 1: The first stage of Design Thinking is Empathise, where practitioners immerse themselves in the perspective of the end-user. This involves conducting interviews, observations, and surveys to gain a comprehensive understanding of the user's challenges, motivations, and aspirations. By cultivating empathy, designers can unearth insights that lay the foundation for innovative solutions.

Stage 2: The Define stage follows, where the gathered insights are synthesized to define the core problem or opportunity. This step involves distilling the complexities identified during the Empathise stage into a clear and actionable problem statement. The framing of the problem is crucial, as it sets the stage for the subsequent ideation and solution phases.

Stage 3: Ideation, the third stage, encourages the generation of a wide range of creative solutions. Design Thinking fosters a culture of brainstorming and free expression, encouraging participants to think beyond conventional boundaries. Techniques such as mind mapping, brainstorming sessions, and ideation workshops are employed to stimulate a diverse array of ideas.

Stage 4: Prototyping marks the fourth stage, where selected ideas are transformed into tangible representations. These prototypes can take various forms, from paper sketches to interactive models, allowing designers to test and refine concepts rapidly. The iterative nature of prototyping enables continuous feedback and refinement, leading to solutions that are more refined and aligned with user needs.

Stage 5: The fifth stage is Test, where prototypes are evaluated by users to gather feedback on their functionality and usability. This testing phase is pivotal in validating assumptions and refining the solution further. Designers closely observe user interactions, allowing them to identify any potential issues and make necessary adjustments.

Stage 6: The final stage, Implement, involves bringing the refined solution to life. This phase extends beyond the design team to encompass the broader organisational context, considering factors such as scalability, feasibility, and sustainability. Successful implementation requires collaboration and coordination across different departments and stakeholders.

Design Thinking's strength lies in its iterative and user-centric nature, fostering continuous refinement and innovation. This methodology is not confined to product design; it has found application in diverse fields such as business strategy, healthcare, and education. By prioritising empathy, collaboration, and creative problem-solving, Design Thinking offers a holistic approach to addressing complex challenges and delivering solutions that resonate with the end-users' needs.

Tools to Assist Design Thinking:

Miro: Miro is an online collaborative whiteboard platform that supports various activities in the Design Thinking process, such as ideation, prototyping, and user journey mapping. It enables teams to work together in real-time, fostering collaboration and creativity.

IDEO Method Cards: IDEO, a pioneer in Design Thinking, offers a set of Method Cards that provide prompts and insights for each phase of the process. These cards serve as a handy tool during workshops and brainstorming sessions, offering inspiration for approaching design challenges.

Sketch and Figma: Tools like Sketch and Figma are popular for digital prototyping and design. They facilitate the creation of interactive prototypes and wireframes, allowing designers to visualise and test ideas before implementation. Figma, in particular, supports real-time collaboration among team members.

UserTesting: UserTesting is a platform that enables designers to conduct remote usability testing. It allows teams to gather valuable insights from real users interacting with prototypes, helping to validate design decisions and identify areas for improvement.

MindMeister: MindMeister is an online mind mapping tool that aids in the ideation phase of Design Thinking. It allows teams to collaboratively create mind maps to explore and organise ideas, fostering a structured approach to ideation and problem definition.

Sticky Notes (Mural, Stormboard): Digital sticky note platforms like Mural and Stormboard replicate the analog experience of using sticky notes during brainstorming sessions. These tools enable teams to virtually post ideas, organise them, and facilitate group discussions.

Useful Online Resources:

IDEO U (ideo.com): IDEO U, the online learning platform from IDEO, offers courses and resources on Design Thinking. It provides in-depth insights into the methodology, case studies, and practical tips for applying Design Thinking principles.

Stanford d.school (dschool.stanford.edu): Stanford University's d.school (Design School) offers a wealth of resources on Design Thinking. Their website provides a toolkit, case studies, and guides that can aid individuals and teams in implementing the methodology.

Nielsen Norman Group (nngroup.com): The Nielsen Norman Group is a renowned UX research and consulting firm. Their website includes a variety of articles, reports, and usability guidelines that align with the principles of Design Thinking, offering valuable insights into user-cantered design.

Coursera - "Design Thinking and Innovation" (coursera.org): Coursera hosts online courses, and some institutions offer courses specifically on Design Thinking. The "Design Thinking and Innovation" course, for instance, provides a comprehensive overview of the methodology and its applications.

Harvard Business Review (hbr.org): Harvard Business Review regularly features articles and case studies related to Design Thinking and innovation. Exploring their Design Thinking section can provide valuable perspectives and real-world examples.

Interaction Design Foundation (interaction-design.org): The Interaction Design Foundation offers an extensive library of articles and courses on various aspects of design, including Design Thinking. It is a valuable resource for individuals looking to deepen their understanding of design methodologies.

Case Study: Barclays Bank - A Design Thinking Approach to Digital Banking Transformation

Barclays Bank, a prominent UK-based financial institution, undertook a transformative journey by applying Design Thinking principles to enhance its digital banking services. Facing the challenge of staying competitive in an increasingly digitised industry, Barclays sought to reimagine its customer experience and develop innovative solutions to address evolving user needs.

1. Empathise: Barclays initiated its Design Thinking process by empathising with its diverse customer base. This involved extensive user research, including in-depth interviews, surveys, and observations to gain a profound understanding of customer behaviours, pain points, and aspirations in the context of digital banking.

2. Define: With insights gathered from the empathise phase, Barclays defined the core challenges and opportunities. The bank identified a need to simplify complex digital interactions, enhance personalisation, and provide more intuitive financial management tools. The framing of these challenges set the foundation for the subsequent ideation and solution phases.

3. Ideate: Barclays fostered a culture of creativity and innovation during the ideation phase. Cross-functional teams participated in brainstorming sessions and workshops using Design Thinking tools, generating a multitude of ideas to address the defined challenges. Concepts ranged from improved mobile banking apps to novel features for personalised financial planning.

4. Prototype: Selected ideas were transformed into tangible prototypes during the next phase. Barclays employed digital prototyping tools to create interactive models and mock-ups of new features and interfaces. These prototypes were tested internally and refined iteratively, ensuring alignment with customer expectations and usability standards.

5. Test: Barclays conducted extensive user testing to gather feedback on the developed prototypes. Real customers interacted with the prototypes, providing valuable insights into the functionality, user experience, and overall satisfaction. This testing phase allowed Barclays to validate assumptions, identify potential issues, and make data-driven refinements.

6. Implement: The final implementation phase extended beyond the design team. Barclays collaborated with its technology and business units to seamlessly integrate the refined digital solutions into its existing infrastructure. The implementation was staged to ensure minimal disruption to ongoing banking services.

Results: Barclays' application of Design Thinking principles resulted in a significant transformation of its digital banking services. The redesigned mobile app, informed by user-centric design principles, offered a more intuitive and personalised experience. Features such as real-time spending insights, personalised budgeting tools, and simplified fund transfers were well-received by customers.

The impact was not only felt in customer satisfaction but also in business outcomes. The user-friendly interfaces led to increased app engagement, a rise in digital transactions, and improved customer retention. Barclays' adoption of the Design Thinking methodology not only addressed immediate challenges but positioned the bank as a forward-thinking innovator in the competitive landscape of digital banking.

Lessons Learned: Barclays' case underscores the importance of embracing Design Thinking as a strategic approach to digital transformation. The focus on empathy, iteration, and collaboration enabled the bank to create solutions that not only met customer expectations but exceeded them, ultimately contributing to the bank's success in a rapidly evolving financial ecosystem.

8.2 User-Centric Design

User-Centric Design is not merely a design philosophy; it's a transformative approach that places the end-user at the core of business strategy and innovation. Recognising the pivotal role of user experience (UX) and engagement in the success of products and services, businesses increasingly embrace user-centric design as a key component of their transformation journey.

1. Understanding User Needs and Pain Points:

User-Centric Design starts with a deep understanding of user needs and pain points. Conducting user research, surveys, and interviews are essential tools to gather qualitative and quantitative data about user preferences and challenges. For instance, tools like UserZoom and Hotjar facilitate the collection of user feedback and behaviour analytics, providing valuable insights into user experiences.

2. Prototyping and Iterative Design:

Prototyping is a crucial step in user-centric design, allowing businesses to create tangible representations of their solutions for user testing and feedback. Tools such as Adobe XD, Figma, or InVision enable designers to rapidly prototype and iterate based on user input. This iterative process ensures that the final product aligns closely with user expectations and preferences.

3. User Personas and Empathy Mapping:

Creating user personas and empathy mapping are tools that help businesses visualise their target audience's characteristics, motivations, and pain points. These tools facilitate a human-centred approach, ensuring that solutions resonate with the diverse needs of the user base. UXPin and Miro are examples of platforms that support the creation of user personas and empathy maps.

4. Accessibility and Inclusive Design:

User-Centric Design extends its focus to inclusivity by considering the needs of users with diverse abilities and backgrounds. Tools like AXE Accessibility Checker and Colour Oracle help designers assess and address accessibility concerns, ensuring that the designed solutions are usable by everyone, regardless of physical or cognitive abilities.

5. User Journey Mapping:

Understanding the user's journey across different touchpoints is critical in delivering a seamless and enjoyable experience. User journey mapping tools, such as Smaply or UXPressia, allow businesses to visualise and analyse the entire user experience, identifying pain points and opportunities for improvement.

6. Usability Testing and Feedback Loops:

Usability testing tools play a vital role in the user-centric design process by allowing businesses to observe how users interact with prototypes or existing products. Tools like UsabilityHub and UserTesting.com facilitate remote and in-person testing, providing valuable insights for refining the user experience. Implementing feedback loops ensures that user input continues to influence design decisions throughout the transformation journey.

User-Centric Design stands as a key component of business transformation by prioritising user experience and engagement. From understanding user needs to iterative design processes and usability testing, the tools available empower businesses to create solutions that resonate with their audience, fostering customer loyalty and driving success in an increasingly user-centric marketplace.

Case Study: Energise Resources

In the midst of an extensive business transformation, Energise Resources, a prominent UK-based company supporting organisations in the not-for-profit sector, embraced

User-Centric Design as a foundational element to revamp its digital services and amplify stakeholder engagement. Acknowledging the critical role of user experience (UX) in the distinctive landscape of not-for-profit initiatives, Energise Resources initiated a profound shift towards prioritising the needs and preferences of its diverse user base. The company commenced by conducting extensive user research, utilising tools such as UserZoom and Hotjar to gather qualitative and quantitative insights into stakeholder preferences and challenges.

Armed with a nuanced understanding of user needs, Energise Resources engaged in prototyping and iterative design using platforms like Figma. This iterative process allowed for swift adjustments based on user feedback, ensuring that the final digital solutions closely aligned with stakeholder expectations. As part of fostering a user-centric culture, the company implemented tools such as UXPin and Miro for the development of user personas and empathy mapping. This facilitated a holistic and human-centered approach, ensuring that the designed solutions resonated with the diverse needs of the not-for-profit sector.

One significant outcome of Energise Resources' User-Centric Design approach was the redesign of its digital platform tailored to support not-for-profit organisations. The company employed Tableau for data visualisation, creating intuitive reports and dashboards that provided insights into stakeholder interactions and preferences. The redesign, informed by user journey mapping using tools like Smaply, resulted in a seamless and engaging digital experience for stakeholders. This user-centric digital transformation not only enhanced stakeholder satisfaction but also significantly contributed to the effectiveness of not-for-profit initiatives, solidifying Energise Resources as a forward-thinking and user-focused partner in the sector.

Chapter 9.

Product Ownership and Product Management

1. **Introduction to Product Management and Ownership:** *An overview of the critical roles of product management and ownership in steering a company's product development initiatives, emphasising the collaborative relationship between these functions.*

2. **Defining Product Vision and Strategy:** *Guidance on crafting a clear and compelling product vision and aligning it with the overall business strategy.*

3. **Responsibilities of a Product Manager:** *Exploration of the day-to-day tasks and long-term responsibilities of a product manager, including market research, feature prioritisation, and maintaining a cohesive product roadmap.*

4. **The Role of a Product Owner:** *An examination of the specific responsibilities of a product owner, focusing on backlog management, sprint planning, and ensuring the development team's understanding of product requirements.*

5. **The role of the Scrum Master:** *How this role fits into Product Development*

6. **Tools for Productivity and Best Practice:** *Tools of the trade that can aid success.*

7. **Collaboration between Product Management and Development:** *Insights into fostering effective communication and collaboration between product management and development teams to ensure a unified approach towards product goals.*

8. **Scaling Product Management for Growth:** *Strategies for scaling product management practices to align with the company's growth, including considerations for expanding product portfolios and managing multiple product lines.*

9. **Overcoming Challenges in Product Ownership:** *Identification and resolution of common challenges faced by product owners, such as balancing competing priorities, managing stakeholder expectations, and adapting to changing market conditions.*

10. **Measuring Product Success and KPIs:** *Defining key performance indicators (KPIs) and metrics for evaluating the success of product initiatives, with a focus on continuous improvement and adapting to market dynamics.*

11. **Case Studies and Best Practices:** *Examining real-world case studies and industry best practices that illustrate successful product management and ownership strategies, offering actionable insights for CTOs.*

9.1 Introduction to Product Management and Ownership

The foundational pillars of product management and ownership intricately guide a company's trajectory in product development. Product management, situated at the strategic helm, orchestrates the complete lifecycle of a product, steering it from conceptualisation through to delivery and beyond. This multifaceted role encompasses delineating a clear product vision, conducting exhaustive market research, and astutely prioritising features that not only resonate with customer needs but also align with the overarching objectives of the business.

Conversely, product ownership assumes the tactical responsibility of translating strategic directives into tangible actions for the development team. It involves meticulous management of the product backlog, facilitation of sprint planning, and serves as the linchpin between various stakeholders and the development team. While product management sets the stage with visionary guidance, product ownership becomes the executor, ensuring that the development process aligns seamlessly with the defined product roadmap. This tandem effort, akin to a well-choreographed dance, underscores the interdependence and cohesiveness required to navigate the complexities of product development.

The collaborative relationship between product management and ownership is pivotal. Product managers articulate the 'what' and 'why' of a product, outlining the broader strategic goals, market positioning, and user value. In tandem, product owners delve into the 'how' and 'when', meticulously breaking down strategic objectives into actionable tasks, facilitating agile methodologies, and maintaining a fluid communication channel with the development team. This collaborative relay of responsibilities ensures that the torch of vision and strategy is smoothly passed from product management to ownership, maintaining clarity and unity throughout the product development journey.

Understanding the intricacies of this collaborative synergy is foundational for managing successful product development initiatives. The forthcoming chapters will delve into the distinct responsibilities of each role, offering strategies to nurture effective collaboration, and providing insights into optimal utilisation of this dynamic partnership. As the narrative unfolds, it will shed light on how this cohesive interplay propels innovation, meets evolving market demands, and ultimately aligns product development with the overarching business objectives of the company.

9.2 Defining Product Vision and Strategy: Crafting a Clear and Compelling Direction for Success

The process of defining a robust product vision and strategy serves as the cornerstone for successful development initiatives. This involves crafting a clear and compelling narrative that outlines the purpose and goals of the product. A well-defined product vision provides a shared understanding among team members and stakeholders about the intended impact of the product on its users and the broader market.

Aligning the product vision with the overall business strategy is a critical aspect of this process. The product should be viewed not in isolation but as a strategic component contributing to the overarching goals of the business. This alignment ensures that the product is developed with a clear understanding of how it fits into the company's broader objectives, market positioning, and growth trajectory.

Crafting a compelling product vision requires foresight and the ability to anticipate market trends, customer needs, and technological advancements. It involves considering the long-term impact of the product on the target audience and the industry. By foreseeing potential challenges and opportunities, product managers can shape a vision that resonates with stakeholders and inspires the development team to work towards a shared goal.

Setting the direction for successful product development involves translating the product vision into a tangible strategy. This includes defining the key features, functionalities, and user experiences that will bring the vision to life. A well-crafted strategy outlines the steps needed to achieve the product vision, considering factors such as market positioning, competition, and technological feasibility.

Moreover, the product strategy serves as a guide for making informed decisions throughout the development lifecycle. It helps in prioritising features, allocating resources efficiently, and adapting to changes in the market landscape. A carefully defined product strategy provides a roadmap for the development team, ensuring that their efforts are directed towards the most impactful and strategic aspects of the product.

In conclusion, the process of defining a product vision and strategy is an essential foundation for successful product management. It involves crafting a narrative that not only inspires but also aligns with the broader business objectives. The resulting strategy serves as a practical guide, directing the development team towards the realisation of the product vision while remaining adaptable to the dynamic nature of the market. By intricately weaving the product vision into the fabric of the business strategy, organisations can set the stage for meaningful and impactful product development.

Vision, Strategy & Roadmap

1. Vision: A vision serves as the overarching, high-level statement that defines the long-term aspirations and purpose of an organisation or a product. It answers the fundamental question of "why" an entity exists and provides a shared understanding of the desired future state. A well-crafted vision is inspirational, guiding the entire team towards a common goal. It is often enduring and doesn't change frequently, encapsulating the core values and aspirations that drive decision-making. In essence, the vision articulates the desired impact and position in the market, offering a north star that aligns the efforts of various stakeholders.

2. Strategy: While a vision outlines the destination, a strategy delineates the path to reach that destination. It is a comprehensive plan that identifies the key objectives and approaches to achieving the vision. Strategy involves making informed choices about where to compete and how to win in the marketplace. It considers factors such as target audience, differentiation, competitive positioning, and resource allocation. Strategies are more concrete and action-oriented than visions, providing a roadmap for decision-making and resource deployment. Strategies can evolve more frequently in response to changing market conditions, competition, or internal capabilities.

3. Roadmap: A roadmap is a tactical, time-bound plan that details the specific steps and milestones necessary to execute a strategy. Unlike the long-term and inspirational nature of a vision, a roadmap is more focused on the short to medium term. It breaks down the strategic initiatives into actionable tasks, defining a sequence of activities and deliverables. Roadmaps are dynamic and can be adjusted based on real-time feedback, emerging challenges, or shifting priorities. They provide a visual representation of the journey towards the strategic goals, helping teams coordinate efforts and stay aligned with the broader strategy.

4. Relationships and Dependencies: The three elements—vision, strategy, and roadmap—are interconnected components of an effective organisational or product planning framework. The vision inspires and sets the ultimate direction, guiding the formulation of a strategy. The strategy, in turn, informs the creation of a roadmap, translating high-level goals into actionable tasks. While a vision is relatively stable, strategies and roadmaps are adaptable, allowing organisations to respond to changing circumstances. The relationship is one of cascading focus, with each level providing a more detailed and actionable perspective, ensuring that day-to-day activities contribute to the realisation of the overarching vision.

One notable example of a company that has clearly articulated its vision, strategy, and roadmap is Microsoft.

Case Study: Microsoft

1. Vision: Microsoft's vision is to empower every person and every organisation on the planet to achieve more. This vision is aspirational, emphasising the company's commitment to leveraging technology to enhance the capabilities and productivity of individuals and organisations globally. It sets the tone for Microsoft's overarching goal and influences its strategic decisions.

2. Strategy: Microsoft's strategy revolves around a shift towards a cloud-first, mobile-first world. Under the leadership of CEO Satya Nadella, the company has focused on cloud computing, artificial intelligence, and other emerging technologies. The strategy involves leveraging the Azure cloud platform, promoting collaboration through products like Microsoft 365, and ensuring accessibility across a wide range of devices. This strategic direction guides Microsoft's investments, partnerships, and product development efforts.

3. Roadmap: Microsoft's roadmap reflects its commitment to ongoing innovation and the delivery of specific products and features. For instance, in the context of Microsoft 365, the roadmap outlines planned updates, enhancements, and new functionalities. The roadmap provides a timeline for the release of features such as collaboration tools, security enhancements, and improvements to user experiences across various devices. This detailed plan allows stakeholders, including customers and partners, to anticipate and prepare for upcoming changes.

Microsoft's transparency in communicating its vision, strategy, and roadmap not only aligns internal teams but also serves as a guide for external stakeholders, including customers, investors, and developers. The company regularly updates these elements to reflect evolving market conditions, technological advancements, and user needs.

Visioning Workshop

Product Vision and Strategy Workshop Agenda with Facilitator's Notes:

1. Introduction (30 minutes):

- **Facilitator's Notes:**

    - Welcome participants and set a positive tone for the workshop.

    - Provide a brief overview of the agenda and the workshop's objectives.

    - Emphasise the importance of collaborative efforts in defining a robust product vision and strategy.

2. Current State Analysis (60 minutes):

- **Facilitator's Notes:**

    - Introduce the session by highlighting the significance of understanding the current market dynamics.

    - Present a concise analysis of current industry trends, emphasising relevant data points.

    - Guide participants through a SWOT analysis to identify internal and external factors affecting the organisation's product landscape.

    - Open the floor for discussions and insights from the C-suite and Senior Management Team.

3. Defining the Product Vision (90 minutes):

- **Facilitator's Notes:**

    - Engage participants in a brainstorming session, encouraging them to envision the future state of the industry.

    - Facilitate a group activity focused on crafting a compelling product vision statement.

    - Ensure that the vision reflects both aspirational goals and realistic market expectations.

    - Foster an open dialogue to achieve a unified understanding of the desired impact.

4. Aligning with Business Objectives (45 minutes):

- **Facilitator's Notes:**

    - Present the overall business strategy and goals, linking them to the product vision.

    - Guide a group discussion on how the product vision aligns with broader business objectives.

    - Encourage participants to identify key success metrics and KPIs that will measure progress towards both product and business goals.

5. Break (15 minutes):

- **Facilitator's Notes:**

    - Provide an opportunity for participants to network, engage in informal discussions, and recharge.

6. Crafting the Product Strategy (120 minutes):

- **Facilitator's Notes:**

    - Lead a workshop on translating the established vision into a tangible product strategy.

    - Facilitate a prioritisation exercise to determine key features and functionalities.

    - Consider market positioning, target audience, and competitive landscape in strategy development.

    - Encourage collaboration and cross-functional discussions during the strategy crafting process.

7. Tools and Technologies Showcase (30 minutes):

- **Facilitator's Notes:**

    - Present a brief overview of various tools and technologies that can aid in product strategy development.

    - Conduct a live demonstration of collaboration platforms, data analytics tools, and project management software.

    - Allow time for a Q&A session to address any queries regarding tool integration and implementation.

8. Breakout Sessions (60 minutes):

- **Facilitator's Notes:**

    - Organise small group breakout sessions with dedicated facilitators for each group.

    - Assign specific aspects of the product strategy to each group for focused development.

    - Encourage participants to actively contribute and collaborate during the breakout sessions.

9. Reporting and Feedback (45 minutes):

- **Facilitator's Notes:**

    - Bring groups back together to present key findings and components of their product strategies.

    - Facilitate an open forum for questions, clarifications, and initial feedback.

    - Use this session to capture diverse perspectives and insights from the participants.

10. Next Steps and Action Planning (30 minutes): - Facilitator's Notes: - Summarise workshop outcomes and key takeaways. - Discuss immediate next steps and assign responsibilities for further research and analysis. - Collaboratively outline actionable items for the product vision and strategy implementation.

11. Closing Remarks (15 minutes): - Facilitator's Notes: - Express gratitude to participants for their active engagement. - Extend invitations for follow-up sessions and continuous collaboration. - Provide any necessary logistical information or reminders for post-workshop activities.

By following this agenda and facilitator's notes, the workshop aims to provide a structured and collaborative environment for developing a clear and compelling product vision and strategy with the C-suite and Senior Management Team.

9.3 Responsibilities of a Product Manager:

Exploration of the day-to-day tasks and long-term responsibilities of a product manager, including market research, feature prioritisation, and maintaining a cohesive product roadmap.

Product managers play a pivotal role in the success of a product, serving as the driving force behind its development and evolution. One of the fundamental day-to-day tasks of a product manager involves conducting extensive market research. This entails staying abreast of industry trends, understanding customer needs, and analysing competitor landscapes. By gathering and synthesising this information, product managers gain crucial insights that inform strategic decisions throughout the product lifecycle.

Another key responsibility of a product manager is feature prioritisation. In a dynamic business environment, where resources are finite, it becomes essential to identify and prioritise the most impactful features that align with both user requirements and overall business objectives. Product managers employ various frameworks and methodologies to assess feature importance, taking into consideration factors such as user feedback, market demand, and technological feasibility. This ensures that the development team focuses on delivering features that provide maximum value to users and contribute to the product's success.

Maintaining a cohesive product roadmap is a long-term responsibility that demands strategic vision and effective communication. The product roadmap serves as a visual representation of the product strategy, outlining the planned features and enhancements over time. Product managers need to balance short-term objectives with long-term goals, ensuring that the roadmap aligns with the overall vision. They collaborate with cross-functional teams, including development, marketing, and sales, to communicate the roadmap, fostering a shared understanding and coordinated effort across the organisation.

In addition to these core responsibilities, product managers act as advocates for the product, championing its value proposition both internally and externally. Internally, they collaborate with development teams to ensure a clear understanding of product goals. Externally, product managers engage with customers and stakeholders, gathering feedback to refine the product strategy. Overall, the role requires a versatile skill set, encompassing analytical thinking, strategic planning, and effective communication, making product managers instrumental in guiding a product from ideation to market success.

Sample Job Description

Job Title: Product Manager

Job Description: We are seeking a skilled and motivated Product Manager to join our dynamic team. The ideal candidate will take ownership of our product portfolio, from conception to delivery, ensuring that it aligns with company goals and resonates with our target audience. The Product Manager will play a key role in shaping the product strategy, conducting market research, and collaborating with cross-functional teams to deliver innovative and high-quality products.

Responsibilities:

1. **Market Research:**

   a) Conduct in-depth market research to understand industry trends, customer needs, and competitive landscapes.

   b) Analyse market data to identify opportunities for product enhancements and innovation.

2. **Product Strategy:**

   a) Develop and execute a comprehensive product strategy that aligns with overall business objectives.

   b) Define product roadmaps, prioritising features based on user feedback, market demand, and technological feasibility.

3. **Cross-Functional Collaboration:**

   a) Collaborate with development teams to communicate product goals and ensure the timely delivery of features.

   b) Work closely with marketing, sales, and customer support to align product messaging and strategies.

4. **Stakeholder Engagement:**

   a) Act as the primary point of contact for stakeholders, gathering and incorporating feedback into the product development process.

   b) Conduct regular meetings with executive leadership to provide updates on product performance and strategy.

5. **Product Advocacy:**

   a) Serve as the product's advocate both internally and externally.

   b) Engage with customers to understand their needs and communicate how the product addresses those needs.

SMART Objectives:

Specific: Increase user engagement by 20% within the next quarter through the implementation of key product enhancements identified through market research.

Measurable: Achieve a 15% improvement in product launch timelines by implementing more efficient project management processes and tools within the next six months.

Achievable: Launch two major product updates within the next year, incorporating features prioritised based on customer feedback and market demand.

Relevant: Enhance customer satisfaction by 25% within the next year through improved user experiences and responsiveness to customer feedback.

Time-Bound: Develop and execute a comprehensive marketing and communication plan for the upcoming product release within the next three months.

Key Relationships:

Internal:
- Collaborate closely with development teams, ensuring a shared understanding of product goals and priorities.
- Engage with marketing and sales teams to align product messaging and strategies.

External:
- Establish and nurture relationships with key customers to gather valuable feedback.
- Act as a liaison with industry partners and stakeholders to stay informed about market trends and potential collaboration opportunities.

9.4 Responsibilities of a Product Owner:

An examination of the specific responsibilities of a product owner, focusing on backlog management, sprint planning, and ensuring the development team's understanding of product requirements.

Within an Agile framework, the role of a Product Owner is key, acting as the bridge between business stakeholders and the development team. One of the primary responsibilities is backlog management, a process involving the prioritisation and organisation of user stories, features, and tasks. The Product Owner meticulously curates the product backlog, ensuring that it aligns with the overall product vision and strategy. This involves constant refinement, where new insights, changing market conditions, or feedback from users are incorporated to optimise the backlog for maximum value delivery.

Sprint planning is another critical aspect of the Product Owner's role. In collaboration with the Scrum Master and the development team, the Product Owner plays a key role in determining the scope of work for the upcoming sprint. This involves selecting high-priority items from the backlog, breaking them down into actionable tasks, and defining acceptance criteria. The Product Owner's input is essential during sprint planning to provide clarity on business priorities and to guide the team towards delivering increments of tangible value within each sprint.

Ensuring the development team's comprehensive understanding of product requirements is fundamental to the Product Owner's responsibilities. The Product Owner serves as the primary source of information about the product, conveying the vision, goals, and specifics of user stories to the development team. Clear communication is crucial to avoid ambiguity and facilitate a shared understanding. The Product Owner actively participates in ceremonies like Sprint Reviews and Sprint Retrospectives, providing feedback and clarifications, fostering an environment of continuous improvement.

Ultimately, the Product Owner's role extends beyond these specific responsibilities. They act as the voice of the customer, advocating for user needs and ensuring that development efforts align with the overall business strategy. The Product Owner embodies a proactive mindset, constantly seeking feedback, collaborating with stakeholders, and adapting strategies to meet evolving requirements. In essence, the Product Owner is a linchpin in the Agile development process, contributing significantly to the success of the product and the satisfaction of end-users.

Job Title: Product Owner

Job Overview: As a Product Owner, you will play a pivotal role in the Agile development process, serving as the key liaison between business stakeholders and the development team. Your primary responsibility is to ensure the successful delivery of high-quality products that align with the company's vision and meet customer needs. This position requires a strategic thinker with excellent communication skills, a deep understanding of the market, and the ability to prioritise and manage product backlogs effectively.

Job Description

Digital Product Owner

Job Overview: As a Product Owner, you will play a pivotal role in the Agile development process, serving as the key liaison between business stakeholders and the development team. Your primary responsibility is to ensure the successful delivery of high-quality products that align with the company's vision and meet customer needs. This position requires a strategic thinker with excellent communication skills, a deep understanding of the market, and the ability to prioritise and manage product backlogs effectively

Responsibilities:

1. **Backlog Management:**

    a) Curate and prioritise the product backlog based on business priorities, customer feedback, and market trends.

    b) Continuously refine and groom the backlog to ensure it reflects the evolving needs of the business and end-users.

    c) Collaborate with stakeholders to gather and incorporate feedback into backlog prioritisation.

2. **Sprint Planning:**

    a) Work closely with the Scrum Master and development team to plan and define the scope of work for each sprint.

    b) Select user stories and features from the backlog for inclusion in sprints, ensuring alignment with business goals.

    c) Break down high-level features into actionable tasks and define acceptance criteria for development.

3. **Requirement Communication:**

   a) Clearly articulate product vision, goals, and specific requirements to the development team.

   b) Act as the primary point of contact for the team regarding clarifications, feedback, and adjustments to requirements.

   c) Ensure a shared understanding of product objectives among all team members.

4. **Stakeholder Collaboration:**

   a) Foster strong relationships with business stakeholders, understanding their needs and expectations.

   b) Collaborate with cross-functional teams, including marketing, sales, and customer support, to gather insights and align product development with overall business strategy.

   c) Act as the voice of the customer within the development team.

Key Relationships:

- **Development Team:** Work closely with developers, QA engineers, and other team members to ensure smooth execution of development tasks.

- **Scrum Master:** Collaborate on sprint planning, backlog refinement, and overall process improvement.

- **Business Stakeholders:** Engage regularly with executives, marketing, sales, and customer support teams to align product development with business objectives.

- **End Users:** Act as an advocate for end-users, gathering feedback and ensuring their needs are prioritised in product development.

SMART Objectives:

1. **Specific:** Clearly define and communicate the top three business priorities to be addressed in the upcoming product release.

2. **Measurable:** Improve sprint velocity by 15% over the next two quarters through efficient backlog management and improved sprint planning.

3. **Achievable:** Reduce customer-reported bugs by 20% by implementing a more rigorous acceptance criteria framework in collaboration with the QA team.

4. **Relevant:** Enhance user satisfaction scores by 10% through the successful delivery of features identified as high-impact by customer feedback.

5. **Time-bound:** Deliver a minimum viable product (MVP) for a key market segment within the next quarter, enabling early market penetration and user feedback.

This role offers a dynamic opportunity to shape the development of cutting-edge products, contributing to the company's growth and success. The successful candidate will combine strategic thinking with hands-on collaboration to drive product excellence.

9.5 Responsibilities of the Scrum Master:

How this role fits into Product Development

Scrum Methodology:

Scrum is an Agile project management framework that promotes iterative and incremental development. It is structured around small, cross-functional teams that work in fixed-length iterations called sprints, usually lasting two to four weeks. At the core of Scrum is the product backlog, a Prioritised list of features and user stories maintained by the Product Owner. Each sprint begins with a sprint planning meeting where the team selects a set of items from the backlog to work on. Daily stand-up meetings keep the team synchronised, highlighting progress and addressing any obstacles. At the end of each sprint, a sprint review is conducted to showcase the completed work, and a retrospective is held to reflect on the process and identify improvements. Scrum provides a flexible and adaptive approach, allowing teams to respond to changing requirements and deliver incremental value throughout the development process.

Kanban:

Kanban is another Agile methodology that focuses on visualising work, limiting work in progress, and maximising flow. Unlike Scrum, Kanban does not prescribe fixed iterations or specific roles. Work items are represented on a Kanban board, which typically consists of columns such as "To Do," "In Progress," and "Done." Teams pull work from the backlog based on capacity, and the goal is to maintain a smooth and continuous flow of work through the system. Work in progress (WIP) limits are set for each column to prevent bottlenecks and overburdening team members. Kanban emphasises continuous improvement, and teams regularly review their processes to optimise efficiency. It is particularly effective for teams with varying workloads or those engaged in continuous delivery rather than fixed-length projects.

Comparing Scrum and Kanban:

While both Scrum and Kanban are Agile methodologies, they have distinct differences. Scrum is characterised by fixed-length sprints, a defined set of roles (Product Owner, Scrum Master, and Development Team), and specific ceremonies (sprint planning, daily stand-ups, sprint review, and retrospective). In contrast, Kanban provides more flexibility regarding roles, time frames, and ceremonies. Scrum encourages teams to commit to a fixed amount of work for each sprint, fostering predictability. Kanban, on the other hand, promotes a continuous flow of work with an emphasis on optimising efficiency and adapting to changes in real-time. Both methodologies share the foundational Agile principles of collaboration, customer feedback, and delivering value iteratively. The choice between Scrum and Kanban often depends on the nature of the project, team dynamics, and the organisation's preferences for structure and adaptability.

Role of a Scrum Master:

A Scrum Master plays a crucial role in facilitating and enabling the Agile development process within a Scrum team. While not a traditional project manager, the Scrum Master is responsible for ensuring that the Scrum framework is understood and implemented effectively. They act as a servant-leader, supporting the team and removing impediments to enhance productivity. The Scrum Master also fosters a culture of continuous improvement, enabling the team to self-organise and deliver high-quality products iteratively.

Responsibilities:

1. **Facilitating Scrum Events:**
    1. Organise and facilitate key Scrum events, including sprint planning, daily stand-ups, sprint review, and sprint retrospective.
    2. Ensure that these events are effective, time-boxed, and focused on achieving their objectives.

2. **Removing Impediments:**
    1. Identify and eliminate obstacles that hinder the team's progress.
    2. Act as a shield for the team, preventing external interruptions and distractions.

3. **Coaching and Training:**
    1. Provide guidance and coaching to the Scrum Team on Agile principles and Scrum practices.
    2. Encourage a culture of continuous learning and improvement.

4. **Ensuring Team Empowerment:**
    1. Foster a self-organising team culture where members collaborate, make decisions collectively, and take ownership of their work.
    2. Empower the team to continuously improve its processes and performance.

5. **Collaborating with Product Owner:**
    1. Collaborate closely with the Product Owner to ensure a well-maintained product backlog and effective communication of product goals.
    2. Assist in refining and prioritising backlog items.

6. **Shielding the Team:**
    1. Protect the team from external disruptions and unnecessary interventions, allowing them to focus on delivering value.

Typical Day of a Scrum Master:

A typical day for a Scrum Master involves a dynamic mix of collaboration, facilitation, and proactive support. In the morning, they may kick off the day with a daily stand-up meeting, during which team members share updates on their work, discuss any impediments, and plan the day's tasks. After the stand-up, the Scrum Master might spend time refining the backlog with the Product Owner, ensuring that it aligns with the current priorities.

Throughout the day, the Scrum Master is often engaged in removing impediments for the team. This could involve coordinating with other teams, addressing technical challenges, or resolving conflicts within the team. They may facilitate collaboration during sprint planning or conduct a retrospective to reflect on the completed sprint and identify areas for improvement.

The Scrum Master also serves as a continuous coach, offering guidance on Agile practices and helping team members overcome challenges. They may schedule one-on-one sessions with team members to discuss individual concerns, provide feedback, and support personal development.

Additionally, a significant portion of the day may be dedicated to observing the team's dynamics, identifying areas for improvement, and fostering a positive and collaborative work environment. The Scrum Master ensures that the Scrum values and principles are upheld, promoting a culture of transparency, inspection, and adaptation within the team.

9.6 Tools and Resources for Product Managers, Product Owners, and Scrum Masters:

1. Product Management Tools:

1. *Jira Software (Atlassian):* Jira is a versatile tool that supports end-to-end product management, from backlog creation and prioritisation to sprint planning and tracking. It enables collaboration among cross-functional teams and provides detailed insights into project progress.

2. *Aha!:* Aha! is a comprehensive product management platform that facilitates strategic planning, roadmapping, and feature prioritisation. It helps Product Managers align product development with organisational goals and communicate a clear product strategy.

3. *Productboard:* This tool streamlines product discovery and prioritisation. It allows Product Managers to gather and centralise customer feedback, Prioritise feature development, and create a visual product roadmap.

2. Product Ownership and Collaboration Tools:

1. *Miro:* Miro is a collaborative online whiteboard platform that is beneficial for Product Owners. It supports activities like backlog grooming, user story mapping, and collaborative sprint planning sessions, fostering effective communication and ideation.

2. *Confluence (Atlassian):* Confluence is a team collaboration platform that aids Product Owners in creating and sharing documentation. It's valuable for maintaining a knowledge base, recording decisions made during sprint planning, and ensuring transparency.

3. *Lucidchart:* Lucidchart is a diagramming tool that can assist Product Owners in creating visual representations of product features, user flows, and process maps. It enhances clarity and communication within the team.

3. Scrum Master Tools and Resources:

1. *Trello:* Trello is a simple and flexible project management tool that Scrum Masters can use for visualising and managing tasks. It's particularly effective for Kanban-style boards, facilitating the tracking of work in progress.

2. *Scrum.org:* Scrum.org provides valuable resources, including training materials, webinars, and articles that can enhance a Scrum Master's understanding of Scrum principles and practices. The Scrum Guide, available on the website, serves as a foundational reference.

3. *Mural:* Mural is a digital workspace that supports collaboration, brainstorming, and retrospective activities. Scrum Masters can use it to facilitate remote or distributed teams in creative problem-solving and continuous improvement sessions.

These tools and resources empower Product Managers, Product Owners, and Scrum Masters to streamline their workflows, enhance collaboration, and ensure effective product development within the Agile framework. Choosing the right combination of tools depends on the specific needs and preferences of the team and the organisation.

9.7 Collaboration between Product Management and Development

The synergy between Product Management and Development teams is paramount for success. Collaboration not only accelerates the delivery of high-quality products but also ensures alignment with strategic business goals. This section delves into key insights and strategies for fostering effective communication and collaboration between these two crucial facets of the product development process, promoting a unified approach towards shared objectives.

Understanding Divergent Perspectives:

At the core of successful collaboration is the recognition and appreciation of the distinct perspectives held by Product Managers and Development teams. Product Managers focus on market demands, user needs, and strategic goals, while Developers concentrate on technical feasibility and implementation details. Bridging this gap requires a mutual understanding of each other's roles and responsibilities. Regular cross-functional training sessions, job-shadowing opportunities, and joint workshops contribute to developing empathy and insight into the challenges faced by both sides.

Establishing Clear Communication Channels:

Clear and open communication is the bedrock of effective collaboration. Establishing transparent channels for communication ensures that information flows seamlessly between Product Management and Development. Regular meetings, such as sprint planning sessions, backlog grooming, and sprint reviews, create forums for discussion and alignment. Leveraging collaboration tools, such as project management platforms and communication apps, enhances real-time information sharing and keeps all stakeholders informed and engaged.

Shared Ownership of Goals:

A fundamental shift towards a unified approach involves instilling a sense of shared ownership of product goals. This entails involving Development teams early in the product planning phase, seeking their input on technical feasibility and potential challenges. Product Managers, in turn, participate in sprint reviews and retrospectives to understand the intricacies of the development process. This shared ownership fosters a collective responsibility for the product's success and encourages a culture of collaboration rather than siloed operations.

Agile Practices for Seamless Integration:

Adopting Agile practices provides a framework for seamless integration between Product Management and Development. Embracing iterative development, regular feedback loops, and adaptive planning, Agile methodologies such as Scrum or Kanban enhance collaboration. Scrum ceremonies, such as sprint planning, daily stand-ups, and retrospectives, offer structured opportunities for interaction, ensuring that the entire team is aligned towards common objectives.

Embracing a Culture of Continuous Improvement:

Collaboration is an evolving process, and fostering a culture of continuous improvement is instrumental in its long-term success. Regular retrospectives, where both Product Management and Development teams reflect on what worked well and what could be improved, provide valuable insights. Establishing a feedback loop, both formal and informal, enables ongoing adjustments to collaboration strategies, ensuring that they remain dynamic and responsive to the evolving needs of the product and the team.

Conclusion:

By acknowledging and addressing the unique challenges each team faces, establishing clear communication channels, embracing shared ownership, adopting Agile practices, and fostering a culture of continuous improvement, organisations can cultivate a collaborative environment where Product Managers and Developers work hand-in-hand towards a common vision. This unity of purpose not only accelerates product delivery but also fuels innovation and ensures that the end result aligns seamlessly with customer expectations and business objectives.

9.8 Chapter: Scaling Product Management for Growth

Scaling Product Management practices is a critical endeavour that requires strategic foresight, adaptability, and a commitment to maintaining product excellence. This section explores strategies for effectively scaling Product Management to align with a company's growth trajectory, emphasising considerations for expanding product portfolios and managing multiple product lines.

Understanding the Growth Landscape:

As a company expands, its product landscape often evolves to encompass a broader array of offerings. Scaling Product Management begins with a comprehensive understanding of the growth landscape, including market dynamics, customer needs, and emerging trends. Product Managers must continuously analyse how the company's products fit into the overall market and strategise ways to meet evolving customer demands.

Strategies for Scaling Product Management:

1. **Cross-Functional Collaboration:** Foster stronger collaboration between Product Management and other departments, including Development, Marketing, and Sales. Cross-functional teams ensure that all aspects of product development align with broader business goals.

2. **Specialised Product Teams:** Consider forming specialised product teams dedicated to specific product lines or customer segments. Each team can have its Product Manager focused on the unique requirements and opportunities within their domain.

3. **Agile Frameworks and Scrum of Scrums:** Implement Agile frameworks such as Scrum and introduce the concept of Scrum of Scrums for larger-scale coordination. This enables efficient collaboration and communication among multiple Product Management teams.

4. **Portfolio Management:** Develop robust portfolio management practices to Prioritise and allocate resources effectively. This involves evaluating the performance of each product line, assessing market dynamics, and making data-driven decisions to optimise the overall product portfolio.

5. **Unified Product Vision:** Ensure that all product teams share a unified product vision and strategy. Regular alignment sessions and communication forums help synchronise efforts and maintain a cohesive narrative across diverse product lines.

6. **Scalable Processes and Tools:** Invest in scalable processes and tools that can accommodate the increased complexity of managing multiple products. This includes adopting product management platforms, collaboration tools, and project management systems that facilitate streamlined workflows.

7. **Talent Development and Recruitment:** Invest in the continuous development of Product Management talent. As the product landscape expands, hiring skilled Product Managers with expertise in different domains becomes essential. Implement training programs and mentorship initiatives to foster a culture of continuous learning.

9.9 Considerations for Managing Multiple Product Lines:

Segmented Customer Personas:

Develop segmented customer personas for each product line to ensure that product features and marketing strategies resonate with specific target audiences.

Resource Allocation:

Implement a dynamic resource allocation strategy that considers the unique needs and potential of each product line. This may involve adjusting team sizes, budgets, and development timelines based on individual product priorities.

Iterative Market Analysis:

Conduct iterative market analyses for each product line to stay abreast of changing market conditions, competition, and customer preferences. Regularly revisit and adapt product strategies based on these analyses.

Unified Branding and Positioning:

Maintain a unified brand image and positioning across all product lines. While each product may address distinct needs, aligning branding elements ensures a coherent and recognisable company identity.

Scaling Product Management for growth requires a strategic blend of adaptability, collaboration, and structured processes. By implementing these strategies and considering the nuances of managing multiple product lines, companies can navigate the complexities of expansion while maintaining a focus on delivering value to customers and achieving sustainable growth. As the product landscape evolves, a scalable and well-coordinated Product Management function becomes the linchpin for orchestrating successful product development journeys.

9.10 Overcoming Challenges in Product Ownership

Product Owners play a pivotal role in steering the development of successful products. However, this role is not without its challenges. This chapter explores the identification and resolution of common challenges faced by Product Owners, providing insights and strategies for overcoming hurdles related to balancing competing priorities, managing stakeholder expectations, and adapting to changing market conditions.

Balancing Competing Priorities:

One of the primary challenges for Product Owners is the delicate balance between competing priorities. This can manifest as conflicting user needs, technical debt, or resource limitations. To address this challenge, Product Owners must employ effective prioritisation techniques. Utilising methods such as MoSCoW (Must-haves, Should-haves, Could-haves, and Won't-haves) or value-driven frameworks helps align priorities with strategic goals. Regular communication with stakeholders and a transparent approach to decision-making assist in managing expectations and fostering understanding among team members.

Managing Stakeholder Expectations:

Stakeholder management is a complex facet of product ownership. Different stakeholders may have diverse expectations and interests. A key strategy for managing this challenge is proactive and open communication. Regularly engaging with stakeholders to gather feedback, providing realistic timelines, and involving them in the decision-making process fosters a collaborative environment. Setting clear expectations from the outset, coupled with transparent reporting on progress and challenges, helps build trust and mitigates the risk of misaligned expectations.

Adapting to Changing Market Conditions:

The dynamic nature of markets presents a significant challenge for Product Owners. Rapid changes in customer preferences, emerging technologies, or unforeseen market disruptions can impact product strategies. To navigate this challenge, Product Owners should embrace agility and responsiveness. Regularly reassessing the product roadmap, staying attuned to market trends, and leveraging user feedback for real-time adjustments are crucial. Adopting Agile methodologies, such as Scrum, facilitates a more adaptive approach, allowing Product Owners to pivot quickly in response to changing market conditions.

Navigating Technical Debt:

Accumulation of technical debt, often a result of expedited development to meet deadlines, poses a persistent challenge. Technical debt can hinder future development and compromise product quality. To address this challenge, Product Owners should work collaboratively with development teams to allocate time for refactoring and addressing technical debt. Prioritising technical tasks alongside user stories ensures a balance between short-term feature development and long-term product sustainability. Regularly reassessing the impact of technical debt on the product roadmap helps maintain a healthy balance.

Dealing with Last-Minute Changes from C-Suite:

A unique challenge faced by Product Owners is the introduction of last-minute changes or shifting priorities initiated by C-Suite members. This dynamic can disrupt the product development process and strain timelines (and drive you crazy!), Mitigating this challenge involves establishing clear communication channels with the C-Suite, ensuring that they are informed about the implications of late-stage changes. It's essential to convey the potential impact on timelines, resources, and overall project goals. Encouraging a collaborative decision-making process and providing insights into the trade-offs involved helps align the C-Suite with the product development strategy.

Conclusion:

Recognising the dynamic nature of some of the challenges here is the first step towards effective mitigation. By employing prioritisation techniques, maintaining transparent communication with stakeholders, embracing agility in response to market changes, addressing technical debt proactively, and navigating last-minute changes from the C-Suite, Product Owners can navigate these challenges with resilience. The ability to adapt, collaborate, and maintain a customer-centric focus is key to overcoming hurdles and steering products towards sustained success in an ever-evolving landscape.

9.11 Measuring Product Success and KPI's

Defining key performance indicators (KPIs) and metrics for evaluating the success of product initiatives, with a focus on continuous improvement and adapting to market dynamics.

Also see Paragraphs 5.3 Measuring Success in Digital Landscape and 4.7 Key KPI's in The Technical Roadmap

Measuring Product Success and KPIs

The ability to measure and gauge the success of product initiatives is crucial for informed decision-making and sustained growth. This section explores the significance of defining key performance indicators (KPIs) and metrics tailored to evaluate product success. Emphasis is placed on the iterative nature of measuring success, fostering continuous improvement, and adapting strategies to dynamic market forces.

Defining Product Success:

Product success extends beyond mere financial gains. While revenue and profit are essential components, a comprehensive understanding of success encompasses customer satisfaction, market impact, and the achievement of strategic goals. Defining success requires alignment with the overarching vision, and this alignment forms the basis for establishing relevant KPIs.

Selecting Key Performance Indicators (KPIs):

KPIs serve as the compass for product success, providing quantifiable and objective measures against predefined goals. The selection of KPIs should be strategic, reflecting the unique objectives of the product and the organisation. Common KPIs include:

1. **Revenue Metrics:** Sales, profit margins, and customer lifetime value.

2. **Customer Satisfaction:** Net Promoter Score (NPS), customer feedback, and user engagement.

3. **Market Impact:** Market share, competitive positioning, and penetration in target segments.

4. **Product Health:** Defect rates, uptime, and reliability.

5. **Innovation Metrics:** Speed of delivery, feature adoption, and time-to-market.

Continuous Improvement through Iterative Measurement:

Measuring success is not a one-time event but a continuous and iterative process. Regularly revisiting KPIs enables product teams to adapt strategies in response to changing market dynamics and customer needs. Continuous improvement involves learning from both successes and failures, adjusting goals, and refining KPIs to reflect evolving priorities.

Adapting to Market Dynamics:

Markets are dynamic, influenced by technological advancements, shifting consumer preferences, and competitive landscapes. Successful products are those that demonstrate adaptability to these changes. KPIs should be flexible enough to accommodate adjustments in response to market dynamics, allowing teams to pivot when necessary and seize new opportunities.

Balancing Short-Term and Long-Term Metrics:

While short-term metrics such as sprint velocity and feature adoption are valuable for immediate feedback, long-term success requires a focus on strategic indicators. Balancing both short-term and long-term metrics ensures that product teams address immediate needs while also building a foundation for sustained growth and innovation.

Communicating Success Metrics:

Effective communication of success metrics is essential for aligning cross-functional teams, stakeholders, and leadership. Visualisation tools, dashboards, and regular reporting sessions facilitate transparent communication, enabling a shared understanding of progress and areas for improvement.

Advantages of Daily Dashboards for Communicating KPIs and Success:

Leveraging regular, daily dashboards for communicating key performance indicators (KPIs) and success metrics offers a myriad of advantages. This real-time approach to reporting fosters transparency, enhances collaboration, and empowers teams with the insights needed for timely decision-making. Here are some key advantages:

1. Immediate Visibility:

Daily dashboards provide immediate visibility into the current state of key metrics. Team members, stakeholders, and leadership can access up-to-date information on performance, allowing them to stay informed about progress and potential challenges.

This immediacy enhances the team's responsiveness to emerging issues or opportunities.

2. Agile Adaptation:

In an Agile environment, where iterative development and continuous improvement are paramount, daily dashboards enable teams to adapt rapidly. The ability to assess KPIs on a daily basis facilitates quick adjustments to strategies, ensuring that the team remains aligned with objectives and can promptly address any deviations.

3. Enhanced Decision-Making:

Regular access to comprehensive dashboards equips decision-makers with the data needed for informed choices. Whether it's adjusting priorities, allocating resources, or pivoting strategies, decision-makers can rely on real-time KPIs to guide their actions. This minimises the lag between data collection and decision implementation.

4. Increased Accountability:

Daily dashboards instil a sense of accountability among team members. When everyone has visibility into their individual and collective contributions to KPIs, there is a heightened awareness of performance expectations. This transparency promotes a culture of ownership and responsibility, fostering a commitment to achieving and exceeding targets.

5. Continuous Improvement Culture:

Regular reporting of KPIs encourages a culture of continuous improvement. Teams can analyse trends, identify patterns, and learn from both successes and setbacks. This iterative process of reflection and adjustment is essential for refining strategies, optimising workflows, and driving sustained success.

6. Cross-Functional Alignment:

Daily dashboards facilitate cross-functional alignment by providing a shared platform for communication. Whether it's the product team, development team, or stakeholders from different departments, a centralised dashboard ensures that everyone is on the same page regarding performance metrics. This alignment contributes to a cohesive and collaborative working environment.

7. Proactive Issue Identification:

Timely access to KPI data allows teams to proactively identify and address issues before they escalate. If a particular metric begins to trend in an undesirable direction, the team can quickly investigate the root cause and implement corrective measures, preventing potential setbacks.

8. Motivational Impact:

The visibility of daily progress can have a motivational impact on team members. Celebrating small wins and milestones on a daily basis fosters a positive work environment and reinforces a sense of achievement. This motivation is crucial for sustaining high levels of productivity and engagement.

By providing immediate visibility, enabling agile adaptation, enhancing decision-making, fostering accountability, promoting a culture of continuous improvement, ensuring cross-functional alignment, facilitating proactive issue identification, and delivering motivational impact, regular dashboards become a cornerstone for effective communication of KPIs and success metrics.

Conclusion:

Measuring product success through well-defined KPIs is not a static exercise but a dynamic journey. It requires a holistic perspective that incorporates financial, customer-centric, and strategic dimensions. By continuously refining and adapting KPIs to align with changing market dynamics, product teams can navigate the complexities of product development, foster a culture of continuous improvement, and ensure that success is not just a destination but a continuous pursuit of excellence.

9.12 Case Studies and Best Practices:

Examining real-world case studies and industry best practices that illustrate successful product management and ownership strategies, offering actionable insights for CTOs.

Case Study Dyson - Transforming Household Appliances with Innovation

Background: Dyson, a UK-based technology company, has revolutionised the household appliances industry with its commitment to innovation and cutting-edge design. One of Dyson's standout products is the Dyson Supersonic hair dryer, introduced in 2016.

Product Management and Ownership Strategies:

1. **Engineering Excellence:** Dyson's product development strategy is anchored in engineering excellence. The Supersonic hair dryer was designed with advanced technology to deliver fast drying while protecting hair from extreme heat. The focus on technological innovation sets Dyson products apart in the market.

2. **Customer-Centric Design:** Dyson places a strong emphasis on customer satisfaction and usability. The Supersonic hair dryer's design is based on extensive research into user preferences and needs. Ergonomics and ease of use are integral elements of Dyson's customer-centric approach.

3. **Brand Identity and Quality:** Dyson has established itself as a brand synonymous with quality and innovation. The Supersonic hair dryer exemplifies the brand's commitment to creating high-performance products that redefine industry standards. Dyson's dedication to quality is reflected in its premium positioning in the market.

Outcomes:

1. **Innovation Leadership:** Dyson's Supersonic hair dryer has positioned the company as a leader in innovation within the household appliances sector. The product's advanced technology and design have set new benchmarks for performance and user experience.

2. **Global Recognition:** Dyson's commitment to engineering excellence and customer-centric design has earned the brand global recognition. The Supersonic hair dryer, along with other Dyson products, has become a symbol of quality and innovation, contributing to the brand's international success.

3. **Diversification and Expansion:** The success of the Supersonic hair dryer has fueled Dyson's diversification into other product categories. Dyson's expansion

into areas such as air purifiers, vacuums, and lighting showcases the brand's ability to apply its product management strategies across various sectors.

Actionable Insights for CTOs:

- **Invest in Engineering Excellence:** Prioritise technological innovation and engineering prowess to create products that redefine industry standards.

- **Prioritise Customer-Centric Design:** Conduct thorough research into user preferences and needs, placing a strong emphasis on creating products that enhance customer satisfaction and usability.

- **Build a Strong Brand Identity:** Establish a brand identity that emphasises quality and innovation. Consistent delivery of high-performance products contributes to brand recognition and success.

Case Study: Apple Inc. - Revolutionising the Consumer Electronics Landscape

Background: Apple Inc., a global tech giant, has consistently demonstrated exemplary product management and ownership strategies, influencing the consumer electronics industry. A standout example is the launch of the iPhone in 2007, transforming mobile communication and setting new standards for smartphones.

Product Management and Ownership Strategies:

1. **User-Centric Design:** Apple's product management philosophy revolves around user-centric design. The iPhone was developed with a deep understanding of user needs, providing an intuitive and seamless experience. This strategy places the user at the center of the product development process.

2. **Ecosystem Integration:** Apple strategically integrated its product ecosystem. The iPhone seamlessly connects with other Apple devices, creating a unified experience. This interconnected ecosystem enhances user loyalty and encourages customers to invest in multiple Apple products, strengthening the brand's market presence.

3. **Iterative Product Development:** Apple employs an iterative approach to product development. Each iPhone release builds upon the successes and learns from the shortcomings of its predecessors. Regular updates, both in terms of hardware and software, ensure continuous improvement and relevance in a dynamic market.

Outcomes:

1. **Brand Loyalty:** Apple's user-centric design and ecosystem integration have cultivated unparalleled brand loyalty. Customers often stay within the Apple ecosystem, upgrading to newer products, and becoming brand advocates.

2. **Market Leadership:** The iPhone's success catapulted Apple to a leadership position in the smartphone market. Apple's ability to set trends and influence consumer preferences has been a driving force in the industry.

3. **Innovation Standard:** Apple's iterative development model and commitment to innovation have set a standard for the tech industry. Competitors often benchmark against Apple, making the company a trendsetter in product management and ownership.

Actionable Insights for CTOs:

- **User-Centric Focus:** Prioritise user needs and experiences in product development to enhance customer satisfaction and loyalty.

- **Ecosystem Integration:** Explore opportunities to create a cohesive product ecosystem that encourages users to invest in multiple offerings.

- **Iterative Development:** Embrace an iterative approach, learning from each product cycle to enhance future releases continually.

Case Study: Atlassian - Transforming Collaboration with Software Solutions

Background: Atlassian, an Australian software company, has made significant contributions to the tech industry through its collaboration and productivity software. One of its standout products is Jira, a project management and issue tracking tool, exemplifying successful product management and ownership strategies.

Product Management and Ownership Strategies:

1. **Agile Development Practices:** Atlassian embodies the principles of Agile development in its product management approach. Jira itself is designed to support Agile methodologies, allowing teams to manage projects iteratively and adapt to changing requirements seamlessly.

2. **Customer-Driven Innovation:** Atlassian Prioritises customer feedback and needs in its product development. Regular updates to Jira incorporate features and enhancements requested by users, ensuring that the software remains aligned with the evolving requirements of project teams and organisations.

3. **Platform Expansion:** Atlassian has strategically expanded its product portfolio to create an ecosystem of tools that complement each other. This approach, with products like Confluence and Bitbucket, allows Atlassian to provide end-to-end solutions for collaborative work, fostering a comprehensive and interconnected user experience.

Outcomes:

1. **Global Adoption:** Jira has gained widespread global adoption, becoming a go-to solution for project management and issue tracking. Its user-friendly interface, adaptability to different project methodologies, and continuous updates have contributed to its success on a global scale.

2. **Enterprise Collaboration:** Atlassian's suite of products, including Jira, has become integral to enterprise collaboration. Organisations globally rely on Atlassian's tools to streamline workflows, enhance communication, and improve overall project efficiency.

3. **Community Engagement:** Atlassian has built a strong community around its products. Forums, user groups, and developer resources encourage collaboration and knowledge-sharing among users. This engagement fosters a sense of community and contributes to the ongoing success and improvement of Atlassian's products.

Actionable Insights for CTOs:

- **Embrace Agile Methodologies:** Implement Agile development practices to enable iterative and adaptable product development.

- **Customer-Centric Approach:** Prioritise customer feedback and incorporate user-driven innovation into product development cycles.

- **Expand Product Ecosystem:** Explore opportunities to create a suite of products that complement each other, providing users with a comprehensive and interconnected platform.

This case study of Atlassian and its product Jira illustrates how successful product management and ownership strategies can lead to global adoption, enterprise collaboration, and the creation of a thriving user community. CTOs can draw actionable insights from Atlassian's approach to enhance their organisations' product development practices.

Case Study: Nubank - Revolutionising Banking in Brazil

Background: Nubank, a Brazilian fintech startup, has disrupted the traditional banking landscape in South America by offering innovative financial services through its mobile app. One of its flagship products, the Nubank Credit Card, exemplifies successful product management and ownership strategies.

Product Management and Ownership Strategies:

1. **Customer-Centric Financial Solutions:** Nubank's approach revolves around providing customer-centric financial solutions. The Nubank Credit Card was designed to address pain points associated with traditional banking, offering transparent terms, lower fees, and a user-friendly mobile app interface.

2. **Data-Driven Decision-Making:** Nubank leverages data analytics to inform its product decisions. The company Utilises customer data to understand spending patterns, preferences, and behaviours, allowing for personalised offerings and continuous improvement of its financial products.

3. **Mobile-First Strategy:** Recognising the growing trend of mobile banking, Nubank adopted a mobile-first strategy. The Nubank app provides users with a seamless and intuitive interface for managing their finances, including credit card transactions, payments, and customer support.

Outcomes:

1. **Financial Inclusion:** Nubank has played a crucial role in promoting financial inclusion in Brazil. The accessibility of its mobile app and transparent financial products has empowered individuals who were previously underserved by traditional banking institutions.

2. **Market Disruption:** Nubank's disruptive approach has reshaped the Brazilian banking landscape. The company has gained significant market share, challenging established banks and encouraging them to reassess their offerings in response to changing customer expectations.

3. **Brand Trust:** Nubank has built a strong reputation for transparency, customer-centricity, and innovation. The Nubank Credit Card, along with other financial products, has become synonymous with a new era of banking in Brazil, fostering trust among users.

Actionable Insights for CTOs:

- **Prioritise Customer-Centric Solutions:** Design products that directly address customer needs and pain points, providing transparent and user-friendly solutions.

- **Utilise Data for Personalisation:** Leverage data analytics to gain insights into customer behaviour, enabling the personalisation of products and services.

- **Embrace Mobile Technology:** Recognise the significance of mobile technology and adopt a mobile-first strategy to cater to the evolving preferences of users.

This case study of Nubank and its Nubank Credit Card demonstrates how successful product management and ownership strategies can lead to financial inclusion, market disruption, and the establishment of a trusted brand. CTOs can draw actionable insights from Nubank's approach to enhance their organisations' product development practices, especially in the context of evolving financial technologies.

Chapter 10.

Project Management: Getting things done.

This chapter on Project Management covers methodologies, roles, case studies, tools, and provide valuable resources for project managers within the technology sector.

1. **Introduction to Project Management:**

    1. Overview of the importance of effective project management within technology focussed projects.

    2. The role of project management in achieving strategic objectives.

2. **Project Management Methodologies:**

    1. Overview of traditional methodologies (Waterfall, PRINCE2) and Agile methodologies (Scrum, Kanban).

    2. Pros and cons of each methodology and considerations for selecting the right approach.

3. **Roles and Responsibilities:**

    1. Definition of key project management roles (Project Manager, Scrum Master, Product Owner).

    2. Responsibilities of team members, stakeholders, and leadership in project success.

4. **Project Initiation:**

    1. Defining project scope, objectives, and deliverables.

    2. Conducting feasibility studies and risk assessments.

    3. Stakeholder identification and communication planning.

5. **Planning and Scheduling:**

    1. Developing project plans, Gantt charts, and timelines.

    2. Resource allocation and management.

    3. Risk mitigation and contingency planning.

6. **Project Monitoring and Control:**

    1. Establishing key performance indicators (KPIs) for project success.

    2. Continuous monitoring of progress, milestones, and budget.

    3. Implementing change control processes.

7. **Project Closure:**

    1. Conducting project reviews and post-implementation assessments.

    2. Documenting lessons learned for continuous improvement.

    3. Handover and knowledge transfer.

8. **Emerging Trends in Project Management:**

    1. Exploring trends such as AI in project management, remote collaboration tools, and the impact of industry 4.0.

9. The PMO: Project Management Office

    1. Responsibilities and organisation structure

10.1 Introduction to Project Management in Technology:

This chapter delves into the intricacies of project management where precision, adaptability, and timely delivery are paramount. As innovation in technology continues to shape industries, the need for robust project management methodologies becomes increasingly crucial. Whether it's the development of cutting-edge software, the implementation of complex infrastructure, or the launch of revolutionary products, project management serves as the guiding force that ensures these ventures align with organisational goals and industry standards.

The Crucial Role of Project Management:

Project management plays a central role in steering technology-focused projects towards success by serving as a conduit between vision and execution. At its core, project management provides the structured framework necessary to navigate the complexities inherent in technological endeavours. Beyond merely overseeing timelines and resources, project management is the driving force that ensures projects align with broader strategic objectives. In the realm of technology, where innovation and efficiency are paramount, effective project management becomes a strategic enabler. It is the instrument through which organisations turn ambitious technological visions into tangible and successful outcomes, influencing not only the bottom line but also shaping the trajectory of the company within the competitive tech landscape.

Achieving Strategic Objectives through Project Management:

By establishing clear project goals, delineating resource requirements, and fostering efficient collaboration, project management becomes the catalyst for translating strategic visions into tangible results. Through meticulous planning, continuous monitoring, and proactive risk management, project management not only safeguards against potential pitfalls but also maximises the potential for achieving strategic milestones in a technology-driven environment.

Navigating the Technological Landscape:

In the context of technology-focused projects, the need for effective project management becomes particularly pronounced. The rapidly evolving nature of technology demands a nimble and adaptive approach to project execution. This chapter will explore the methodologies, best practices, and tools that project managers in the technology sector can leverage to navigate the technological landscape successfully. As we delve deeper, we will uncover case studies, draw insights from industry leaders, and provide practical guidance for project managers tasked with steering technology-driven initiatives. Emphasising the symbiosis between project management and strategic objectives, this chapter serves as a foundational guide for anyone seeking to understand and excel in the realm of project management within technology-focused projects.

10.2 Project Management Methodologies:

This section provides an overview of both traditional and Agile methodologies, highlighting the distinctive approaches offered by traditional methodologies like Waterfall and PRINCE2, as well as the dynamic methodologies of Agile, represented by Scrum and Kanban.

Traditional Methodologies: Waterfall and PRINCE2

Waterfall: Waterfall is a linear and sequential approach to project management, where tasks progress through distinct phases such as initiation, planning, execution, monitoring, and closure. Each phase must be completed before moving on to the next, making it particularly suitable for well-defined projects with stable requirements. However, its rigid structure can become a limitation when faced with evolving project needs.

PRINCE2 (Projects IN Controlled Environments): PRINCE2 is a process-driven project management methodology that provides a flexible framework applicable to various types of projects. It emphasises dividing projects into manageable stages, each with defined roles, responsibilities, and control points. PRINCE2 is particularly effective in promoting governance and control throughout the project lifecycle.

Agile Methodologies: Scrum and Kanban

Scrum: Scrum is an iterative and incremental Agile framework that operates within fixed-length cycles called sprints. It encourages cross-functional teams to collaborate closely, delivering small, functional increments at the end of each sprint. Scrum's adaptability makes it well-suited for projects where requirements are subject to change, allowing teams to adjust their focus and priorities regularly. (see section on Scrum Chapter 6.5)

Kanban: Kanban is a visual and flow-based Agile methodology that emphasises continuous delivery. It uses a Kanban board to visualise work stages and limits work in progress (WIP) to enhance efficiency. Kanban is particularly effective for projects with fluctuating priorities and a constant flow of incoming tasks, providing a real-time view of the project's status.

Pros and Cons, Considerations for Selection:

Waterfall and PRINCE2: Pros:

- Clear project structure and defined stages.
- Suitable for projects with stable requirements.
- Emphasises documentation and control.

Cons:
- Limited adaptability to changes.
- Lengthy project timelines.
- Strict phase dependencies.

Considerations:
- Well-defined project scope.
- Stable and predictable requirements.
- A need for comprehensive documentation.

Scrum and Kanban: Pros:
- Adaptability to changing requirements.
- Continuous delivery of increments.
- Enhanced team collaboration and flexibility.

Cons:
- Less emphasis on documentation.
- May require more frequent communication.
- May not suit projects with highly stable requirements.

Considerations:
- Dynamic or evolving project requirements.
- Frequent client or stakeholder involvement.
- Emphasis on delivering value incrementally.

In selecting the right methodology, project managers should carefully evaluate the project's nature, requirements, and the level of flexibility needed. While traditional methodologies provide structure and control, Agile methodologies offer adaptability and responsiveness, allowing project managers to tailor their approach to the specific demands of the project at hand.

10.3 Roles and Responsibilities within Projects:

This section delves into the crucial roles that contribute to project success, with a particular focus on stakeholders and business owners. Understanding and defining these roles are pivotal to creating a harmonious and efficient project environment.

1. Stakeholders: Navigating the Web of Influence

Stakeholders, individuals or groups with a vested interest in the project, play a pivotal role in shaping its trajectory. From internal team members to external clients, their influence and expectations must be carefully managed.

- *Project Sponsor:* The champion of the project, often a senior executive, responsible for securing resources, ensuring alignment with organisational goals, and resolving high-level issues.

- *Clients and End Users:* External stakeholders whose needs and expectations must be understood and met. Their feedback is integral to project success.

- *Project Team:* Internal stakeholders responsible for executing tasks, managing timelines, and achieving project goals. This includes project managers, developers, designers, and other specialists.

Understanding the unique needs and expectations of each stakeholder group is essential for effective communication and collaboration.

2. Business Owners: Steering the Ship

Business owners hold a distinctive role, bridging the gap between project execution and overarching organisational objectives. This section delineates their responsibilities:

- *Defining Project Objectives:* Business owners articulate the project's overarching goals, ensuring alignment with the organisation's strategic vision.

- *Resource Allocation:* Allocating human and financial resources strategically to maximise project efficiency and success.

- *Risk Management:* Identifying potential risks and working with the project team to mitigate them, safeguarding both short-term and long-term interests.

Business owners must provide clear direction, balancing the project's immediate needs with long-term business goals.

3. Project Manager: None of the glory – all the blame!

At the heart of project execution lies the project manager, a multi-faceted leader responsible for coordinating various moving parts.

- *Project Planning:* Developing comprehensive project plans that include timelines, resources, and risk mitigation strategies.

- *Team Leadership:* Motivating and leading the project team, fostering collaboration and effective communication.

- *Problem Solving:* Navigating challenges and obstacles, ensuring the project stays on course.

The project manager is the linchpin, translating the vision set by Business and Product Owners into actionable plans and ensuring successful project delivery.

4. Collaboration and Communication: The Glue that Binds

Effective collaboration and communication are integral to the success of any project. This section explores how roles intersect and emphasises the importance of transparent communication channels:

- *Regular Updates:* Establishing regular communication channels for status updates, addressing concerns, and sharing progress.

- *Conflict Resolution:* Navigating conflicts between stakeholders and team members to maintain a positive and productive work environment.

- *Documentation:* Ensuring thorough documentation of decisions, plans, and changes to provide a clear project history.

Transparent communication fosters trust, mitigates risks, and enhances overall project efficiency.

5. Project Board: Joint Responsibility

- Executive:
    - Responsibilities: Approves project initiation, authorises key decisions, ensures value for money and business alignment.
    - Success: Project delivers within budget, time, and quality while meeting business objectives.
- Senior User:
    - Responsibilities: Defines user needs, approves deliverables, represents user interests.
    - Success: Project delivers usable and beneficial outputs that meet user requirements.
- Senior Supplier:
    - Responsibilities: Manages supplier agreements, resource availability, risk mitigation.
    - Success: Project deliverables successfully completed within agreed-upon scope and cost.

6. Project Management Team:

- Project Manager:
    - Responsibilities: Plans, directs, controls, and motivates the project team, reports to Project Board.
    - Success: Project achieves objectives defined in the Project Initiation Document (PID).
- Team Manager(s):
    - Responsibilities: Leads and supervises project team members, manages individual tasks, reports progress to Project Manager.
    - Success: Team members deliver their assigned tasks on time and to quality standards.
- Project Support:
    - Responsibilities: Provides administrative and technical support to the project team, manages documents and records.
    - Success: Project team has efficient access to information and resources to complete their tasks.

7. **Additional Roles:**

- Change Analyst:
  - Responsibilities: Identifies, assesses, and manages changes to the project scope, plan, and budget.
  - Success: Changes are effectively managed with minimal impact on project goals.
- Risk Manager:
  - Responsibilities: Identifies, analyses, and manages project risks, develops mitigation strategies.
  - Success: No major risks materialise, and any identified risks are effectively managed.
- Quality Manager:
  - Responsibilities: Develops and implements quality management plans, monitors and assures project deliverables meet quality standards.
  - Success: Project outputs meet agreed-upon quality standards and user requirements.
- Product Manager:
  - Responsibilities: Defines and manages the product requirements, monitors product development, ensures product meets user needs.
  - Success: Product successfully delivered to meet user requirements and project objectives.

Sample Job Description of a Project Manager

Responsibilities:

Project Planning and Execution:

- Develop and execute comprehensive project plans, ensuring alignment with organisational objectives and timelines.
- Manage all phases of the project life cycle, from initiation to closure, with a focus on technology integration and implementation.

Stakeholder Communication:

- Act as the primary point of contact for project stakeholders, ensuring clear and effective communication throughout the project.
- Collaborate with cross-functional teams, including development, quality assurance, and operations, to ensure seamless project delivery.

Risk Management:

- Identify and assess project risks, developing mitigation strategies to ensure successful project outcomes.
- Proactively address challenges and obstacles, maintaining flexibility to adapt to changing project requirements.

Resource Management:

- Allocate and manage project resources efficiently, ensuring optimal utilisation of technology teams and external vendors.
- Monitor project budgets and timelines, making adjustments as needed to maintain project health.

Quality Assurance:

- Implement robust quality assurance processes to guarantee the delivery of high-quality technology solutions.
- Conduct regular reviews and assessments to ensure adherence to project specifications and standards.

Technology Integration:

- Oversee the seamless integration of technology components, collaborating with technical teams to ensure compatibility and functionality.
- Stay abreast of industry trends and emerging technologies, incorporating relevant advancements into project strategies.

Personal Attributes:

- **Leadership:** Demonstrate strong leadership skills, motivating and guiding project teams to achieve project goals.
- **Adaptability:** Possess the ability to adapt to changing project requirements and proactively address challenges.
- **Communication:** Excellent communication and interpersonal skills, with the ability to convey technical concepts to both technical and non-technical stakeholders.
- **Problem-solving:** Exhibit strong problem-solving abilities and a strategic mindset in addressing project issues.
- **Organisation:** Display exceptional organisational skills, ensuring efficient resource allocation and project timeline adherence.
- **Initiative:** Take initiative in identifying areas for improvement and driving continuous enhancements in project processes.

Sample Job Description for the Business Sponsor

Responsibilities:

Strategic Leadership:
- Provide strategic leadership and vision for technology-focused projects, ensuring they align with the broader organisational objectives.
- Collaborate with executive leadership to define project goals in line with the company's strategic direction.

Stakeholder Engagement:
- Act as a primary bridge between project teams and executive leadership, ensuring transparent communication of project progress, challenges, and opportunities.
- Engage with stakeholders to gather insights, address concerns, and ensure project outcomes are closely aligned with business priorities.

Resource Management:
- Collaborate with Project Managers to strategically allocate resources, optimising the utilisation of technology teams and external partners.
- Review and endorse project budgets, making informed decisions that support successful project delivery.

Risk Oversight:
- Work closely with project teams to identify and evaluate project risks, offering strategic guidance on mitigation strategies.
- Proactively address challenges and contribute to decision-making processes that safeguard project success.

Quality Assurance:
- Oversee the implementation of robust quality assurance processes, guaranteeing the delivery of high-quality technology solutions.
- Conduct periodic reviews and assessments to ensure adherence to project specifications and organisational standards.

Personal Attributes:
- Strategic Visionary: Demonstrate a strategic mindset, contributing to the development and execution of technology strategies aligned with organisational goals.
- Effective Communicator: Possess excellent communication and interpersonal skills, adept at conveying complex technical concepts to diverse stakeholders.
- Decision-making Authority: Make informed and strategic decisions, considering both project requirements and the broader organisational objectives.

- Collaborative Leadership: Foster a culture of collaboration and cross-functional teamwork to maximise project success.
- Adaptability: Exhibit adaptability and responsiveness to changing project dynamics and business needs.

10.4 Project Initiation:

Project initiation marks the genesis of every successful endeavour, setting the stage for what lies ahead. In this critical phase, project managers lay the groundwork by defining the project's scope, objectives, and deliverables. The scope acts as the project's boundary, delineating what is included and excluded, while objectives and deliverables provide the roadmap for project success. This section delves into the intricacies of these foundational elements, emphasising the importance of clarity and precision in articulating the project's purpose and desired outcomes.

Conducting feasibility studies and risk assessments is an indispensable facet of project initiation. This involves a meticulous examination of the project's viability, considering technical, operational, and financial factors. Feasibility studies ensure that the project aligns with organisational goals and can be realistically accomplished. Simultaneously, risk assessments identify potential obstacles and uncertainties, allowing project managers to proactively plan for contingencies. These evaluations in the initiation phase are essential to mitigate potential challenges and enhance the project's chances of success.

Stakeholder identification and communication planning play a pivotal role in ensuring that the project aligns with organisational objectives and meets the expectations of those involved. Identifying stakeholders involves mapping out individuals or groups with a vested interest in the project's outcome. This includes internal team members, external clients, and anyone else affected by the project. Communication planning then addresses how information will be shared, ensuring a transparent and efficient flow of communication. Establishing clear channels for engaging stakeholders and keeping them informed is a hallmark of effective project initiation.

Differences in project initiation between PRINCE2 and Scrum reflect the inherent distinctions in their methodologies. PRINCE2, with its structured approach, places a significant emphasis on defining project scope, objectives, and deliverables in a detailed Project Initiation Document (PID). Feasibility studies and risk assessments are integral components of this document, providing a comprehensive foundation for project governance. In contrast, Scrum, with its Agile philosophy, takes a more iterative approach to project initiation. While the project's overall vision is established, Scrum allows for flexibility in adapting to changing circumstances, with detailed planning occurring incrementally throughout the project's lifecycle. Stakeholder engagement and communication planning in Scrum are ongoing processes, with regular interactions fostering collaboration and adaptability. The key distinction lies in the adaptability of Scrum versus the structured documentation of PRINCE2 during the initiation phase. Both approaches, however, underscore the critical nature of thorough initiation processes in achieving project success.

Useful Tools and Resources

During the project initiation phase, various online resources can prove valuable for project managers and teams. These resources offer guidance, templates, and tools that streamline the initiation process. Here are some recommended online resources:

1. **ProjectManagement.com:**

    1. Website: ProjectManagement.com

    2. Description: ProjectManagement.com offers a wealth of resources, including articles, templates, webinars, and forums. The site covers various project management topics, and its templates can be particularly useful during the initiation phase for developing project plans and documentation.

2. **PRINCE2 Official Website:**

    1. Website: PRINCE2 Official Website

    2. Description: For those following the PRINCE2 methodology, the official website provides guidance, documentation, and templates specific to the PRINCE2 framework. This includes details on creating a Project Initiation Document (PID) and other key PRINCE2 artifacts.

3. **PMI (Project Management Institute):**

    1. Website: PMI.org

    2. Description: PMI is a globally recognised organisation in project management. Their website offers a range of resources, including standards, publications, and guides. PMI's templates and best practices can be beneficial during the initiation phase.

4. **MindTools - Project Initiation Checklist:**

    1. Website: MindTools Project Initiation Checklist

2. Description: MindTools provides a comprehensive project initiation checklist, guiding project managers through essential steps. This checklist covers key aspects such as defining objectives, identifying stakeholders, and conducting risk assessments.

5. **Smartsheet - Project Management Templates:**

   1. Website: Smartsheet Project Management Templates

   2. Description: Smartsheet offers a collection of free project management templates, including Gantt charts, project plans, and timelines. These templates can be customised for the project initiation phase, aiding in planning and documentation.

6. **Lucidchart - Stakeholder Analysis Template:**

   1. Website: Lucidchart Stakeholder Analysis Template

   2. Description: Lucidchart provides a Stakeholder Analysis template, a crucial component of project initiation. This tool helps identify and analyse stakeholders, ensuring effective communication planning.

7. **Project Smart:**

   1. Website: Project Smart

   2. Description: Project Smart offers articles, templates, and guides on project management. Their resources cover various aspects of project initiation, providing insights into best practices and considerations.

8. **Microsoft Project - Project Initiation Templates:**

   1. Website: Microsoft Project Templates

   2. Description: Microsoft Project provides a range of project initiation templates within its Office templates collection. These templates can be used with Microsoft Project or adapted for other tools.

9. **Trello:**

   1. Website: Trello

   2. Description: Trello is a widely-used Kanban board tool that allows teams to visualise tasks and workflow. During project initiation, Trello can be employed to create a Kanban board to outline and manage initial project tasks, objectives, and stakeholders.

10. **JIRA Software:**

    1. Website: JIRA Software

    2. Description: JIRA Software, part of the Atlassian suite, is a powerful Agile project management tool. It supports Scrum and Kanban methodologies and facilitates project initiation by providing tools for backlog creation, sprint planning, and real-time collaboration.

11. **Kanbanize:**

    1. Website: Kanbanize

    2. Description: Kanbanize is a Kanban project management tool that enables visual project tracking and collaboration. It offers features like workflow automation and analytics, making it useful during project initiation to set up the initial Kanban board and define workflows.

12. **VersionOne:**

    1. Website: VersionOne

    2. Description: VersionOne is an Agile project management tool designed to support various Agile methodologies, including Scrum and Kanban. It provides capabilities for backlog management, release planning, and collaboration, facilitating the initiation phase of Agile projects.

13. **LeanKit:**

    1. Website: LeanKit

    2. Description: LeanKit is a visual project management tool that supports the Kanban methodology. It allows teams to visualise work on virtual Kanban boards, making it helpful during project initiation for setting up workflows and defining initial project elements.

14. **Agile Alliance:**

    1. Website: Agile Alliance

    2. Description: Agile Alliance is a non-profit organisation supporting the Agile community. Their website offers a wealth of resources, including articles, guides, and case studies that can be valuable during the initiation phase of Agile projects.

15. **Kanban University:**

    1. Website: <u>Kanban University</u>

    2. Description: Kanban University provides training and resources related to the Kanban methodology. Their materials can be beneficial for project managers and teams looking to understand and implement Kanban during project initiation.

16. **Agile Manifesto:**

    1. Website: <u>Agile Manifesto</u>

    2. Description: The Agile Manifesto outlines the guiding values and principles of Agile development. Project managers and teams can refer to the manifesto during project initiation to ensure alignment with Agile principles and foster a collaborative mindset.

When using online resources, it's essential to ensure that the information aligns with the specific methodology and requirements of the project. Additionally, project managers may find value in industry-specific forums and communities for insights tailored to their field.

10.5 Planning and Scheduling:

Planning and Scheduling: Mastering the Blueprint of Success

Developing Project Plans, Gantt Charts, and Timelines:

The cornerstone of effective project management lies in the meticulous development of project plans. This involves outlining the project's scope, objectives, and deliverables in a comprehensive document that serves as a roadmap for the entire team. Gantt charts, a visual representation of tasks against time, and timelines further refine project planning, providing a clear and intuitive overview of project progression.

Resource Allocation and Management:

Resource allocation can be time consuming but it has to a high priority as if you get this wrong it will have implications throughout the life of the project. This involves judiciously assigning human resources, budgetary provisions, and technological assets to various project tasks. Effective resource management requires not only a keen understanding of each team member's skills and strengths but also a strategic approach to balancing workloads

Risk Mitigation and Contingency Planning:

Uncertainties and risks are inevitable. Identifying potential risks, whether they be technological challenges, resource constraints, or external factors, is the first step. Developing proactive strategies to mitigate these risks ensures that the project remains resilient in the face of adversity. Contingency planning involves establishing alternative courses of action, allowing the project team to navigate unforeseen challenges without compromising timelines or deliverables. By meticulously addressing potential pitfalls, project managers can foster a proactive and adaptive project environment.

Advantages of Having a Risk Register:

Maintaining a comprehensive risk register is a cornerstone of effective project management, offering a structured approach to identifying, assessing, and mitigating potential risks. One of the primary advantages is enhanced risk visibility, providing project managers and teams with a clear understanding of potential challenges that may impact project success. A well-maintained risk register serves as a proactive tool, allowing teams to anticipate, analyse, and prioritise risks, fostering a culture of preparedness. Additionally, having a centralised repository of risks enables efficient communication among team members and stakeholders, ensuring that everyone is aware of potential threats and the corresponding mitigation strategies. This not only minimises the element of surprise but also empowers the team to respond promptly and effectively, ultimately contributing to successful risk management and project outcomes.

Online Resources for Creating a Risk Register:

Creating a robust risk register requires a structured approach and access to valuable resources. Project managers can leverage online tools and templates to streamline the process. Websites such as ProjectManagement.com offer a variety of downloadable templates and guides for creating effective risk registers. The Project Management Institute (PMI) provides insights and best practices through articles and webinars, guiding project managers in developing and maintaining comprehensive risk registers. Additionally, software platforms like Asana, Trello, or Microsoft Excel can be customised to serve as digital risk registers, offering collaborative features for team members to contribute to risk identification and mitigation efforts. These resources empower project managers to implement industry best practices and tailor risk registers to their specific project needs.

- ProjectManagement.com Templates - https://www.projectmanagement.com/Templates

- Project Management Institute (PMI) - https://www.pmi.org

- Asana - Risk Management Templates - https://asana.com

- Trello - Risk Register Template - https://trello.com

- Microsoft Excel - Risk Register Template - https://create.microsoft.com

Seamless Integration for Holistic Project Success:

A well-crafted project plan serves as the foundation upon which resources are allocated strategically. Simultaneously, the risk mitigation strategies embedded in the project plan anticipate potential challenges and inform resource allocation decisions. This holistic approach ensures that planning, resource management, and risk mitigation operate in tandem, creating a resilient and adaptable project ecosystem. By understanding the interconnectedness of these elements, project managers can steer projects with precision, anticipating challenges, and optimising resources for successful outcomes.

Planning and scheduling are not mere administrative tasks; they are the linchpin that holds the project's structure together. This section provides a comprehensive guide to crafting meticulous project plans, optimising resource allocation, and navigating the unpredictable terrain of risk, establishing a robust foundation for project success.

10.6 Implementing Change Control and Monitoring Progress:

Implementing change control and monitoring progress are twin pillars that uphold the integrity of project management, ensuring that projects remain adaptable to evolving circumstances while maintaining a vigilant eye on predefined objectives.

1. Implementing Change Control: The Adaptive Backbone

Change is an inherent aspect of project dynamics, and effective change control mechanisms act as the adaptive backbone that accommodates and manages alterations to the project scope, requirements, or timelines. This process involves establishing a formalised change control board, comprising key stakeholders, to evaluate proposed changes rigorously. This board assesses the impact of changes on project objectives, timelines, and resources before making informed decisions. By implementing a structured change control process, project managers can ensure that alterations are aligned with strategic goals and avoid potential disruptions to the project's trajectory.

2. Defining Change Protocols: A Proactive Approach

A proactive approach to change control involves defining clear protocols for how changes are proposed, assessed, and implemented. This includes establishing a transparent communication channel for change requests, detailing the information required for evaluation, and delineating the criteria for approving or rejecting proposed changes. Furthermore, a well-defined change management plan encompasses communication strategies to disseminate information about approved changes to the relevant stakeholders. By setting clear guidelines and protocols, project managers instill a sense of order and predictability into the change control process, fostering an environment where modifications can occur seamlessly without compromising project stability.

3. Monitoring Progress: The Ongoing Vigilance

Monitoring progress is an ongoing and dynamic process that requires vigilant oversight to ensure that the project stays on course. This involves regularly tracking key performance indicators (KPIs), milestones, and deliverables against the predetermined project plan. Utilising project management tools such as Gantt charts, progress dashboards, and collaborative platforms, project managers can gain real-time insights into the project's health. Regular status meetings and progress reports serve as checkpoints, allowing the project team to address challenges promptly and make informed decisions to keep the project on track.

4. Iterative Feedback Loops: Fine-Tuning the Trajectory

An effective strategy for monitoring progress involves establishing iterative feedback loops that facilitate continuous improvement. By gathering feedback from team members, stakeholders, and end-users, project managers can identify areas of improvement, assess the effectiveness of implemented changes, and adjust the project plan accordingly. These feedback loops not only contribute to refining the project's trajectory but also create a culture of adaptability, where the project team actively contributes to the ongoing evolution of project processes.

By implementing robust change control mechanisms, defining proactive change protocols, vigilantly monitoring progress, and incorporating iterative feedback loops, project managers can navigate the complexities of project evolution with finesse, ensuring that projects remain agile, adaptable, and ultimately successful.

10.7 Practical Insights on Project Closure:

Project closure marks the culmination of a collective effort, requiring meticulous attention to detail and strategic reflection. This section provides practical insights into the crucial phase of project closure, emphasising the importance of tying up loose ends, conducting comprehensive reviews, and celebrating achievements with an eye toward continuous improvement.

As the project nears completion, it becomes imperative to conduct a systematic review of all deliverables and objectives. This involves ensuring that all project tasks have been completed satisfactorily, and any outstanding issues or discrepancies are addressed promptly. By conducting a comprehensive audit of the project's status, project managers can ascertain the level of success achieved, identify areas for improvement, and facilitate a smooth transition to the closure phase.

The formal closure of a project involves more than ticking off a checklist; it is an opportunity for reflection and learning. Project managers should facilitate a project closure meeting or workshop where team members can share their insights, lessons learned, and feedback on the overall project process. This collaborative reflection not only reinforces a sense of shared accomplishment but also lays the groundwork for continuous improvement in future projects.

Addressing the administrative aspects of closure, such as finalising contracts, releasing project resources, and ensuring all documentation is complete, is equally crucial. This process requires meticulous attention to detail to avoid any lingering issues that may arise post-closure. Moreover, disseminating project documentation, including lessons learned and best practices, contributes to organisational knowledge and serves as a valuable resource for future projects.

A thoughtful and celebratory closure is the final touch to a well-executed project. Acknowledging and appreciating the efforts of the project team fosters a positive work culture and motivates team members for future challenges. Whether through a formal closing ceremony, team outing, or a simple thank-you note, expressing gratitude for the hard work and dedication invested in the project contributes to a sense of accomplishment and camaraderie among the project team.

In essence, project closure is not merely the end of a project but a transition point that sets the stage for future successes. By conducting thorough reviews, facilitating open communication, addressing administrative tasks with precision, and celebrating achievements, project managers can ensure that project closure becomes a valuable stepping stone in the ongoing journey of organisational growth and excellence.

Agile Project Closure and Retrospectives: Nurturing Continuous Improvement

In Agile methodologies, the project closure process is a crucial phase that not only signifies the completion of deliverables but also emphasises continuous improvement through retrospectives. Here's a detailed breakdown of the Agile project closure process:

Completion of Deliverables: The first step in Agile project closure involves ensuring that all planned deliverables have been completed and meet the acceptance criteria. The Agile team conducts a final review to confirm that the product increment aligns with the initial project goals and customer expectations. Any outstanding work or incomplete features are addressed, and the product is brought to a state of readiness for release.

Product Release and Deployment: Once the product is deemed ready, the Agile team proceeds with the release and deployment phase. This involves deploying the product into the live environment, making it accessible to end-users. Continuous integration and continuous delivery (CI/CD) practices are often leveraged in Agile projects to automate the release process, ensuring a smooth and efficient deployment.

Retrospective Meeting: A key component of Agile project closure is the retrospective meeting. This is a dedicated session where the Agile team reflects on the entire project lifecycle, from planning to execution. The retrospective aims to identify what went well, areas for improvement, and action items for future projects. It provides a forum for open and honest communication, fostering a culture of continuous learning and adaptation.

Key Elements of Retrospectives:

- *Start, Stop, Continue:* The team evaluates activities they should start doing, stop doing, and continue doing in future projects.

- *SWOT Analysis:* A Strengths, Weaknesses, Opportunities, and Threats (SWOT) analysis helps identify internal and external factors affecting project performance.

- *Timeline Review:* A timeline review allows the team to pinpoint specific events or sprints that had a significant impact, positive or negative, on the project.

Action Items and Improvement Plans: Based on the insights gained from the retrospective, the Agile team collaboratively identifies action items and improvement plans. These can range from adjusting team processes and communication strategies to adopting new tools or techniques. The emphasis is on implementing tangible changes that contribute to enhanced efficiency and effectiveness in future projects.

Documentation and Knowledge Sharing: Agile project closure includes thorough documentation of the retrospective findings and improvement plans. This information is valuable not only for the current team but also for future teams within the organisation. Lessons learned, best practices, and challenges faced are documented to contribute to organisational knowledge and inform future decision-making.

Celebrating Achievements: Agile project closure isn't just about addressing challenges; it's also an opportunity to celebrate achievements. Recognising the hard work, collaboration, and successes of the team fosters a positive and motivating work culture. This celebration can take various forms, from a simple acknowledgment during the retrospective to a dedicated team event.

By integrating retrospectives into the Agile project closure process, teams embrace a culture of continuous improvement, where each project serves as a stepping stone for refining processes, enhancing collaboration, and delivering greater value in subsequent endeavours.

10.8 Emerging Trends in Project Management

AI in Project Management: Transformative Tools and Practical Benefits

Artificial Intelligence (AI) has emerged as a game-changer in project management, revolutionising traditional practices and enhancing project efficiency. A range of AI-powered tools has been developed to automate routine tasks, provide predictive insights, and streamline decision-making processes. One notable example is **Monday.com's Automations**, which employs AI algorithms to automate repetitive tasks, such as task assignments and status updates. By leveraging AI, project managers can significantly reduce manual workload, allowing teams to focus on strategic aspects of project delivery.

Benefits of AI in Project Management:

1. **Automated Task Management:** AI-powered tools automate mundane and repetitive tasks, such as data entry, status updates, and notifications. This not only saves time but also minimises the risk of human error, ensuring that routine activities are executed consistently and accurately.

2. **Predictive Analytics for Resource Allocation:** Tools like **LiquidPlanner** utilise AI-driven predictive analytics to forecast project timelines and resource needs. By analysing historical data and project trends, these tools help project managers

make informed decisions about resource allocation, preventing bottlenecks and optimising team efficiency.

3. **Smart Risk Management:** AI contributes to proactive risk management by analysing historical project data and identifying potential risks before they escalate. Tools like **Proggio** use AI algorithms to assess project risks and suggest mitigation strategies, allowing project managers to address issues before they impact project timelines.

4. **Natural Language Processing (NLP) for Communication:** AI-driven communication tools, such as **Slack**, employ Natural Language Processing to enhance team collaboration. These tools can understand and interpret natural language, facilitating seamless communication and information retrieval within project teams. Integrating NLP-powered chatbots can also provide instant responses to common queries, improving overall team efficiency.

Practical Implementation Strategies:

1. **AI Training and Integration:** To harness the benefits of AI in project management, it's essential for project managers and team members to undergo training on AI tools and their functionalities. Integrating AI seamlessly into existing workflows ensures a smooth transition and maximises the tools' impact.

2. **Continuous Monitoring and Optimisation:** AI tools often improve over time through machine learning. Regularly monitoring and optimising these tools based on project feedback and evolving requirements ensure that they adapt to the specific needs of the project team.

3. **Customisation for Project Specifics:** AI tools should be customised to align with the unique requirements of each project. This involves configuring algorithms, setting parameters, and tailoring the tool's functionalities to address the specific challenges and goals of the project at hand.

4. **User Feedback and Adaptation:** Encouraging user feedback and actively seeking insights from the project team fosters a collaborative approach to AI implementation. This iterative feedback loop allows project managers to fine-tune AI tools based on real-world experiences, ensuring they align with the evolving dynamics of the project environment.

The practical integration of AI in project management, exemplified by tools such as Monday.com, LiquidPlanner, Proggio, and Slack, offers tangible benefits. From automating tasks and predicting resource needs to smart risk management and enhanced communication, AI empowers project managers to lead projects with greater efficiency and agility.

10.9 The Project Management Office (PMO)

A Project Management Office (PMO) is a centralised entity within an organisation that is responsible for defining and maintaining project management standards and practices. Its primary function is to ensure that projects are executed efficiently, consistently, and align with the organisation's strategic goals. One of the key functions of a PMO is to establish and communicate project management methodologies, standards, and best practices. This includes defining project life cycles, creating templates for project documentation, and establishing guidelines for project planning, execution, and monitoring.

The organisational structure of a PMO can vary based on the size and needs of the organisation. Generally, there are three common types of PMO structures: supportive, controlling, and directive. In a supportive PMO, the focus is on providing templates, training, and support to project managers. In a controlling PMO, there is a higher level of control and oversight, with the PMO actively managing projects and ensuring compliance with standards. In a directive PMO, the PMO takes full control of projects and manages them directly. The choice of structure depends on the organisation's culture, project complexity, and the level of control required.

Another crucial function of a PMO is portfolio management. PMOs often oversee the organisation's project portfolio, ensuring that projects align with strategic objectives and deliver value. This involves prioritising projects, allocating resources effectively, and monitoring the overall health of the project portfolio. By providing a holistic view of all projects, a PMO helps organisations make informed decisions about resource allocation and project prioritisation.

Additionally, a PMO plays a vital role in fostering a culture of continuous improvement. Through the collection and analysis of project data and metrics, the PMO identifies areas for improvement in project management processes. It then initiates corrective actions, shares lessons learned, and updates methodologies to enhance project performance. This focus on continuous improvement contributes to increased project success rates and overall organisational efficiency.

A PMO serves as a central hub for project management excellence within an organisation. Its functions encompass defining and maintaining project management standards, establishing methodologies, overseeing project portfolios, and driving continuous improvement. The organisational structure of a PMO can vary, but its ultimate goal is to enhance project success.

Chapter 11.

Data-Driven Decision-Making:

11.1 Leveraging Analytics for Transformation

Data-driven decision-making stands as a linchpin, offering organisations a powerful tool to navigate the complexities of re-engineering processes. This chapter delves into the pivotal role of data in driving informed decisions during the re-engineering phase and explores how analytics can be harnessed for transformative outcomes.

The Foundation of Data-Driven Decision-Making

At the heart of data-driven decision-making lies the recognition that data is not merely a byproduct of operations but a strategic asset. Before embarking on the journey of re-engineering processes, organisations must establish a robust data infrastructure. This includes implementing advanced analytics tools, ensuring data quality, and fostering a data-centric culture across the organisation.

Case Study: Netflix's Data-Driven Success Netflix's data-driven approach to content recommendation and production exemplifies the power of leveraging analytics for transformation. By analysing viewer data, Netflix tailors its content offerings, enhancing user satisfaction and retention. This strategic use of data has played a pivotal role in making Netflix a global streaming giant. (Reference: Davenport, T. H. (2014). Big Data at Work: Dispelling the Myths, Uncovering the Opportunities)

Data as a Catalyst for Process Re-Engineering

During the re-engineering phase, data serves as a catalyst for identifying inefficiencies, bottlenecks, and areas ripe for improvement. Analysing historical and real-time data enables organisations to pinpoint aspects of their processes that require enhancement or complete overhaul. This insight-driven approach ensures that re-engineering efforts are targeted and aligned with actual performance data.

Example: British Airways' Operational Efficiency British Airways, a leader in the aviation industry, utilises data analytics to drive operational efficiency during re-engineering initiatives. By analysing data on flight schedules, passenger loads, and aircraft maintenance, British Airways optimises its operations, reducing costs and improving overall service quality. (Reference: British Airways. (n.d.). Sustainability)

Predictive Analytics for Informed Decision-Making

Incorporating predictive analytics into the re-engineering process enables organisations to foresee potential challenges and proactively address them. Predictive models, built on historical data and machine learning algorithms, empower decision-makers to make informed choices that anticipate future trends and outcomes.

Example: Walmart's Supply Chain Optimisation Walmart, a retail giant, employs predictive analytics in re-engineering its supply chain processes. By analysing data on consumer purchasing patterns, inventory levels, and external factors, Walmart optimises its supply chain, ensuring products are available when and where customers need them. This data-driven approach has been instrumental in maintaining Walmart's position as a leader in the retail industry. (Reference: Davenport, T. H., & Harris, J. (2007). Competing on Analytics: The New Science of Winning)

Real-Time Decision-Making with Data Analytics

Real-time data analytics empowers organisations to make decisions on the fly. By harnessing the power of real-time data, decision-makers can respond swiftly to changing circumstances, ensuring that the re-engineering process remains agile and adaptive.

Case Study: Uber's Dynamic Pricing Uber's dynamic pricing model is a prime example of real-time decision-making through data analytics. During peak demand, Uber uses real-time data on ride requests, traffic patterns, and driver availability to dynamically adjust prices. This data-driven approach not only maximises revenue for the company but also ensures efficient allocation of resources during periods of high demand. (Reference: Chen, L., Cheng, Y., & Liu, W. (2015). Dynamic Pricing and Demand Information Dissemination in Transportation Networks)

Ensuring Data Privacy and Security

As organisations leverage data for decision-making during re-engineering, it becomes imperative to prioritise data privacy and security. Ensuring compliance with relevant regulations and implementing robust cybersecurity measures safeguards sensitive information and builds trust with stakeholders.

Global Example: GDPR Compliance at Siemens, a multinational conglomerate, prioritises data privacy during its re-engineering efforts. In compliance with the General Data Protection Regulation (GDPR), Siemens has implemented stringent data protection measures, ensuring that customer and employee data is handled with the utmost care. This commitment to data privacy aligns with Siemens' values and fosters trust with its global customer base. (Reference: Siemens. (2024). Data Privacy)

Driving Insight from Data

Deriving meaningful insights from data is a pivotal aspect of data-driven decision-making, and various techniques and tools contribute to this process. A fundamental technique involves data visualisation, employing tools like Tableau, Power BI, or Python's matplotlib. These tools present data through charts, graphs, and dashboards, enhancing comprehension and enabling stakeholders to discern trends, correlations, and outliers, ultimately supporting more informed decision-making.

Machine learning and statistical analysis represent powerful techniques for extracting insights from data. Utilising tools such as Python's scikit-learn, R, or TensorFlow facilitates the implementation of machine learning algorithms to identify patterns and make predictions based on historical data. Statistical methods, including regression analysis and hypothesis testing, provide quantitative insights into relationships within the data, helping to uncover trends, predict future outcomes, and understand the impact of different variables on business metrics.

Furthermore, tools for exploratory data analysis (EDA), such as Python's Pandas and R's tidyverse, are crucial in the initial stages of understanding and summarising datasets. These tools facilitate data manipulation, summarisation, and profiling, aiding data analysts and scientists in comprehending the structure and characteristics of the data. By combining these techniques and tools, organisations can transform raw data into actionable insights, empowering decision-makers to formulate strategies based on a comprehensive understanding of their data landscape.

Measuring the Impact of Data-Driven Re-Engineering

The effectiveness of data-driven decision-making in the re-engineering process can be measured through key performance indicators (KPIs) and metrics. Analysing the impact of changes, both quantitatively and qualitatively, allows organisations to refine their approach continuously and ensure that the re-engineering efforts align with overarching strategic goals.

In conclusion, the integration of data-driven decision-making into the re-engineering process is transformative. By treating data as a strategic asset, utilising predictive and real-time analytics, ensuring data privacy, and measuring impact, organisations can navigate the complexities of re-engineering with precision and agility.

10-Point Plan for Effective Data-Driven Decision-Making

1. **Establish a Data-Centric Culture:** Cultivate a workplace environment that values and prioritises data. Ensure that employees across all levels understand the importance of data in decision-making and encourage a culture of continuous learning about data analytics.

2. **Define Clear Objectives:** Clearly articulate the objectives and goals that your organisation aims to achieve through data-driven decision-making. Align these objectives with the overall strategic vision to ensure a cohesive integration of data into the decision-making process.

3. **Invest in Data Infrastructure:** Build a robust data infrastructure that includes advanced analytics tools, secure data storage, and efficient processing capabilities. Ensure that the infrastructure is scalable to accommodate the growing volume and complexity of data.

4. **Educate and Train Employees:** Provide comprehensive training to employees on data analytics tools, methodologies, and best practices. Invest in upskilling initiatives to empower teams to independently extract meaningful insights from data.

5. **Prioritise Data Quality:** Establish data quality standards to ensure the accuracy, completeness, and reliability of data. Regularly audit and clean datasets to maintain the integrity of the information used for decision-making.

6. **Develop Key Performance Indicators (KPIs):** Define and monitor KPIs that align with your organisation's objectives. These indicators should be measurable, relevant, and directly linked to the outcomes you want to achieve through data-driven decision-making.

7. **Implement Predictive Analytics:** Incorporate predictive analytics models to anticipate trends and future outcomes. Leverage historical data to build models that can inform decision-makers about potential scenarios and guide proactive strategies.

8. **Ensure Data Privacy and Security:** Prioritise data privacy by adhering to relevant regulations such as GDPR. Implement robust cybersecurity measures to safeguard sensitive information and build trust with stakeholders.

9. **Encourage Cross-Functional Collaboration:** Foster collaboration between departments to break down data silos. Encourage cross-functional teams to share insights and collaborate on projects that require a holistic understanding of the organisation's data.

10. **Establish Continuous Improvement Processes:** Implement feedback loops and regularly review the impact of data-driven decisions. Use insights gained from these reviews to refine strategies, update processes, and continuously improve the effectiveness of data-driven decision-making.

11.2 Data-Driven Decision-Making Infrastructure: Building Blocks and Components

In the realm of data-driven decision-making, a robust and scalable infrastructure is pivotal for efficiently storing, processing, and analysing vast amounts of data. Key components such as data warehouses, data lakes, and emerging concepts like lake houses play crucial roles in shaping a comprehensive infrastructure.

1. Data Warehouses:

Data warehouses are central repositories designed for querying and reporting. They store structured data in a format optimised for analysis. Amazon Redshift, for example, is a cloud-based data warehouse service that enables organisations to run complex queries across large datasets with high performance. Its columnar storage architecture and parallel processing capabilities make it well-suited for data-driven decision-making.

2. Data Lakes:

Data lakes serve as vast storage reservoirs capable of handling structured and unstructured data at scale. AWS Glue is a serverless data integration service that plays a crucial role in populating and maintaining data lakes. It simplifies the process of discovering, preparing, and transforming data, ensuring that the information stored in data lakes is accessible and meaningful for analytical purposes.

3. Lake Houses:

Lake houses represent an integration of data warehouses and data lakes, combining the best of both worlds. They offer the structured querying capabilities of data warehouses along with the flexibility and scalability of data lakes. Delta Lake, for instance, is a storage layer that brings ACID (Atomicity, Consistency, Isolation, Durability) transactions to data lakes, providing reliability and ensuring data consistency.

4. Amazon Redshift:

Amazon Redshift, a fully managed data warehouse service, stands out for its ease of use and scalability. It allows organisations to Analyse large datasets with high performance, making it an ideal choice for data-driven decision-making. With features such as automatic optimisation and seamless integration with other AWS services, Redshift streamlines the analytical process.

5. AWS Glue:

AWS Glue is a comprehensive data integration service that simplifies the process of preparing and loading data for analysis. It automates much of the ETL (Extract, Transform, Load) process, making it easier for organisations to ensure data quality and consistency in their data lakes. By automating these tasks, AWS Glue enables faster and more reliable data-driven decision-making.

6. Delta Lake:

Delta Lake is an open-source storage layer that brings ACID transactions* to Apache Spark and big data workloads. It plays a crucial role in the lake house architecture by ensuring data consistency and reliability. Delta Lake's capabilities enable organisations to build data pipelines that support real-time analytics and data-driven decision-making.

Constructing an effective infrastructure for data-driven decision-making involves a careful orchestration of components like data warehouses, data lakes, and the emerging concept of lake houses. Services such as Amazon Redshift, AWS Glue, and Delta Lake provide the necessary tools to streamline data storage, integration, and analysis, empowering organisations to extract meaningful insights and make informed decisions based on their data. This integrated approach is crucial for navigating the complexities of modern data environments and deriving actionable intelligence from vast datasets.

*The incorporation of ACID transactions in a storage layer denotes the application of Atomicity, Consistency, Isolation, and Durability principles not only at the database level but extends these guarantees to the underlying storage infrastructure. This approach ensures that data transactions are treated as indivisible units, preserving the integrity and consistency of the data. Isolation prevents interference between concurrent transactions, and durability guarantees the permanence of committed changes even in the face of system failures. Whether within a database management system or a distributed storage environment, implementing ACID transactions at the storage layer provides a robust foundation for reliable and secure data management across various computing scenarios.

11.3 Useful Summary of Data Workflow

The workflow of data flowing between data warehouses and data lakes involves a systematic process that includes extraction, transformation, loading (ETL), and the use of connectors to bring in data from various systems. Let's break down this workflow step by step:

1. Data Extraction:

The process begins with data extraction from various source systems, which can include transactional databases, external APIs, logs, and more. Connectors are instrumental at this stage, acting as interfaces that facilitate the extraction of data from source systems. These connectors are designed to understand the structure and format of the source data, ensuring a smooth extraction process.

2. Transformation:

Once the data is extracted, it often undergoes a transformation process to make it suitable for analysis. Transformation involves cleaning, aggregating, and structuring the data according to the requirements of the target data warehouse or data lake. Connectors play a role here by facilitating the seamless transfer of transformed data to the destination system.

3. Loading into Data Warehouse:

The transformed data is then loaded into the data warehouse, which is optimised for structured querying and reporting. In this step, connectors specific to the chosen data warehouse come into play. These connectors are tailored to the data warehouse's architecture, ensuring efficient loading of data while maintaining performance.

4. Loading into Data Lake:

Simultaneously, the transformed data can be loaded into the data lake, especially if there's a need to store raw or unstructured data for more flexible and scalable analysis. Connectors designed for data lakes, like those compatible with Apache Hadoop or cloud-based data lake storage, facilitate the seamless transfer of data into this repository.

5. Lake House Architecture:

In scenarios where a lake house architecture is employed, connectors are crucial for managing data consistency between the data warehouse and data lake. Delta Lake, as a storage layer for lake houses, ensures ACID transactions and reliability in data lakes.

Connectors facilitate the coordination of data updates and changes between the two environments, maintaining a unified and reliable dataset.

6. Data Analysis:

With the data now residing in both the data warehouse and data lake, organisations can perform analyses based on their specific needs. Analysts and data scientists can run complex queries on the structured data in the data warehouse, while also leveraging the flexibility of the data lake for exploratory analysis. This duality provides a comprehensive approach to data-driven decision-making.

7. Connectors for Continuous Integration:

As new data becomes available in source systems, connectors play a pivotal role in ensuring continuous data integration. They monitor changes in source systems, trigger the extraction process, and facilitate the seamless flow of new data into the data warehouse and data lake. This continuous integration is crucial for maintaining the relevance and timeliness of the data used in decision-making.

The workflow of data flowing between data warehouses and data lakes is a well-orchestrated process that involves connectors at multiple stages. These connectors act as bridges between source systems, data warehouses, and data lakes, ensuring a smooth and efficient flow of data. As organisations strive for more comprehensive and unified data environments, connectors play a crucial role in enabling seamless integration and empowering data-driven decision-making.

11.4 Data Analysis – Extracting Value

In the dynamic landscape of data analytics, organisations leverage powerful tools to extract valuable insights from vast and varied datasets within data lakes or lake houses. These tools play a crucial role in not only analysing data but also creating comprehensive reports, intuitive dashboards, and business intelligence to guide informed decision-making. Additionally, many tools incorporate artificial intelligence (AI) components to enhance the depth and accuracy of insights. Among these, Microsoft Power BI stands out as a prominent player in facilitating data-driven decision-making.

Data Analysis Tools:

Apache Spark:

Apache Spark is a distributed computing framework that is highly effective for processing large-scale data in a parallel and fault-tolerant manner. Within data lakes, Spark can handle complex data processing tasks, including data cleaning, transformation, and analysis. Its versatility and efficiency make it a popular choice for organisations dealing with extensive datasets.

Databricks:

Databricks provides a unified analytics platform built on Apache Spark, enabling seamless collaboration between data scientists, engineers, and analysts. It facilitates interactive data exploration, machine learning, and data engineering, making it a comprehensive tool for leveraging data within a lake house environment.

Reporting and Dashboard Creation:

Tableau:

Tableau is a robust data visualisation tool that excels in creating interactive and insightful reports and dashboards. It connects to various data sources, including data lakes, allowing organisations to visually represent complex data in an easily understandable format. Tableau's drag-and-drop interface makes it accessible for users across different skill levels.

Looker:

Looker is a business intelligence platform that provides a centralised hub for creating and sharing reports and dashboards. It can directly connect to data lakes and lake houses, offering a semantic layer for data exploration. Looker's strength lies in its ability to create a single source of truth for data, promoting consistency and accuracy in reporting.

Business Intelligence (BI) Tools:

Microsoft Power BI:

Power BI is a comprehensive business analytics tool by Microsoft that integrates seamlessly with various data sources, including data lakes and lake houses. It allows users to connect, transform, and visualise data, enabling the creation of interactive reports and dashboards. Power BI's AI capabilities include features like natural language queries, automated insights, and predictive analytics, making it a powerful ally in deriving meaningful business intelligence.

AI Components in Data Analytics Tools:

Amazon SageMaker:

For organisations leveraging AI within data lakes, Amazon SageMaker offers an end-to-end platform for building, training, and deploying machine learning models. It integrates with data stored in lakes to extract patterns and insights, enhancing the analytical capabilities of the organisation.

Google Cloud AI Platform:

Google Cloud AI Platform provides tools and services for implementing machine learning models on cloud infrastructure. It can seamlessly integrate with data lakes, offering advanced AI capabilities for predictive analytics, anomaly detection, and other data-driven insights.

The tools available for data analysis, reporting, and business intelligence within a data lake or lake house environment are diverse and powerful. These tools empower organisations to harness the potential of their data, uncover actionable insights, and make informed decisions. Microsoft Power BI, with its integration capabilities and AI features, exemplifies the evolution of tools towards more intelligent and comprehensive solutions for data-driven decision-making.

Chapter 12:

The Many Faces of the CTO

Leadership roles have become increasingly diverse and intricate. This chapter explores key technology leadership positions, shedding light on the roles of Chief Technology Officer (CTO), Chief Product and Technology Officer (CPTO), Chief Information Officer (CIO), Chief Digital Officer (CDO), Chief Innovation Officer(CIO), and Chief Data Officer(CIO). Each role contributes uniquely to an organisation's technological ecosystem, with nuanced responsibilities that often intersect and evolve.

We also explore the Consulting CTO, The Fractional CTO, The CTO as a Salesman, the list is ever growing.

12.1 Chief Technology Officer (CTO): The Technological Maestro

The CTO is the visionary technologist, responsible for crafting and implementing the organisation's technological vision. They navigate the complexities of emerging technologies, ensuring that the company stays ahead in an ever-competitive landscape. From spearheading innovation to overseeing the technical aspects of product development, the CTO is the driving force behind the organisation's technological prowess.

At the helm of technological innovation stands the Chief Technology Officer (CTO), a role that transcends mere technical expertise to embody the essence of a visionary technologist. This strategic leader is entrusted with the monumental task of shaping and executing the organisation's technological vision, a compass guiding the company through the dynamic landscapes of the digital era. The CTO doesn't merely react to technological trends but anticipates them, strategically positioning the organisation to stay not just current but ahead in an ever evolving and fiercely competitive environment.

Navigating the Technological Landscape

The CTO is akin to a digital navigator, skillfully steering the organisation through the intricate web of emerging technologies. In this multifaceted role, they don the hat of a technological trendspotter, consistently scanning the horizon for innovations that can be harnessed for strategic advantage. This proactive approach ensures that the company remains at the forefront of technological advancements, ready to leverage new tools and methodologies before they become industry norms. The CTO is not only a custodian of current technologies but a pioneer, leading the charge into uncharted territories that promise competitive differentiation.

Catalyst for Innovation and Product Development

Spearheading innovation is a cornerstone of the CTO's responsibilities. Beyond staying abreast of the latest technologies, they foster a culture of creativity and experimentation within the organisation. This involves not only identifying opportunities for disruptive innovation but also creating an environment where team members are encouraged to ideate and explore unconventional solutions. The CTO is intricately involved in the lifecycle of product development, providing technical oversight that goes beyond routine management. They act as a catalyst, infusing the development process with strategic technological insights to ensure that products align seamlessly with the overarching technological vision.

Driving Technological Prowess

At the heart of the CTO's role is the responsibility for driving and maintaining the organisation's technological prowess. This encompasses a spectrum of tasks, from setting technical standards and ensuring the scalability of systems to optimising infrastructure and fostering a culture of continuous improvement. The CTO isn't merely a guardian of existing technologies; they are architects of the future, building and fortifying the technological foundations that support the organisation's growth. Their strategic decisions reverberate through the entire technological ecosystem, shaping not just the present but laying the groundwork for sustained innovation and success.

In essence, the CTO is the glue connecting technology, innovation, and organisational strategy. Their role extends beyond the realm of a traditional technologist; they are the architects of the technological future, the innovators propelling the organisation forward, and the stewards of a dynamic and ever-evolving technological landscape.

12.2 Chief Product and Technology Officer: Bridging Tech and Product Strategy

This role combines technology leadership with a keen focus on product strategy. The Chief Product and Technology Officer plays a pivotal role in aligning technological initiatives with product development goals. By merging technical expertise with a deep understanding of market needs, this role ensures that technology serves as a catalyst for product innovation and business growth.

The role of the Chief Product and Technology Officer stands as a dynamic confluence, intertwining the realms of technology and product strategy into a seamless narrative of innovation. At its core, this role embodies a visionary approach, orchestrating the fusion of technological advancements with strategic product development. This leadership position goes beyond conventional boundaries, establishing a symbiotic relationship between technology and product, a synergy that propels the organisation into new dimensions of growth.

Strategic Alignment at the Nexus of Technology and Product Development

In the combination of technology and product strategy, the Chief Product and Technology Officer assumes the role of a strategic choreographer. This involves more than mere alignment; it necessitates a deep dive into understanding the intricate threads of both realms. With a nuanced understanding of market dynamics, consumer behaviour, and emerging trends, this leader crafts a roadmap where technological initiatives seamlessly align with the organisation's product development goals. It's a delicate balance of foresight and adaptability, ensuring that the technological landscape is not just navigated but actively shaped to drive product innovation.

Catalyst for Innovation: Nurturing Technological Creativity

Within this multifaceted role, the Chief Product and Technology Officer emerges as a catalyst for innovation. Beyond the structured alignment of tech and product, this leader fosters an environment where technological creativity flourishes. By cultivating a culture that encourages experimentation and embraces calculated risks, they empower teams to push the boundaries of what's possible. This goes beyond the routine; it's about infusing a spirit of creativity that transforms technological solutions into visionary products, setting the stage for the organisation's distinctive imprint on the market.

Driving Business Growth Through Holistic Technological Leadership

The Chief Product and Technology Officer's role extends beyond individual projects or products; it becomes a driving force for overall business growth. The fusion of technical expertise with a profound understanding of market needs ensures that each technological endeavour is not just an isolated innovation but a strategic leap forward in the broader context of the business landscape. In essence, this role represents a holistic approach, where technological leadership becomes synonymous with steering the entire organisation towards sustained growth and market leadership.

12.3 Chief Information Officer (CIO): Architect of Information Ecosystems

The CIO is entrusted with managing the organisation's information systems and technology infrastructure. They play a crucial role in ensuring the seamless flow of information across departments, safeguarding data integrity, and implementing technologies that enhance operational efficiency. The CIO's focus extends beyond technology to encompass the strategic use of information for organisational success.

At the helm of an organisation's technological landscape, the Chief Information Officer (CIO) assumes the pivotal role of an architect, crafting and overseeing the intricate web of information systems and technology infrastructure. Charged with a broad mandate, the CIO is not merely a guardian of servers and databases; they are the orchestrators of an expansive and dynamic information ecosystem that fuels every facet of the business.

Strategic Oversight and Seamless Information Flow

Central to the CIO's responsibilities is the facilitation of a seamless flow of information across departments. Beyond the day-to-day management of IT systems, the CIO acts as a strategic visionary, aligning technological capabilities with organisational objectives. They meticulously assess the unique information needs of each department, ensuring that data traverses the organisational landscape effortlessly, fostering collaboration and informed decision-making.

Safeguarding Data Integrity and Security

A core facet of the CIO's role is the commitment to safeguarding data integrity. In an era where data is a paramount asset, the CIO implements robust measures to fortify the organisation against cyber threats, unauthorised access, and data breaches. By deploying state-of-the-art security protocols and staying abreast of evolving cyber threats, the CIO ensures that the organisation's data remains an impenetrable fortress, instilling confidence among stakeholders and customers alike.

Enhancing Operational Efficiency through Technological Innovation

The CIO is not solely a custodian of existing technologies; they are at the forefront of driving technological innovation to enhance operational efficiency. This involves a strategic evaluation of emerging technologies, understanding their potential impact on the organisation, and orchestrating their integration where beneficial. From implementing advanced analytics for data-driven decision-making to embracing automation for streamlined processes, the CIO leverages technology as a catalyst for transformative efficiency gains.

Strategic Use of Information for Organisational Success

Beyond the realm of technology, the CIO's focus extends to the strategic use of information as a key driver for organisational success. They collaborate closely with executive leadership to align information initiatives with broader business strategies. Whether it's identifying market trends through data analytics or harnessing customer insights for product development, the CIO ensures that information becomes a strategic asset, propelling the organisation forward in a competitive landscape.

The role of the Chief Information Officer encompasses strategic vision, meticulous data stewardship, and the judicious application of technology to foster a resilient, agile, and information-driven organisation. As the architect of information ecosystems, the CIO is instrumental in shaping the technological future of the organisation, ensuring that it not only adapts to the digital age but thrives in it.

12.4 Chief Digital Officer: Pioneering Digital Transformation

In the era of digitalisation, the Chief Digital Officer leads the charge in transforming traditional business models into digital enterprises. This role is focused on leveraging digital technologies to enhance customer experiences, optimise business processes, and drive overall digital strategy. The Chief Digital Officer is a catalyst for organisational agility in the digital age.

The CDO undertakes a multifaceted role that extends beyond mere technological integration, delving into the realms of strategic innovation and customer-centric evolution. At the heart of their mandate is the relentless pursuit of leveraging cutting-edge digital technologies to fundamentally reshape how businesses operate, compete, and deliver value.

Strategic Leverage of Digital Technologies

The role of the Chief Digital Officer is intricately woven with the strategic leverage of digital technologies across the entire organisational spectrum. From harnessing the power of artificial intelligence and data analytics to embracing the Internet of Things (IoT) and immersive technologies, the CDO strategically navigates the digital landscape. This extends to a meticulous orchestration of technological resources to enhance customer experiences, ensuring that every digital touchpoint becomes an avenue for engagement and satisfaction. By crafting a technology-infused strategy, the CDO aims not only to meet but to surpass customer expectations in an increasingly digital-centric marketplace.

Optimisation of Business Processes and Operational Dynamics

Beyond customer-facing initiatives, the Chief Digital Officer is deeply entrenched in the optimisation of internal business processes and operational dynamics. This involves a comprehensive evaluation of existing workflows, identification of bottlenecks, and the seamless integration of digital solutions to streamline operations. Automation, data-driven decision-making, and the implementation of innovative technologies like robotic process automation (RPA) become strategic tools in the CDO's arsenal, fostering efficiency, agility, and adaptability. The overarching goal is to create an organisational framework that responds nimbly to market shifts and technological advancements.

Catalyst for Organisational Agility in the Digital Age

At the core of the Chief Digital Officer's mission is the establishment of organisational agility. This involves not only the adoption of digital technologies but a cultural transformation that embraces change, innovation, and continuous improvement. The CDO serves as a catalyst for cultivating a digital mindset within the organisation, encouraging experimentation, and fostering a culture where adaptability is not just encouraged but ingrained. Through strategic leadership, the Chief Digital Officer navigates the complexities of digital transformation, steering the organisation towards resilience and competitiveness in an era defined by rapid technological evolution.

The Chief Digital Officer stands as a visionary leader shaping the future trajectory of the organisation in the digital era. Beyond the technical dimensions of digitalisation, the role encompasses strategic vision, operational excellence, and cultural evolution.

12.5 Chief Innovation Officer: Cultivating a Culture of Creativity

The Chief Innovation Officer is at the forefront of fostering a culture of creativity and experimentation within the organisation. They identify opportunities for disruptive innovation, drive collaboration across departments, and champion initiatives that push the boundaries of traditional thinking. The Chief Innovation Officer ensures that innovation becomes a core component of the organisational DNA.

The role of the Chief Innovation Officer (CIO) is pivotal in steering organisations towards a dynamic future marked by creative vitality. Positioned at the forefront of strategic leadership, the CIO serves as the chief architect in fostering a culture of innovation and experimentation. Their mandate extends beyond mere ideation; the CIO is charged with identifying untapped opportunities for disruptive innovation that can propel the organisation into new frontiers.

Identifying Opportunities for Disruptive Innovation: The CIO as Visionary Strategist

At the heart of the CIO's responsibilities is the ability to identify and capitalise on opportunities for disruptive innovation. This involves a keen understanding of market trends, emerging technologies, and potential gaps in the industry landscape. The CIO serves as a visionary strategist, leveraging their insights to propose transformative initiatives that challenge the status quo. By aligning innovation with strategic objectives, the CIO ensures that the organisation remains agile and responsive to the ever-evolving business environment.

Driving Collaboration Across Departments: Orchestrating Cross-Functional Synergy

Collaboration is a cornerstone of innovation, and the CIO orchestrates this collaborative ethos across departments. Acting as a catalyst for cross-functional synergy, the CIO breaks down silos, fostering an environment where diverse teams can converge to exchange ideas and perspectives. This collaborative spirit extends not only within the organisation but also involves external partnerships, industry networks, and ecosystems. Through this expansive collaboration, the CIO ensures that innovation is a collective endeavour, drawing from a rich pool of insights and expertise.

Championing Initiatives that Push Boundaries: The CIO as Trailblazer

A true trailblazer, the CIO champions initiatives that push the boundaries of traditional thinking. This involves navigating the fine line between calculated risk and strategic advancement. By advocating for projects that challenge existing norms, the CIO instils a sense of boldness within the organisation. Whether it's exploring new business models, embracing emerging technologies, or reimagining customer experiences, the CIO sets the tone for a culture that values creativity as a driver of progress. Their role as a champion extends to providing the necessary resources, support, and advocacy for initiatives that have the potential to redefine the organisation's trajectory.

Ensuring Innovation as Organisational DNA: A Legacy of Creative Excellence

Ultimately, the Chief Innovation Officer ensures that innovation becomes ingrained in the organisational DNA. This goes beyond sporadic bursts of creativity; the CIO seeks to embed a lasting legacy of creative excellence. Through strategic initiatives, continuous learning programmes, and a commitment to fostering a culture of curiosity, the CIO shapes an environment where innovation is not just a buzzword but a fundamental aspect of how the organisation operates. This legacy ensures that the organisation is not merely responsive to change but is actively shaping the future landscape through continuous and transformative innovation.

12.6 Chief Data Officer: Steward of Data Assets

The Chief Data Officer is responsible for managing and leveraging the organisation's data as a strategic asset. This role involves ensuring data quality, security, and compliance, while also exploring ways to extract insights that drive business decision-making. The Chief Data Officer plays a critical role in transforming raw data into actionable intelligence.

Interactions and Synergies: Weaving a Collaborative Tapestry

These roles are not siloed; instead, they form a collaborative tapestry where interactions are key. The CTO may collaborate closely with the Chief Product and Technology Officer to align technological innovation with product strategy. The Chief Digital Officer may partner with the Chief Information Officer to integrate digital technologies seamlessly into existing information systems. In some organisations, roles may be amalgamated, with individuals assuming responsibilities that span multiple traditional roles.

Emerging Trends and Future Outlook: The Evolution of Tech Leadership

As technology continues to advance, these roles are not static; they are subject to continuous evolution. The emergence of new technologies such as artificial intelligence, blockchain, and the Internet of Things is reshaping the responsibilities of technology leaders. The convergence of roles, where a single leader may embody aspects of CTO, CIO, and Chief Digital Officer, reflects an adaptive response to the dynamic technological landscape.

12.7 Evolution of CTO Roles: Navigating Modern Challenges

The Chief Technology Officer (CTO) role has expanded into diverse dimensions, each catering to unique organisational needs. This section explores three nuanced aspects of the CTO role: The Consulting CTO, The Fractional CTO, and The CTO as a Salesman. These dynamic roles reflect the adaptability and multifaceted nature of modern technology leadership.

1. The Consulting CTO: Strategic Technological Guidance

The Consulting CTO epitomises the strategic advisor within an organisation, offering expert technological guidance without the constraints of a full-time executive position. This role is characterised by an emphasis on consultancy, leveraging deep technical knowledge to provide strategic insights. Competencies required for a Consulting CTO include strong analytical skills, the ability to translate technical complexities into business language, and a comprehensive understanding of industry trends. In modern times, the Consulting CTO role has seen growth due to the increasing demand for specialised technological expertise without the need for a full-time executive presence. Organisations often engage Consulting CTOs for specific projects or to guide them through pivotal technological decisions.

2. The Fractional CTO: Balancing Leadership Across Endeavours

The Fractional CTO role emerges as a solution for organisations seeking high-level technological leadership without committing to a full-time executive. This part-time CTO collaborates with organisations on a fractional basis, bringing strategic insights and technical leadership to the table. Competencies required for a Fractional CTO include adaptability, effective project management skills, and the ability to seamlessly integrate into diverse organisational cultures. The growth of remote work and project-based engagements has propelled the Fractional CTO role into prominence, allowing organisations to access top-tier technology leadership tailored to their specific needs.

3. The CTO as a Salesman: Bridging Technology and Business Goals

In the evolving landscape, the CTO is not only a technological architect but also a bridge between technology and business objectives. The CTO as a Salesman excels in articulating the value of technological solutions to both internal stakeholders and external clients. Competencies required for this role include excellent communication skills, a deep understanding of market needs, and the ability to align technological offerings with business goals. The growth of this role is driven by the recognition that effective communication and collaboration between technical and non-technical teams are crucial for successful project implementation and achieving business objectives. The CTO as a Salesman is instrumental in building strong client relationships and driving innovation with a market-centric approach.

Modern Dynamics and Future Trajectories

The growth of these nuanced CTO roles reflects the dynamic nature of technology leadership in modern times. As organisations navigate the complexities of the digital age, the need for strategic guidance, flexible leadership structures, and effective communication between technical and business realms becomes increasingly paramount. The Consulting CTO, Fractional CTO, and CTO as a Salesman represent not only adaptations to modern challenges but also opportunities for organisations to access specialised expertise and drive innovation in an ever-evolving technological landscape.

12.8 Motivation and Recruitment

Why would we cover CTO motivation in this book? Because it will not only be CTO's reading it! This section isn't true for all CTO's as at the end of the day we are not clones, but there are some basic traits here.

Motivating a Chief Technology Officer:

Involves understanding their unique role and recognising that their motivations may differ from those of a salesperson, a CFO or other C Suite Members. A CTO is primarily driven by a passion for technology, innovation, and the strategic impact of their decisions on the organisation's technical landscape. Unlike a salesperson who may be motivated by financial incentives and commissions, a CTO seeks challenges, opportunities to innovate, and a sense of accomplishment derived from successful technological advancements. Interview your CTO when relaxed, sitting at the bar or when relaxed with a cup of tea or coffee and ask them about their early thoughts, their dreams as a child, about what inspired them in their career direction. I would bet, as I have spent a career recruiting technology talent into my teams, that their early motivation wasn't money or becoming rich.

To effectively motivate a CTO, it's crucial to align their goals with the organisation's mission and vision. Provide them with a clear understanding of how their technical leadership contributes to the company's overall success. Acknowledge their expertise, involve them in strategic decision-making processes, and allow them the autonomy to shape the technological direction of the company. Recognise and celebrate milestones achieved through technical innovation, fostering a culture that values and acknowledges the CTO's contributions.

Creating a bonus structure for a CTO requires careful consideration. Traditional sales-oriented bonus structures may not resonate well with CTOs, as their success is often measured in long-term technical advancements, new concepts and transformational ideas rather than short-term sales figures. It's essential to design a bonus structure that aligns with the CTO's objectives and offers achievable goals within a reasonable timeframe. Avoiding structures where scope changes or external factors beyond the CTO's control can render the bonus unattainable is crucial. A demotivating bonus structure is worse than not having one at all and can diminish the CTO's enthusiasm and commitment to pursuing ambitious technological goals.

In addition to a well-crafted bonus structure, fostering a supportive work environment is key to motivating a CTO. Provide opportunities for professional growth, encourage continuous learning, and offer a collaborative atmosphere where the CTO can lead a team of talented technologists. Recognise the significance of their role in shaping the company's technological future and ensure that their contributions are valued and acknowledged at all levels of the organisation. By understanding the unique motivations of a CTO and tailoring motivational strategies accordingly, organisations can inspire their technology leaders to drive innovation and contribute meaningfully to the company's success.

Recruiting and being Recruited as a CTO

How many interviews have you sat through where you have felt you were being asked the wrong questions? A recruitment agency or headhunter may have asked a series of questions that got you to that short list of candidates. You shake the hand of the CEO or HR director before you sit down and then the questions begin.

The Marlow Business school is often asked to help organisations prepare the CEO or HR Director to interview for senior technology positions, or interview candidates on their behalf. These are a good set of questions to prepare for your interview.

Technical Expertise and Innovation

When recruiting a Chief Technology Officer (CTO), technical competence is paramount. A successful CTO must possess a good understanding of current and emerging technologies relevant to the company's industry. Look for a candidate with a proven track record of driving innovation, introducing cutting-edge solutions, and staying abreast of industry trends. Their ability to align technology strategies with business goals, foster a culture of innovation, and lead technical teams towards impactful outcomes should be key considerations.

"Can you share an example from your career where your technical expertise directly contributed to driving innovation within a company? How did your understanding of emerging technologies align with the business goals, and what were the outcomes of this initiative?"

Why ask this: This question evaluates the candidate's practical experience in applying technical expertise to foster innovation. It assesses their ability to align technology strategies with business objectives and showcases their impact on driving meaningful outcomes.

"How do you stay informed about current and emerging technologies relevant to our industry? Can you provide an instance where staying abreast of industry trends directly influenced a strategic decision or initiative under your leadership?"

Why ask this: This question delves into the candidate's commitment to continuous learning and staying updated on industry trends. It assesses their proactive approach to technology awareness and how this awareness translates into strategic decisions for the benefit of the company.

Leadership and Strategic Vision

Effective leadership is a cornerstone competency for a CTO. Seek candidates who have demonstrated the ability to lead and inspire technical teams. A successful CTO should have a strategic vision that aligns technology initiatives with the overall business strategy. Their capacity to articulate a clear technological roadmap, make informed decisions, and adapt to evolving industry landscapes is crucial. Strong communication skills are essential to convey technical complexities in a comprehensible manner to both technical and non-technical stakeholders.

"Can you share an example from your recent experience where you successfully led and inspired a technical team to achieve a significant milestone or overcome a challenge? How did your leadership contribute to the team's success?"

Why this question is effective: It prompts the candidate to provide a real-world scenario, demonstrating their practical application of leadership skills and their ability to drive positive outcomes within a technical team.

"Describe a situation where you had to align a technological roadmap with broader business strategies. How did you ensure that the technology initiatives supported and enhanced the overall business objectives? What challenges did you face, and how were they addressed?"

Why this question is effective: It delves into the candidate's strategic thinking and ability to align technology with business goals. Their response will provide insights into their decision-making process, adaptability, and problem-solving skills in the context of evolving industry landscapes.

Team Building and Collaboration

A CTO plays a pivotal role in fostering a collaborative and efficient work environment. Look for candidates who have a history of building and managing high-performing technical teams. Assess their ability to collaborate across departments, as the role often involves working closely with executives, product managers, and other stakeholders. An effective CTO should be adept at creating cross-functional synergy and promoting a culture of knowledge sharing and collaboration.

"Can you share a specific example from your experience where you successfully built and managed a high-performing technical team? What strategies did you employ to enhance collaboration and efficiency within the team?"

Why ask: This question prompts the candidate to provide a real-world scenario that demonstrates their ability to lead and cultivate excellence within a technical team. It also gauges their strategies for fostering collaboration and efficiency.

"In your previous roles, how have you collaborated with executives, product managers, and other stakeholders to ensure the alignment of technology initiatives with broader business goals? Can you share a notable achievement that resulted from effective cross-functional collaboration?"

Why ask: This question assesses the candidate's interpersonal skills and their capacity to collaborate beyond the technical realm. It encourages them to provide concrete examples of successful collaboration with different stakeholders, demonstrating their ability to align technology strategies with overall business objectives.

Problem-Solving and Decision-Making

In the dynamic landscape of technology, challenges and uncertainties are inevitable. A strong CTO should excel in problem-solving and decision-making. Evaluate candidates based on their ability to analyse complex issues, devise effective solutions, and make timely decisions. Their approach to risk management and the ability to balance short-term goals with long-term strategies are critical indicators of their decision-making prowess.

"Can you share a specific example from your experience where you faced a complex technological challenge? How did you approach the problem, devise a solution, and make decisions to overcome it?"

Why this question is effective: It prompts the candidate to provide a real-world scenario, allowing you to assess their problem-solving methodology, decision-making process, and the strategies employed to address a challenging situation.

"In the ever-evolving tech landscape, uncertainties are common. Share an instance where you had to navigate such uncertainties, and describe how you managed risk while ensuring a balance between short-term goals and long-term strategies."

Why this question is effective: It delves into the candidate's ability to handle uncertainties and make decisions that align with both immediate objectives and long-term vision. It provides insights into their strategic thinking and risk management skills.

Business Acumen and Results Orientation

Beyond technical acumen, a CTO must understand the business implications of technology decisions. Seek candidates with a keen sense of business acumen, including financial literacy and an understanding of market dynamics. Assess their results orientation by examining past achievements and the impact of their technical leadership on the company's success. A successful CTO should be results-driven, aiming for tangible outcomes that contribute to the organisation's overall growth and success.

How do you integrate your technical decisions with broader business objectives? Can you provide an example where your understanding of market dynamics influenced a significant technology-related decision, positively impacting the overall business strategy?

Why this question is effective: This question assesses the candidate's ability to align technology decisions with business goals and demonstrates their understanding of the broader market context, showcasing their business acumen.

Share a specific achievement from your past roles where your technical leadership directly contributed to the success and growth of the company. How do you measure and ensure tangible outcomes in your role as a CTO, and can you provide an example of a successful outcome you've driven?

Why this question is effective: This question evaluates the candidate's results-driven approach and their ability to quantify the impact of their technical leadership on the organisation's growth and success. It provides insights into their track record of achieving tangible outcomes.

Industry Recognition

Consider a candidate's standing within the industry. Recognitions, awards, or a notable presence in industry forums can indicate their impact and influence in the field. Have they been a judge for an awards programme or presented key note speeches at industry events. Industry recognition showcases their expertise and the respect they command among peers, further affirming their suitability for a CTO role.

"Have you been recognised for achievement by industry forums outside of your organisation or has your impact and influence in technology or innovation been acknowledged by your peers?"

Why this question is effective: This question wants to identify personal influence and a horizon that sits outside the confines of their workplace.

"Have you ever served as a judge for an awards programme or presented keynote speeches at industry events? If not is it something you have ever considered doing?"

Why this question is effective: This question delves into the candidate's involvement in industry-related activities beyond their immediate role. Serving as a judge or delivering keynote speeches demonstrates a level of expertise and leadership that extends beyond day-to-day responsibilities, providing insights into their broader impact on the industry.

Emotional Intelligence

Emotional intelligence is a crucial aspect often overlooked in technical roles. Assess a candidate's ability to understand and manage their emotions, communicate effectively, and navigate interpersonal relationships. A CTO with high emotional intelligence can foster a positive work environment, resolve conflicts, and lead teams with empathy and understanding.

"Imagine a scenario where a critical project deadline is approaching, and tensions are high among team members due to differing opinions on the project's direction. As a CTO, how would you approach this situation to ensure effective collaboration, manage conflicting emotions, and guide the team towards a successful outcome?"

Why this question is effective: This question presents a realistic scenario that requires the candidate to demonstrate their emotional intelligence in a high-pressure environment. It assesses their ability to understand and manage emotions within a team, navigate through conflicting perspectives, and lead with empathy and resilience.

The response provides insights into their practical application of emotional intelligence in challenging professional scenarios, a crucial aspect for a CTO.

"In a collaborative project involving members of the C-suite, conflicts can naturally arise due to diverse perspectives. Can you share an experience where you effectively navigated a disagreement or difference of opinion with a fellow C-suite executive? How did you approach the situation to maintain a positive working relationship and ensure the project's success?"

Why this question is effective: This question probes into the candidate's emotional intelligence in a leadership context, particularly when working closely with other C-suite executives. It assesses their ability to manage interpersonal dynamics, handle conflicts diplomatically, and contribute to a harmonious collaboration within the highest echelons of the organisation. The response will provide insights into their collaborative skills, emotional resilience, and leadership diplomacy in high-level decision-making scenarios.

Ethics and Integrity

Evaluate a candidate's commitment to ethical practices and integrity. A CTO should uphold high ethical standards in decision-making, especially given the responsibility for handling sensitive data and driving technological advancements. Assess their track record for ethical conduct, adherence to industry regulations, and commitment to fostering a culture of integrity within their teams.

"Can you share a specific example from your career where you had to navigate a challenging ethical decision in a technology-related context? How did you approach it, and what were the outcomes?"

Why this question is effective This question aims to gauge the candidate's real-world experience in handling ethical dilemmas within the technology landscape. It assesses their decision-making process, thoughtfulness in ethical considerations, and the ability to navigate challenges while upholding integrity.

"If you are recruiting a member to your team, how would you asses their integrity?

Why this question is effective This question is valuable because it prompts the candidate to articulate their approach to evaluating the essential quality of integrity in potential team members. It provides insights into the candidate's understanding of the importance of integrity in a team setting and allows them to demonstrate their strategies for assessing and ensuring ethical conduct within the team. The response can reveal their awareness of the significance of integrity in team dynamics and their ability to make informed decisions during the recruitment process.

In the recruitment process, a holistic evaluation of these competencies will enable organisations to identify a CTO who not only possesses technical prowess but also aligns with the company's strategic goals, fosters collaboration, demonstrates effective leadership, and embodies qualities of industry recognition, emotional intelligence, and ethical conduct.

Chapter 13.

The CTO's Guide To Finance and Budgeting.

Financial Fluency for CTOs: Navigating Key Processes and Terminology

This chapter has been written with an aim to simplify a topic which many of us CTO's see as a burden and necessary evil which tends to dull our creativity. However financial acumen is not just a skill; it's a strategic imperative. This chapter unapologetically serves as an introduction guide for Chief Technology Officers (CTOs) to navigate key financial processes and terminologies, empowering them to align technological initiatives with fiscal responsibilities. Some of you reading this will be financial wizzes, and in which case move quickly past, but for many CTO's our expertise lay elsewhere, so it's starting at the beginning.

1. Net Fee Income: The Bottom Line in Technological Investments

Net Fee Income stands as a key metric in evaluating the profitability of technology projects. This section delves into how it is calculated, and its significance in assessing the financial success of technological endeavours.

2.EBITDA: or WTF is that?

EBITDA, or Earnings Before Interest, Taxes, Depreciation, and Amortisation, serves as a vital indicator of operational profitability. This section offers insights into the financial health of technology-driven initiatives.

3.Cash Flow: Ensuring Liquidity for Innovative Ventures

Cash Flow is the lifeblood of any organisation, and for CTOs, understanding how technological projects impact cash flow is paramount. This section explores the intricacies of cash flow management, and the financial ebbs and flows of their technological undertakings.

4. Balance Sheets and P&L: Decoding Financial Health

A helicopter view into Balance Sheets and Profit and Loss (P&L) statements. A holistic view of an organisation's financial health. This section gives CTOs a base overview how to interpret these documents and make informed decisions based on their insights.

5. Bank Covenants: Navigating Financial Agreements with Finesse

Bank covenants are contractual agreements that impact an organisation's financial flexibility. Learn how to navigate these agreements while ensuring that technological projects align with stipulated financial conditions.

6. ROI in Technological Investments: Maximising Returns for Innovation

Return on Investment (ROI) is a cornerstone metric for evaluating the success of technological investments. This section explores how CTOs can calculate and maximise ROI, ensuring that every technological initiative contributes meaningfully to the organisation's financial objectives.

13.1 Understanding Net Fee Income in Simple Terms:

Net Fee Income is like a financial magnifying glass that helps us see how much money we really make from our technology projects. It's a crucial tool, especially for leaders to figure out if our tech projects are making a profit.

How it's Calculated:

Calculating Net Fee Income isn't as complicated as it sounds. First, we look at all the money we make from our technology projects. Then, we subtract the direct costs, like the money we spent on making those projects happen. What's left after this subtraction is our Net Fee Income. It's like figuring out how much we actually keep from the money we make after covering all the project costs. It's pretty obvious really, but I get the impression the finance team can enjoy making it sound much more complicated.

Why it Matters: Getting to the Heart of Financial Success

Why is this Net Fee Income so important? Well, it's not just about counting money; it's about understanding the real success of our tech adventures. Unlike just looking at how much money we bring in (revenue), Net Fee Income tells us the real profit, which means we can make smart decisions. By knowing how much profit our tech projects are really making, we can decide where to invest more, which projects are working well, and how to keep our tech adventures financially healthy.

In Simple Terms: Making Smart Choices for Success

Think of Net Fee Income like looking at your household budget, subtracting what you spent on wine last month (!) and finding out how much money you really have left. It's about making sure our technology projects not only look good on the outside (bringing in money) but are also strong and healthy on the inside (making a real profit). Understanding Net Fee Income helps leaders make smart choices to keep our technology projects successful and financially strong.

13.2 Understanding EBITDA: or WTF is that?

EBITDA stands for Earnings Before Interest, Taxes, Depreciation, and Amortisation. It is a financial metric used to gauge the profitability of a business or, in this context, the success of technological initiatives. Let's break down what it means and why it's important in a way that's easy to understand.

What is EBITDA?

EBITDA is like a financial snapshot that focuses on the core profitability of a business, excluding certain expenses. The acronym helps us remember what's excluded: Earnings Before Interest, Taxes, Depreciation, and Amortisation.

- **Earnings:** This is the money a business makes.

- **Interest:** The cost of borrowing money.

- **Taxes:** The amount a business pays in taxes.

- **Depreciation:** The gradual decrease in the value of physical assets, like machinery.

- **Amortisation*:** The spreading out of the cost of intangible assets, like patents, over time.

How is EBITDA Calculated?

The formula is quite straightforward. Start with the earnings or profit a business has made, then add back the interest, taxes, depreciation, and amortisation. The goal is to focus on the core operating performance without the impact of certain financial factors.

Why is EBITDA Important for Technological Endeavours?

For Chief Technology Officers (CTOs), EBITDA is crucial because it gives a clear picture of how well a technology project is contributing to the overall profitability of the company. By excluding interest, taxes, and other non-core expenses, EBITDA allows us to see the raw earnings generated by our technological initiatives. This helps in making strategic decisions, prioritising projects that not only drive innovation but also contribute positively to the financial success of the organisation.

In simpler terms, EBITDA is like looking at the money a business makes from its main operations without getting into the nitty-gritty of interest payments, taxes, and other financial details. For CTOs, understanding EBITDA is a powerful tool for evaluating the true impact of their technological ventures on the company's bottom line. It's also useful for the CFO to know we understand that.

Amortisation is like spreading the cost of something over time. It's like paying for a big purchase with smaller payments instead of one big lump sum. For example, you might amortise the cost of a car loan by making monthly payments. In accounting, amortisation is used to spread the cost of intangible assets, such as software or patents, over their useful life.

13.3 Understanding Cash Flow and Liquidity

What is Cash Flow and Liquidity?

Cash flow refers to the movement of money in and out of a business over a specific period. It represents the cash generated and used by an organisation in its daily operations. Liquidity, on the other hand, is a measure of how easily an organisation can convert its assets into cash. It reflects the ability to meet short-term financial obligations promptly.

How is Cash Flow Calculated?

Cash flow is calculated by taking the cash generated from operating activities and subtracting cash used in investing and financing activities. The formula is:

Cash Flow=Cash from Operating Activities−(Cash from Investing Activities+Cash from Financing Activities)Cash Flow=Cash from Operating Activities−(Cash from Investing Activities+Cash from Financing Activities)

Liquidity is often assessed using ratios like the current ratio (current assets divided by current liabilities) or the quick ratio (excluding inventory from current assets).

Why is it Important to a CTO?

Understanding cash flow and liquidity is crucial for a Chief Technology Officer (CTO) for several reasons:

1. **Operational Stability:** Cash flow ensures that there's enough money to cover day-to-day operations, from paying employees to acquiring necessary technology resources. It safeguards against disruptions in technological initiatives due to financial constraints.

2. **Strategic Decision-Making:** Cash flow insights aid in making strategic decisions about technology investments. A positive cash flow indicates financial health, providing the confidence to pursue innovation and invest in future projects.

3. **Meeting Financial Obligations:** Liquidity measures the ability to meet short-term obligations. For a CTO, this means having the financial capacity to address unexpected technology issues, procure essential resources promptly, or seize strategic opportunities.

4. **Adaptability to Change:** A healthy cash flow and liquidity position provide flexibility. In the dynamic realm of technology, where changes and disruptions are frequent, having the financial adaptability to pivot or invest in emerging trends is invaluable.

5. **Investor and Stakeholder Confidence:** Positive cash flow and robust liquidity are indicators of financial strength. This instils confidence in investors and stakeholders, showcasing the organisation's ability to sustain and grow its technological initiatives.

In essence, for a CTO, understanding cash flow and liquidity is like having a clear view of the financial runway. It ensures not only the smooth functioning of current technological projects but also the ability to take off into new ventures with confidence and resilience.

13.4 Balance Sheets and P&L: A Simplified Guide

Understanding Balance Sheets:

A Balance Sheet is like a financial snapshot that shows what a company owns (assets), what it owes (liabilities), and what remains for the owners (equity) at a specific point in time. It follows a simple formula: Assets = Liabilities + Equity. Assets can include things like cash, equipment, and buildings, while liabilities encompass debts and obligations. Equity represents the ownership stake of shareholders. For CTOs, the Balance Sheet offers a clear picture of the financial health of the company, helping them gauge its capacity to fund technology projects and manage debts.

Understanding P&L (Profit and Loss):

A Profit and Loss (P&L) statement, also known as an Income Statement, is like a financial report card that outlines a company's revenues, costs, and expenses during a specific period. The basic formula is: Revenue - Expenses = Profit (or Loss). Revenue includes sales and other income, while expenses cover costs like salaries, rent, and utilities. If the result is positive, it's a profit; if negative, it's a loss. For CTOs, the P&L is crucial as it provides insights into how much money the company is making or losing from its operations. It's a key tool for assessing the financial performance of technology projects and their impact on the company's overall profitability.

Why They Matter to CTOs:

1. **Financial Health Assessment:** Balance Sheets and P&Ls help CTOs evaluate the financial health of their companies. This assessment is vital for making informed decisions about the feasibility and sustainability of technology projects.

2. **Resource Allocation:** CTOs need to allocate resources efficiently. Understanding the financial position from Balance Sheets helps in determining available funds, while insights from P&L guide decisions on where to invest for the most significant impact.

3. **Debt Management:** Balance Sheets reveal the company's debts, aiding CTOs in managing financial obligations associated with technology initiatives. It ensures a balanced approach to funding projects without compromising the financial stability of the organisation.

4. **Profitability Analysis:** P&Ls provide a direct view of how technology projects contribute to the company's overall profitability. CTOs can assess which projects generate the most value and align future initiatives with strategic financial goals.

Balance Sheets and P&Ls are essential financial tools that empower us to make well-informed decisions about resource allocation, debt management, and the overall financial health of our organisations. They serve as navigational aids, guiding CTOs in steering their technological ventures towards sustainable growth and success.

13.5 Understanding Bank Covenants: Navigating Financial Agreements

Bank covenants are essentially promises or agreements made by a company to its bank, outlining certain financial conditions that need to be maintained. Think of them as rules that help ensure a company stays financially healthy, especially when it has borrowed money from a bank. For a Chief Technology Officer (CTO), understanding these covenants is like understanding the rules of the financial game.

When it comes to technological projects, aligning them with stipulated financial conditions in bank covenants is crucial. This alignment ensures that the financial promises made to the bank are kept while the company invests in and executes its tech projects. The conditions could involve maintaining a certain level of profitability, managing debt levels, or even having a specific amount of cash on hand.

Calculating bank covenants involves looking at specific financial metrics. For instance, a common covenant might be the Debt-to-Equity ratio, which is calculated by dividing the company's total debt by its shareholders' equity. If a CTO is overseeing a project that involves borrowing money, they need to consider how this project impacts these financial ratios. A significant change in these ratios could affect the company's compliance with the bank covenants.

Why is this important to a CTO? Well, failing to meet the conditions outlined in bank covenants can have serious consequences. It might lead to increased interest rates, restrictions on further borrowing, or even the bank calling back the loan. Understanding and aligning technological projects with these financial conditions ensures that the company can continue to innovate without jeopardising its financial health and relationship with the bank. So, for a CTO, it's not just about tech; it's about keeping the financial house in order to support and sustain technological advancements.

13.6 ROI in Technological Investments:

Return on Investment (ROI) serves as a pivotal measure to help us understand the effectiveness of technological investments for the organisation. Calculating ROI involves examining the gains or benefits in comparison to the costs incurred in implementing a technology project. It's akin to determining whether the money and effort invested in a project are yielding sufficient value, or more simply, is it worth it?

To calculate ROI, a straightforward formula can be applied: ROI = (Net Gain from

Investment / Cost of Investment) x 100. Put plainly, it assesses how much was gained in relation to how much was spent, expressed as a percentage. A positive ROI indicates a fruitful investment, while a negative ROI suggests that the project may not be delivering the expected returns.

Comprehending ROI is paramount for CTOs as it enables them to make informed decisions regarding resource allocation. Consider having two projects – one exhibiting a high ROI and the other a lower ROI. Armed with this knowledge, a CTO can prioritise the project with a higher ROI, ensuring better returns for the resources invested. This strategic approach ensures that technology initiatives not only align with the organisation's goals but also yield tangible benefits, harmonising innovation with the overall success of the company.

Return on Investment (ROI) extends beyond monetary measures; it encapsulates a broader spectrum of gains that are equally crucial for Chief Technology Officers (CTOs) to consider. In addition to financial returns, ROI can manifest in the form of increased market share, successful expansion into new geographies or markets, and contributions to social or environmental benefits. For instance, a technology project that enhances the company's market presence, helps it establish a foothold in previously untapped regions, or contributes positively to social and environmental causes can all be seen as valuable returns on investment. This holistic understanding of ROI allows CTOs to gauge the comprehensive impact of their technological initiatives, aligning them not only with financial objectives but also with broader strategic and societal goals. It reflects a nuanced perspective that acknowledges the multifaceted nature of success in the ever-evolving technological landscape.

In essence, ROI functions as a report card for technology projects. It provides insight into the performance of investments, indicating whether they are thriving or if adjustments are necessary. By regularly evaluating ROI, CTOs can guarantee that technological endeavours not only keep pace with the latest innovations but also contribute positively to the organisation's financial prosperity.

Book 3 – Innovation

Innovation in technology is the catalyst for industry progress, transforming ideas into user-centric solutions. CTOs play a pivotal role in cultivating a culture of experimentation and learning, driving companies to stay ahead by embracing innovation throughout the product lifecycle. This book sets the stage for a detailed exploration of strategies and best practices for accelerating innovation through the hands of the CTO.

A CTO's Guide to Driving Innovation

Innovation is not just about fancy gadgets; it's the engine driving progress across industries. It's about transforming ideas into practical solutions that work better, smoother, and delight users. Innovation is the move from continuous improvement to a step change or leap in business or technical performance. To harness this power, companies need champions like CTOs who foster a culture of experimentation, risk-taking, and learning. By embracing innovation across its lifecycle, from product development to data-driven insights, companies can stay ahead of the curve and shape the future, not just adapt to it. This chapter lays the foundation for exploring deeper dives into strategies and best practices for leading innovation in the ever-changing tech landscape.

Chapter 14: Innovation in Practice:
- Defining innovation in the context of technology.
- The role of the CTO in fostering a culture of innovation.

Chapter 15: The Innovation Mindset:
- Developing a mindset that encourages creativity and risk-taking.
- Overcoming challenges to instill an innovation-oriented culture.

Chapter 16: Creating an Innovation Lab:
- Designing and implementing an innovation lab within the organisation.
- Best practices for fostering experimentation, collaboration, and breakthrough innovation.

Chapter 17: The 3 Horizon Squad Model

- **Strategic Balance:** 3 Horizon Model aligns short and long-term innovation for success.
- **Agile Autonomy:** Squad methodology empowers teams for independent, efficient collaboration and innovation.

Chapter 18: The Power of Imagination

- Imagination fuels breakthroughs, sparking creative solutions and transformative technological advancements.
- Innovative technologies often originate from imaginative visions, pushing the boundaries of possibility.

Chapter 19: Global Perspectives on Innovation:

- Exploring innovation trends and practices on a global scale.
- Learning from international case studies and success stories.

Chapter 14:

Innovation in Practice

14.1 Defining Innovation in the Context of Technology

Innovation in technology is a dynamic force that propels industries forward, driving progress and shaping the future. It transcends the mere introduction of new gadgets or software; instead, it encapsulates the transformation of ideas into practical solutions that enhance efficiency, effectiveness, and user experience. In the context of technology, innovation is the engine that powers evolution.

One widely accepted definition of innovation in technology comes from the work of Clayton Christensen, a renowned Harvard Business School professor. In his book "The Innovator's Dilemma," Christensen defines disruptive innovation as the process by which smaller companies with fewer resources successfully challenge established incumbent businesses. Disruptive innovation often introduces new technologies or business models that redefine the competitive landscape.

Moreover, the concept of open innovation, popularised by Henry Chesbrough, emphasises the importance of collaboration and knowledge-sharing across organisational boundaries. Open innovation involves leveraging external ideas, as well as sharing internal innovations, to advance technology and create a more interconnected ecosystem.

To illustrate, the evolution of smartphones provides a compelling example of technological innovation. The introduction of touchscreens, app ecosystems, and mobile internet connectivity revolutionised the way individuals communicate, work, and access information. Companies like Apple, with the iPhone, exemplified the transformative power of innovation in reshaping entire industries.

The Role of the CTO in Fostering a Culture of Innovation

The Chief Technology Officer plays a pivotal role in driving and sustaining a culture of innovation within an organisation. A CTO is not merely a technical overseer but a catalyst for change, inspiring teams to explore new horizons and push the boundaries of what is possible.

In the pursuit of fostering innovation, the CTO must create an environment that encourages experimentation and risk-taking. Google's "20% time," where employees were allowed to spend a portion of their work hours on personal projects, is a notable example. This initiative led to the development of products such as Gmail, AdSense and Google Maps, showcasing the power of giving employees the freedom to explore innovative ideas.

Furthermore, the CTO should actively seek external collaborations and partnerships to tap into a broader pool of ideas and expertise. IBM's "Innovation Jams," large-scale online brainstorming sessions involving employees, clients, and partners, exemplify the value of inclusive innovation processes. This started as an internal experiment and now supports tens of thousands of participants, jamming in real time across time zones supported by gamification and AI.

Embracing a mindset of continuous learning and adaptation is essential for a culture of innovation. CTOs can facilitate this by promoting professional development opportunities, encouraging cross-functional collaboration, and fostering a culture that values and learns from both successes and failures.

In conclusion, this chapter establishes the foundation for understanding innovation in technology and highlights the critical role of the CTO in championing a culture that nurtures creativity and advancement. By exploring key concepts and drawing on real-world examples, we lay the groundwork for the subsequent chapters in the CTO Handbook, where we delve deeper into strategies and best practices for leading innovation in the ever-evolving landscape of technology.

In this chapter we will be focusing upon:

Innovation in Product Development
a. Integrating innovation into the product development lifecycle.
b. Balancing innovation with product stability and customer expectations.

Open Innovation and Collaboration
a. Imagination fuels breakthroughs, sparking creative solutions and transformative technological advancements.
b. Innovative technologies often originate from imaginative visions, pushing the boundaries of possibility.

Innovation Key Methodologies
a. Frameworks to accelerate innovation.
b. Overview of methodologies

Frugal Innovation
a. Integrating Frugal Innovation into the product development lifecycle.
b. Innovating on a shoestring

Measuring and Managing Innovation:
a. Key performance indicators for tracking innovation success.
b. Strategies for managing risks associated with innovation initiatives.

Balancing Incremental and Disruptive Innovation:
a. Understanding the spectrum from incremental improvements to disruptive innovations.
b. Developing a portfolio approach to innovation.

Innovation Culture and Employee Engagement:
a. Nurturing a culture where every employee feels empowered to contribute ideas.
b. Employee engagement strategies that fuel innovation.

Intellectual Property and Innovation:
a. Protecting and leveraging intellectual property in the innovation process.
b. Strategies for managing patents, copyrights, and trade secrets.

Ethics in Innovation:
a. Navigating ethical considerations in the pursuit of innovation.
b. Balancing progress with responsible technological development.

Innovation in Legacy Systems:
a. Strategies for injecting innovation into existing systems.
b. Modernisation approaches for legacy technologies.

The Role of Data in Innovation:
a. Leveraging data analytics for informed decision-making.
b. Harnessing big data and machine learning for innovative solutions.

Cybersecurity in Innovative Environments:
a. Integrating cybersecurity measures into the innovation process.
b. Balancing innovation with robust security practices.

14.2 Innovations in Product Development:

Effective product development demands the seamless integration of innovation throughout the lifecycle, balanced with the need for stability and meeting customer expectations. By adopting a strategic approach and learning from industry leaders, companies can navigate the dynamic landscape of innovation in product development.

Integrating Innovation into the Product Development Lifecycle

Innovation is a driving force in the evolution of products, pushing boundaries and creating new market opportunities. To effectively integrate innovation into the product development lifecycle, companies must adopt a holistic approach that spans ideation to launch.

Ideation and Conceptualisation: The first stage of product development sets the tone for innovation. Companies like Apple have excelled in fostering a culture of creativity through internal brainstorming sessions and cross-functional collaboration (Isaacson, 2011). Furthermore, open innovation platforms, such as ***Wazoku*** and ***NineSigma***, allow companies to tap into external expertise, widening the pool of innovative ideas.

Agile Development: Adopting Agile methodologies facilitates the seamless integration of innovation into the development process. Agile promotes iterative development, allowing teams to respond swiftly to changing market conditions and incorporate innovative features (Beck et al., 2001). Online resources like the Agile Alliance provide valuable insights into implementing Agile practices.

Prototyping and Testing: Companies like Google have embraced rapid prototyping and beta testing as key elements in the product development lifecycle. This allows for quick validation of innovative concepts, minimising the risk of investing in ideas that may not resonate with users. See Tom Chi's TED Talk

Continuous Innovation: Innovation doesn't end with product launch. Regular updates and improvements based on user feedback, as seen in the software industry's continuous delivery model, keep products relevant and competitive. Tools such as A/B testing and user analytics contribute to ongoing innovation by providing data-driven insights.

Balancing Innovation with Product Stability and Customer Expectations

While innovation drives progress, maintaining a delicate equilibrium between innovation, product stability, and meeting customer expectations is critical.

Risk Mitigation: Innovation inherently carries risks, and managing these risks is paramount. The concept of the "innovation sandbox," where companies allocate a specific budget and resources for experimental projects, allows for controlled risk-taking without jeopardising the core product stability (Terwiesch & Ulrich, 2009).

Customer-Centric Innovation: Successful product development requires a deep understanding of customer needs. Amazon's customer-centric approach involves continuous engagement, feedback loops, and a commitment to understanding evolving customer expectations (Bezos, 2017). By aligning innovation with customer expectations, companies can ensure that new features resonate with their user base.

Stability in Core Features: While innovation is essential, the core features that define a product should remain stable to provide a consistent user experience. The concept of the "innovation toggle," as discussed by O'Reilly (2014), allows companies to selectively introduce innovative features without disrupting the core functionality.

References:

- Isaacson, W. (2011). Steve Jobs. Simon & Schuster.

- Beck, K., Beedle, M., Van Bennekum, A., Cockburn, A., Cunningham, W., Fowler, M., ... & Kern, J. (2001). Manifesto for Agile Software Development. Agile Alliance.

- Mayer, M. (2013). Google's Product Development Process: 8 Steps to Success. Forbes.

- Humble, J., & Farley, D. (2010). Continuous Delivery: Reliable Software Releases through Build, Test, and Deployment Automation. Addison-Wesley.

- Kohavi, R., Tang, D., & Xu, Y. (2009). Trustworthy Online Controlled Experiments: Five Puzzling Outcomes Explained. Proceedings of the 16th ACM SIGKDD International Conference on Knowledge Discovery and Data Mining.

- Terwiesch, C., & Ulrich, K. T. (2009). Innovation Tournaments: Creating and Selecting Exceptional Opportunities. Harvard Business Press.

- Bezos, J. (2017). 2017 Letter to Shareholders. Amazon.

- O'Reilly, T. (2014). The Innovator's Dilemma. O'Reilly Media.

14.3 Introduction to Rapid Prototyping

Rapid prototyping, a cornerstone in contemporary product development, facilitates the swift transformation of conceptual ideas into tangible models. This iterative approach minimises time-to-market, fosters collaboration among diverse teams, and ensures that end-users' needs are met through continuous refinement.

Key Principles of Rapid Prototyping

1. **Iterative Design:** Emphasising an iterative design process, rapid prototyping enables constant refinement based on user feedback and testing.

2. **User-Centric Focus:** By involving end-users early in the design process, prototypes can be tailored to meet their evolving needs and preferences.

3. **Collaborative Development:** Cross-functional collaboration is enhanced as various teams interact with tangible prototypes, providing valuable insights and fostering a holistic approach to product development.

Rapid Prototyping Techniques

1. 3D Printing

Revolutionising prototyping, 3D printing offers a cost-effective and swift transformation of digital designs into physical objects. Boeing's innovative use of 3D printing for rapid prototyping of aircraft components stands as a testament to its efficiency[1].

2. Virtual Reality Prototyping

The integration of virtual reality (VR) in prototyping allows for the creation of immersive, interactive experiences. Valve Corporation, during the development of the Valve Index virtual reality system, extensively employed VR prototyping to refine and enhance user experiences[2].

3. Paper Prototyping

In the early stages of product development, paper prototyping remains a simple yet effective method. Dropbox, a leader in cloud-based services, utilises paper prototyping to ideate and validate user interfaces[3].

Digital Twins in Product Development

1. Definition and Concept

Digital twins involve the creation of a virtual replica of a physical product or system. This digital representation, often updated in real-time, mirrors the physical entity, offering valuable insights throughout its lifecycle.

2. Applications in Rapid Prototyping

Digital twins play a pivotal role in enhancing rapid prototyping. By creating a virtual counterpart, designers and engineers can simulate and analyse various scenarios, ensuring that the physical prototype meets predefined specifications before production begins[4].

3. Real-Time Feedback and Analysis

The real-time nature of digital twins enables continuous monitoring and analysis. This ensures that any deviations or issues are promptly identified and addressed, streamlining the prototyping process and reducing the likelihood of costly errors in the final product.

Benefits of Rapid Prototyping and Digital Twins

1. **Time Efficiency:** Rapid prototyping, coupled with digital twins, significantly reduces the time required to move from concept to tangible model, enhancing the overall efficiency of product development.

2. **Cost Reduction:** Identifying and addressing design flaws early in the digital twin phase minimises costly changes in later stages of physical prototyping.

3. **Holistic Development:** The synergy between rapid prototyping and digital twins fosters a holistic approach to product development, ensuring that the end product aligns seamlessly with user expectations.

The integration of rapid prototyping and digital twins heralds a new era in product development. By leveraging techniques like 3D printing, virtual reality prototyping, and paper prototyping, coupled with the power of digital twins, companies can navigate the complexities of design, streamline collaboration, and bring high-quality products to market faster than ever before.

References:

Footnotes

1. Boeing. "Boeing Embraces 3D Printing for the Dreamliner." Boeing.

2. Valve Corporation. "Behind the Scenes of Half-Life: Alyx." Valve.

3. Dropbox. "Designing with Dropbox Paper." Dropbox Design.

4. Glaessgen, E. H., & Stargel, D. S. "The digital twin paradigm for future NASA and U.S. air force vehicles." NASA Technical Reports Server.

14.4 Open Innovation and Collaboration

Harnessing the Power of External Collaborations and Partnerships

In the pursuit of innovation, organisations are moving beyond traditional boundaries and tapping into external expertise. Collaborating with external partners, including startups, research institutions, and other companies, provides a diverse pool of knowledge and resources. One prominent example is the collaboration between Apple and Imagination Technologies, where Apple leveraged Imagination's graphics processing units (GPUs) for its iPhones and iPads.

Open innovation fosters a culture of knowledge exchange, allowing companies to access new technologies and ideas. The concept of open innovation was popularised by Henry Chesbrough, who emphasised the importance of leveraging external ideas to drive internal innovation. The open approach has been widely adopted by companies like Procter & Gamble, which actively seeks external innovation through its Connect+Develop programme.

To delve deeper into this topic, readers can refer to "Open Innovation: Researching a New Paradigm" by Chesbrough, H.

b. Building Ecosystems for Shared Innovation

In the quest for sustained innovation, organisations are increasingly embracing the concept of ecosystems, creating collaborative environments where various stakeholders interact and contribute. An illustrative example is the Linux operating system, developed collaboratively by a global community of developers. The Linux ecosystem exemplifies the power of collective innovation, where contributors from diverse backgrounds collaborate to enhance the software.

Building ecosystems for shared innovation goes beyond individual partnerships. It involves creating a network of interconnected entities that collectively drive innovation. Amazon Web Services (AWS) is a prime example, providing a comprehensive ecosystem of cloud services that enables businesses to innovate by leveraging scalable computing power, storage, and other resources.

For a comprehensive exploration of ecosystem-driven innovation, readers can explore "The Business Model Navigator: 55 Models That Will Revolutionise Your Business" by Gassmann, O., Frankenberger, K., & Csik, M.

In conclusion, open innovation and collaboration are pivotal in navigating the complexities of today's business environment. By tapping into external collaborations and building ecosystems for shared innovation, organisations can unlock new opportunities, drive creativity, and stay at the forefront of industry advancements.

Case Study: LEGO Group - Embracing Open Innovation for Creativity

Introduction: The LEGO Group, a Danish toy company, is renowned for its iconic plastic building bricks. In the early 2000s, LEGO faced financial challenges and needed to innovate to stay relevant in a rapidly changing market. The company decided to embrace open innovation, inviting external collaborators to contribute to its creative processes.

Challenges Faced: At the start of the millennium, LEGO was grappling with declining sales and a perception that its traditional brick-based toys were losing appeal in the age of digital entertainment. The company recognised the need to adapt to changing consumer preferences and find new ways to engage its audience.

Open Innovation Initiatives:

1. **LEGO Ideas Platform:** In 2008, LEGO launched the LEGO Ideas platform, an online community where fans could submit their own LEGO set ideas. Today it has over 100,000 current users. Members of the community can vote on their favourite designs, and if a project received enough support, LEGO committed to reviewing it for potential production. This initiative allows LEGO to tap into the creativity of its fanbase and bring new, innovative products to market.

2. **Collaboration with MIT Media Lab:** LEGO collaborated with the MIT Media Lab to explore innovative ways of integrating digital technology with physical play. This partnership resulted in the creation of LEGO Mindstorms, a programmable robotics kit that allowed children to build and program their own robots. This collaboration exemplifies how LEGO leveraged external expertise to enter the robotics and technology education space.

Outcomes:

1. **Crowdsourced Innovation:** The LEGO Ideas platform has become a thriving hub for innovation, with numerous fan-designed sets making it to production. Notable examples include the "NASA Apollo Saturn V" and the "Doctor Who" sets, both of which originated from fan submissions.

2. **Market Resurgence:** Embracing open innovation played a pivotal role in LEGO's resurgence. The company transformed from near bankruptcy to becoming one of the most profitable and innovative toy manufacturers globally. LEGO's revenue and market share increased, showcasing the positive impact of open collaboration on business outcomes.

Key Takeaways:

1. **Engaging the Community:** By actively involving its community, LEGO transformed its relationship with consumers from passive buyers to engaged co-creators. This engagement not only contributed to product development but also built a strong and loyal fanbase.

2. **Diversification through Collaboration:** Partnerships with external entities, such as the MIT Media Lab, allowed LEGO to diversify its product offerings and enter new market segments, demonstrating the power of open collaboration in driving innovation.

3. **Continuous Learning:** LEGO's open innovation journey reflects a commitment to continuous learning and adaptation. The company remains open to new ideas and perspectives, fostering a culture of innovation that has become integral to its success.

This case study illustrates how the LEGO Group successfully navigated industry challenges by embracing open innovation and collaboration, turning adversity into an opportunity for creativity and growth.

Footnotes

1. "Apple's Innovation Strategy, Innovation Process, Insights, Case Study" - Innovation Tactics. (https://innovationtactics.com/apple-innovation-strategy/)

2. Chesbrough, H. "Open Innovation: Researching a New Paradigm." Oxford University Press.

3. "Connect+Develop" - Procter & Gamble. (https://www.pg.com/innovation/connect-develop/)

4. "How Linux was Built" - The Linux Foundation. (https://www.linuxfoundation.org/blog/how-linux-was-built/)

5. "Amazon Web Services (AWS)" - Amazon. (https://aws.amazon.com/)

6. Gassmann, O., Frankenberger, K., & Csik, M. "The Business Model Navigator: 55 Models That Will Revolutionise Your Business." Pearson.

14.5 Innovation Key Methodologies

Design Thinking:

When applied within an Innovation lab, Design Thinking provides a structured and human-centric approach to problem-solving and idea generation. The methodology is characterised by a series of iterative steps that emphasise empathy, collaboration, and experimentation. In the context of an innovation lab, the Design Thinking process typically begins with the identification of a specific challenge or opportunity. This initial phase involves extensive empathetic research, including user interviews, observations, and immersion into the problem space. Innovation labs often leverage interdisciplinary teams to bring diverse perspectives to the forefront, ensuring a holistic understanding of the challenges at hand.

Once insights are gathered, the ideation phase commences, where teams brainstorm and generate a wide array of ideas. The innovation lab environment fosters a culture that encourages wild ideation and unconventional thinking. Rapid prototyping is a central element of the methodology within an innovation lab. Teams create tangible representations of their ideas, enabling quick and low-cost testing and iteration. This hands-on approach encourages experimentation and allows for a faster and more responsive development process. The iterative nature of Design Thinking aligns seamlessly with the dynamic and adaptive environment of an innovation lab, promoting a culture of continuous improvement and learning as ideas evolve through multiple iterations. This methodology in an innovation lab fosters a mindset that values agility, creativity, and user-centricity, driving the development of innovative solutions that address real-world challenges.

Lean Startup:

The Lean Startup methodology, when implemented within an Innovation Lab, revolves around fostering a culture of rapid experimentation and iterative development to efficiently bring innovative ideas to fruition. At its core, the Lean Startup methodology, pioneered by Eric Ries, emphasises a systematic and scientific approach to creating and managing successful startups and projects. Within the context of an innovation lab, this methodology encourages teams to validate assumptions, test hypotheses, and adapt their strategies based on real-time feedback. The process typically begins with the identification of a problem or opportunity, followed by the creation of a minimum viable product (MVP) – a scaled-down version of the envisioned solution. This MVP is then swiftly released to gather user feedback, enabling quick learning and adjustment of the product or service based on market responses.

In an Innovation Lab setting, the Lean Startup methodology is instrumental in mitigating risks associated with untested assumptions. The iterative nature of the process allows for the discovery of potential flaws or improvements early on, preventing the wasteful allocation of resources on concepts that may not resonate with users. Moreover, the methodology encourages a lean and agile mindset, where the emphasis is on validated learning and efficient resource allocation. This approach aligns well with the dynamic environment of an innovation lab, fostering a culture that values adaptability, resilience, and a relentless focus on delivering value to end-users.

Stage-Gate Model:

The Stage-Gate Model is a widely adopted innovation management framework designed to guide the development and launch of new products or services. When applied within an Innovation Lab, the methodology serves as a structured process to systematically advance ideas from conception to execution. The model consists of a series of stages (or gates), each representing a critical juncture in the innovation process. In the context of an Innovation Lab, the Stage-Gate Model becomes a dynamic and iterative framework, fostering creativity and adaptability.

Within the Innovation Lab setting, the methodology typically begins with the identification and generation of innovative ideas through brainstorming sessions, workshops, or collaborative forums. These ideas then enter the initial stages of the Stage-Gate process, where they are carefully assessed, refined, and developed. The Innovation Lab environment, with its emphasis on experimentation and cross-disciplinary collaboration, aligns seamlessly with the iterative nature of the Stage-Gate Model. At each gate, a multidisciplinary team reviews and evaluates the progress of the innovation, considering factors such as feasibility, market potential, and alignment with strategic objectives. The Innovation Lab's flexible structure allows for rapid iteration and refinement, ensuring that promising ideas are nurtured and less viable concepts are discontinued early in the process, ultimately optimising resource allocation and enhancing the overall innovation journey.

TRIZ (Theory of Inventive Problem Solving):

TRIZ, or the Theory of Inventive Problem Solving, is a systematic and structured approach to innovation that originated in the Soviet Union and has since gained global recognition. When applied within an Innovation Lab, TRIZ provides a powerful framework for tackling complex problems and driving creative solutions. The methodology of TRIZ revolves around the identification of inventive principles and patterns that have proven successful in solving problems across various domains. Within the Innovation Lab context, the first step typically involves clearly defining the problem at hand, breaking it down into specific contradictions or challenges.

Once the problem is delineated, TRIZ employs a vast toolkit of inventive principles and methodologies to guide problem-solving. This includes tools such as the 40 Inventive Principles, which offer strategies for resolving contradictions and finding innovative solutions. In an Innovation Lab setting, practitioners of TRIZ collaborate in cross-functional teams to leverage diverse perspectives and expertise. TRIZ encourages thinking beyond conventional solutions by challenging assumptions and fostering creativity. The iterative nature of the TRIZ process aligns well with the dynamic and experimental environment of an Innovation Lab, where rapid prototyping and learning from failures are integral to the innovation process. Overall, TRIZ provides a systematic and repeatable methodology that complements the experimental and collaborative ethos of an Innovation Lab, offering a structured approach to inventive problem-solving and fostering a culture of continuous improvement and innovation.

Kaizen:

Kaizen, a Japanese term meaning "continuous improvement," serves as a powerful methodology within an Innovation Lab, fostering a culture of incremental enhancements and sustainable innovation. In the context of an Innovation Lab, Kaizen involves a systematic approach to problem-solving and creativity, encouraging teams to regularly evaluate processes, identify areas for improvement, and implement changes collaboratively. The methodology hinges on the principle that small, incremental changes, when accumulated over time, lead to significant improvements. Within an Innovation Lab, Kaizen becomes a guiding philosophy, instilling a mindset that values experimentation, feedback, and adaptability.

Teams in an Innovation Lab applying Kaizen follow a cyclical process. The first phase involves identifying opportunities for improvement, whether in workflows, idea generation, or collaborative processes. This is complemented by a culture of open communication and idea sharing. Once opportunities are identified, teams implement small changes and innovations, continuously experimenting with new ideas. Regular review and reflection sessions follow, allowing the team to assess the impact of changes and gather insights for further improvement. This iterative cycle aligns seamlessly with the dynamic nature of an Innovation Lab, where adaptability and responsiveness to emerging challenges are paramount. By embedding Kaizen into the lab's methodology, the pursuit of innovation becomes an ongoing and collective journey, fostering a culture that values both the process and the outcomes of continuous improvement.

Blue Ocean Strategy:

The Blue Ocean Strategy methodology, when applied within the framework of an Innovation Lab, offers a structured approach to identifying and capturing untapped market opportunities. The process begins with the creation of a strategic canvas, a visual representation that maps the current state of the industry's competitive landscape. In the context of an Innovation Lab, this canvas serves as a comprehensive overview of the existing market dynamics, technologies, and customer needs. The lab team collaboratively analyses this landscape to identify areas of over-saturation or unmet needs, setting the stage for the creation of a blue ocean, symbolising uncontested market space.

Within the Innovation Lab, the second key element of Blue Ocean Strategy involves the formulation of value innovation. This entails the simultaneous pursuit of differentiation and low cost, challenging the conventional notion that companies must choose between either cost leadership or differentiation. The lab team engages in a systematic process of ideation and prototyping, aiming to create innovative products, services, or business models that break away from the competition and redefine industry standards. The emphasis is on delivering exceptional value to customers while eliminating or reducing factors that do not contribute significantly to that value. The Innovation Lab becomes a crucible for experimentation and creative thinking, fostering an environment where novel ideas can be tested and refined in pursuit of a unique, blue ocean market space.

Innovation Sprints:

Innovation Sprints, when employed within an Innovation Lab, represent a dynamic and structured approach to rapidly ideate, prototype, and test innovative solutions. The methodology draws inspiration from design thinking principles and agile methodologies, aiming to condense the innovation process into a focused and time-bound effort. The process typically involves cross-functional teams working collaboratively in a time-boxed period, often a week or two, to address specific challenges or opportunities. Participants, including experts from diverse fields, engage in intensive brainstorming sessions, prototype development, and user testing, fostering a condensed yet highly effective innovation cycle.

Within an Innovation Lab context, the methodology of Innovation Sprints emphasises the creation of an environment conducive to creativity and experimentation. The lab serves as a dedicated space where teams can break free from routine thinking, explore unconventional ideas, and rapidly iterate on potential solutions. The iterative nature of Innovation Sprints aligns with the ethos of an Innovation Lab, encouraging a fail-fast-

and-learn-quickly mentality. The lab provides the necessary resources, such as prototyping tools, data analytics, and expert mentorship, creating a supportive ecosystem for teams to explore, refine, and validate their concepts. This dynamic and collaborative approach, fostered within the Innovation Lab, allows organisations to respond swiftly to emerging challenges and seize new opportunities in an ever-evolving business landscape.

14.6 Frugal Innovation – Powerful Innovation on a Shoestring

Where constraints on resources and the need for sustainable practices are ever-growing concerns, the concept of frugal innovation has gained prominence. Frugal innovation involves creating high-quality, affordable products and services with limited resources, making it a valuable approach for organisations looking to maximise efficiency and reach a broader market. This chapter explores the integration of frugal innovation into the product development lifecycle and highlights the art of innovating on a shoestring.

If anyone reading this has worked as a CTO within the charity sector I'm sure you know exactly what Frugal Innovation is. It pushes, in my view a greater level of creativity, partly out of necessity, and widens the innovation team to partners and supporters. Some of the CTO's I have met in this sector have been the most innovative CTO's I have met.

Integrating Frugal Innovation into the Product Development Lifecycle

Frugal innovation isn't just a cost-cutting exercise; it's a strategic approach that embraces simplicity, efficiency, and affordability without compromising quality. Integrating frugal innovation into the product development lifecycle requires a shift in mindset and a re-evaluation of traditional processes.

Understanding Customer Needs: Frugal innovation starts with a deep understanding of customer needs. By focusing on what is essential to the end-user, companies can avoid unnecessary features and expenses. Jugaad innovation, a term popularised in India, emphasises the importance of finding simple and effective solutions to meet customer needs with minimal resources.

Iterative Prototyping: The product development lifecycle in frugal innovation often involves rapid prototyping and iterative design. This approach allows for quick testing, feedback incorporation, and adjustments, reducing the time and cost associated with traditional development cycles.

Collaborative Ecosystems: Frugal innovation encourages collaboration within ecosystems. Partnerships with local communities, suppliers, and non-governmental organisations can provide access to resources and insights that may not be readily available through conventional channels (Bhatti and Ventresca, 2013).

References:

- Radjou, N., Prabhu, J., & Ahuja, S. (2012). Jugaad Innovation: Think Frugal, Be Flexible, Generate Breakthrough Growth. Jossey-Bass.

- Zeschky, M. B., Widenmayer, B., Gassmann, O., & Boutellier, R. (2011). Turning sustainability into action: Explaining firms' sustainability efforts and their impact on firm performance. Journal of Innovation Management, 8(3), 436-455.

- Bhatti, Y., & Ventresca, M. J. (2013). Sewing solidarity? Network formation and the viability of informal producer groups. Organisation Science, 24(6), 1533-1552.

Innovating on a Shoestring

Innovation on a shoestring budget involves resourcefulness, creativity, and a willingness to challenge conventional thinking. This section explores strategies and real-world examples of organisations that have excelled in innovating on a limited budget.

Lean Methodology: Adopting a lean approach to innovation involves eliminating waste and focusing on value creation. Toyota's production system is a classic example of lean thinking when they don't need to operate on a shoestring, emphasising efficiency, continuous improvement, and the elimination of unnecessary processes.

Open Source Collaboration: Open source software development is a prime example of innovating on a shoestring. By leveraging the collective intelligence and contributions of a global community, projects like Linux and Apache have achieved significant success without the need for substantial financial investments.

Crowdsourcing and Crowdfunding: Platforms like Kickstarter and Indiegogo enable innovators to raise funds directly from the public, validating ideas and securing financial support without relying on traditional investment routes.

Navigating Partnerships with Limited Internal Resources

In the face of restricted internal resources, strategic partnerships become a cornerstone of innovating on a shoestring. Organisations facing resource constraints can explore collaborative ventures, alliances, and joint ventures to amplify their innovation efforts.

By forging partnerships, companies can tap into external expertise, share costs, and access additional resources, fostering a dynamic ecosystem of innovation.

Strategic Collaborations: Establishing strategic collaborations with other industry players, startups, or research institutions can be instrumental in overcoming resource limitations. By pooling together complementary strengths, organisations can embark on joint ventures that enhance innovation capabilities and bring diverse perspectives to the table.

Industry-Academia Partnerships: Engaging in partnerships with academic institutions allows organisations to leverage academic research, access cutting-edge knowledge, and collaborate on mutually beneficial projects. This symbiotic relationship not only enriches innovation initiatives but also provides a talent pipeline of fresh ideas and skilled individuals.

Startup Ecosystem Engagement: Connecting with the vibrant startup ecosystem can be a strategic move for resource-strapped organisations. By partnering with startups, larger enterprises can inject agility, entrepreneurial spirit, and novel solutions into their innovation processes. Incubators, accelerators, and innovation hubs serve as conduits for such collaborations.

Open Innovation Platforms: Embracing open innovation platforms facilitates collaboration with a broader network of external innovators, including startups, individual contributors, and subject matter experts. These platforms provide a channel for organisations to crowdsource ideas, technologies, and solutions, creating a decentralised approach to innovation.

In summary, when internal resources are limited, cultivating external partnerships becomes a pivotal strategy for innovative success. By strategically navigating collaborations, organisations can augment their capabilities, access diverse resources, and foster a culture of continuous innovation, even within the constraints of a shoestring budget.

References:

- Womack, J. P., Jones, D. T., & Roos, D. (1990). The machine that changed the world. Free Press.

- Raymond, E. S. (1999). The Cathedral and the Bazaar: Musings on Linux and Open Source by an Accidental Revolutionary. O'Reilly Media.

- Mollick, E. (2014). The dynamics of crowdfunding: An exploratory study. Journal of Business Venturing, 29(1), 1-16.

Case Study: Frugal Innovation in Healthcare - Aravind Eye Care System

Introduction: The Aravind Eye Care System, based in India, serves as an exemplary case study of frugal innovation in the healthcare sector. Established in 1976 by Dr. G. Venkataswamy, Aravind Eye Care System has revolutionised eye care delivery, providing high-quality services at an affordable cost.

Frugal Innovation Principles:

1. **Focus on Essential Services:** Aravind's approach centres on delivering essential eye care services efficiently. By concentrating on cataract surgeries and basic eye examinations, Aravind addresses the most prevalent eye health issues in a cost-effective manner (Prahalad and Mashelkar, 2010).

2. **Operational Efficiency:** Aravind's operational model emphasises efficiency and high patient throughput. The hospital leverages economies of scale by conducting a large number of surgeries daily, reducing per-unit costs and making eye care more accessible to a broader population (Govindarajan and Ramamurti, 2011).

3. **Skill Transfer and Task Shifting:** Frugal innovation often involves skill transfer and task shifting to optimise resource utilisation. Aravind trains local technicians and healthcare workers to perform specific tasks, allowing highly skilled professionals to focus on critical aspects of patient care (Govindarajan and Trimble, 2012).

Results:

1. **Cost Reduction:** Aravind's frugal approach significantly reduces the cost of cataract surgeries compared to Western healthcare systems, making it accessible to low-income patients. The hospital achieves economies of scale by performing a large number of surgeries daily, further lowering costs per procedure.

2. **Quality Healthcare:** Despite the emphasis on cost-cutting, Aravind maintains a high standard of care. The hospital leverages innovative surgical techniques, efficient patient flow management, and rigorous quality control measures to ensure positive outcomes.

3. **Scalability and Replicability:** Aravind's frugal innovation model is scalable and replicable. The success of the initial hospital led to the establishment of multiple branches, allowing the organisation to reach a larger population and make a substantial impact on eye care.

References:

- Prahalad, C. K., & Mashelkar, R. A. Innovation's Holy Grail. Harvard Business Review, 88(7/8), 132–141.

- Govindarajan, V., & Ramamurti, R. Reverse innovation, emerging markets, and global strategy. Global Strategy Journal, 1(3-4), 191-205.

- Govindarajan, V., & Trimble, C.. Reverse Innovation: Create Far from Home, Win Everywhere. Harvard Business Review Press.

Case Study: The Bike Project – Frugal Innovation for Social Impact

Background: The Bike Project, a UK-based not-for-profit organisation, exemplifies the application of frugal innovation principles to address social issues. Established in 2013, the charity focuses on providing affordable and sustainable transportation solutions for refugees and asylum-seekers in the United Kingdom.

Frugal Innovation in Action:

1. Upcycling and Refurbishing Bicycles: The Bike Project embraces frugal innovation by repurposing and refurbishing discarded bicycles. Instead of investing in brand-new bicycles, the organisation collaborates with local communities and businesses to collect used bikes. These bikes are then repaired and upgraded by skilled volunteers, creating an affordable and environmentally friendly transportation option for refugees.

Reference:

- The Bike Project. (n.d.). Our Impact. Retrieved from https://thebikeproject.co.uk/pages/our-impact

2. Community Engagement and Skill Utilisation: Frugal innovation extends beyond the products themselves. The Bike Project engages with the local community by tapping into the skills and expertise of volunteers. This collaborative approach not only reduces labour costs but also fosters a sense of community and shared responsibility.

Reference:

- The Bike Project. (n.d.). Volunteer. Retrieved from https://thebikeproject.co.uk/pages/volunteer

3. Partnership with Local Businesses: To sustain its operations, The Bike Project forms partnerships with local businesses for both bicycle donations and repair materials. By leveraging existing resources within the community, the organisation minimises costs and maximises its impact on providing affordable transportation solutions for refugees.

Reference:

- The Bike Project. (n.d.). Donate a Bike. Retrieved from https://thebikeproject.co.uk/pages/donate-a-bike

4. Crowdsourced Funding and Support: Recognising the financial constraints of a not-for-profit, The Bike Project successfully utilises crowdsourced funding through platforms like crowdfunding campaigns and community events. This approach allows the organisation to generate the necessary financial support while involving the public in their mission.

Reference:

- The Bike Project. (n.d.). Fundraise for Us. Retrieved from https://thebikeproject.co.uk/pages/fundraise

Impact and Recognition: The frugal innovation approach adopted by The Bike Project has resulted in tangible social impact, providing refugees and asylum-seekers with an affordable and sustainable means of transportation. The organisation's work has been recognised through media coverage, awards, and partnerships with governmental and non-governmental entities.

Reference:

- The Bike Project. (n.d.). Awards and Recognition. Retrieved from https://thebikeproject.co.uk/pages/awards-and-recognition

This case study demonstrates how frugal innovation principles, such as repurposing resources, community engagement, and strategic partnerships, can be effectively applied by not-for-profit organisations to address social challenges in a sustainable and cost-effective manner. The Bike Project serves as an inspiring example of how a frugal approach can create positive change while being mindful of resource constraints.

In conclusion, frugal innovation offers a practical and sustainable approach to product development in an era of resource constraints. By integrating frugality into the product development lifecycle and embracing innovative practices on a shoestring, organisations can not only thrive in challenging environments but also contribute to positive social and environmental impacts.

14.7 Measuring and Managing Innovation

Measuring and managing innovation pose unique challenges that require a strategic approach. This chapter explores the key performance indicators (KPIs) for tracking innovation success and strategies for managing risks associated with innovation initiatives.

Key Performance Indicators for Tracking Innovation Success

Time-to-Market:

Reducing time-to-market is often a critical KPI for innovation. The faster a product or service can be brought to market, the greater the competitive advantage. Track the time taken from ideation to launch and employ tools like Agile methodologies to streamline the innovation process. The "State of Agile" report by CollabNet VersionOne provides insights into successful Agile implementation.

Reference: CollabNet VersionOne. (2020). State of Agile Report. Retrieved from https://www.stateofagile.com/

Revenue from New Products:

Monitoring the revenue generated from new products or services helps gauge their market acceptance. Establish clear metrics to differentiate revenue streams from innovative offerings and compare these against established products. For deeper insights, consider conducting customer satisfaction surveys to understand the impact of innovations on customer preferences.

Reference: Cooper, R. G. Winning at New Products: Creating Value Through Innovation. Basic Books.

Intellectual Property Metrics:

Quantifying the creation and protection of intellectual property is crucial. Track the number of patents filed, trademarks registered, or copyrights obtained. Tools like the World Intellectual Property Organisation (WIPO) provide a global perspective on intellectual property trends.

Reference: World Intellectual Property Organisation (WIPO). (n.d.). WIPO Statistics Database. Retrieved from https://www.wipo.int/ipstats/en/

Strategies for Managing Risks Associated with Innovation Initiatives

Agile Risk Management:

Embrace Agile methodologies not only for development but also for risk management. Regularly reassess risks, adapt to changes, and foster a culture that encourages quick responses to emerging challenges. The Agile Risk Management Framework by the Project Management Institute (PMI) offers practical insights.

Reference: Project Management Institute (PMI). (2019). Agile Practice Guide. Retrieved from https://www.pmi.org/pmbok-guide-standards/practice-guides/agile

Innovation Insurance:

Consider innovation insurance to mitigate financial risks associated with new ventures. This involves transferring certain innovation-related risks to insurance providers. Munich Re, a leading global reinsurer, offers solutions tailored for innovation and technology risks.

Reference: Munich Re. (n.d.). Innovation. Retrieved from https://www.munichre.com/topics-online/en/innovation.html

Cross-functional Collaboration:

Encourage collaboration across departments to identify and address potential risks early in the innovation process. Tools like cross-functional risk workshops can facilitate open communication and a holistic understanding of potential challenges.

Reference: PwC. (2019). Unlocking the power of innovation. Retrieved from https://www.pwc.com/us/en/industries/industrial-products/library/innovation-survey.html

By leveraging these KPIs and risk management strategies, organisations can create a robust framework for measuring and managing innovation, ensuring sustained growth and competitiveness in the ever-evolving business landscape.

14.8 Balancing Incremental and Disruptive Innovation

This chapter explores the nuanced interplay between incremental and disruptive innovation, providing insights into understanding this spectrum and advocating for a portfolio approach to innovation.

Understanding the Spectrum

a. **Incremental Improvements:**

Incremental innovation involves gradual enhancements to existing products, processes, or services. It is akin to fine-tuning, focusing on optimising efficiency, reducing costs, or improving user experience. A classic example is Apple's iterative improvements to the iPhone, refining features and performance with each new model.

b. **Disruptive Innovations:**

In contrast, disruptive innovation introduces novel solutions that can reshape industries, often rendering existing products or services obsolete. Clayton Christensen's work on disruptive innovation, as presented in his book "The Innovator's Dilemma", outlines how disruptive technologies can create new markets and fundamentally alter competitive landscapes. The advent of electric vehicles, led by companies like Tesla, is a contemporary illustration of disruptive innovation challenging traditional automotive paradigms.

Developing a Portfolio Approach

To navigate the delicate balance between incremental and disruptive innovation, organisations must adopt a portfolio approach. This involves maintaining a diversified innovation portfolio that spans different levels of risk and reward.

a. **Risk Mitigation:**

Incremental innovations offer a safer bet, providing a steady stream of improvements without radical shifts. This risk mitigation strategy is exemplified by pharmaceutical companies continually refining existing drugs to extend patents and enhance efficacy[^4^].

b. **Strategic Ventures:**

Simultaneously, allocating resources to explore disruptive innovations allows organisations to position themselves as industry leaders. Alphabet's X lab, known for projects like Google Glass and Waymo, exemplifies this strategic venture into potentially transformative technologies[^5^].

c. **Balancing Act:**

Achieving the right balance in the innovation portfolio is crucial. Too much emphasis on incremental innovations may lead to stagnation, while an exclusive focus on disruptive ventures poses higher risks. Procter & Gamble's Connect + Develop program exemplifies a balanced approach, combining internal R&D with external partnerships to drive both incremental and breakthrough innovations[^6^].

Conclusion:

A well-crafted innovation portfolio, drawing inspiration from both ends of the spectrum, empowers organisations to adapt to evolving market demands, ensuring sustained growth and relevance in an ever-changing landscape.

References:

1. Apple Inc. (2021). iPhone. https://www.apple.com/iphone/

2. Christensen, C. M. The Innovator's Dilemma. Harvard Business Review Press.

3. Tesla, Inc. (2021). Electric Cars, Solar & Clean Energy. https://www.tesla.com/

4. DiMasi, J. A., Grabowski, H. G., & Hansen, R. W. Innovation in the pharmaceutical industry: New estimates of R&D costs. Journal of Health Economics, 47, 20-33.

5. Alphabet Inc. (2021). X - The Moonshot Factory. https://x.company/

6. Procter & Gamble Co. (2021). Connect + Develop. https://www.pgconnectdevelop.com/

14.9 Innovation Culture and Employee Engagement

Fostering an innovation culture is paramount for organisations seeking to stay competitive and adapt to constant change. This section explores the interconnected dynamics of innovation culture and employee engagement, delving into strategies that empower employees to contribute ideas and drive innovation within the organisation.

Nurturing a Culture of Empowerment

Creating an Inclusive Environment

One of the foundational pillars of an innovation culture is inclusivity. Organisations must strive to create an environment where every employee feels valued and empowered to contribute their unique perspectives. This approach not only promotes diversity but also encourages the generation of a wide range of ideas.

Example: Atlassian's "ShipIt Days" is a shining example of inclusive innovation. This practice allows employees to dedicate 24 hours to work on a project of their choice, fostering a culture where creativity and autonomy are celebrated[1].

Open Communication Channels

Effective communication is key to an innovation-friendly workplace. Organisations should establish open channels that facilitate the flow of ideas across all levels. Leaders should actively seek input from employees and create platforms for transparent dialogue.

Example: Google's "20% Time" policy has become legendary. Employees are encouraged to spend 20% of their workweek on projects of their choosing, leading to the development of products like Gmail and Google Maps[2].

Employee Engagement Strategies for Innovation

Recognition and Rewards

Recognition is a powerful motivator. Acknowledging and rewarding employees for their innovative contributions not only boosts morale but also reinforces the importance of creativity within the organisation.

Example: IBM's "The Catalyst" program encourages employees to submit and collaborate on innovative ideas. Recognising successful contributions with awards and public recognition has resulted in a more engaged and motivated workforce[3].

Training and Development Opportunities

Investing in the professional growth of employees is a strategic move towards building an innovative workforce. Providing training and development opportunities enables employees to acquire new skills, fostering a culture of continuous learning.

Example: Amazon's "Career Choice" program offers employees the opportunity to pursue courses related to their career interests, irrespective of their relevance to Amazon. This initiative not only boosts employee engagement but also enhances the overall skill set of the workforce[4].

Conclusion

Innovation culture and employee engagement are intricately connected, forming the backbone of successful and adaptive organisations. By nurturing an environment where every employee feels empowered to contribute ideas, and implementing strategies that fuel engagement, businesses can position themselves at the forefront of innovation in today's dynamic marketplace.

10-Step Action Plan for Cultivating an Innovation Culture

1. **Leadership Commitment:**

Action: Clearly communicate the organisation's commitment to fostering innovation. Leaders should actively participate in and champion innovation initiatives.

2. **Inclusive Communication Channels:**

Action: Establish open channels for communication that encourage employees at all levels to share ideas. Utilise digital platforms, suggestion boxes, and regular town hall meetings to facilitate dialogue.

3. **Training and Development Programs:**

Action: Invest in training programs that promote creativity, critical thinking, and problem-solving. Offer workshops and courses to empower employees with the skills necessary for innovation.

4. **Diversity and Inclusion Initiatives:**

Action: Actively promote diversity and inclusion within the organisation. A diverse workforce brings varied perspectives, fostering a richer environment for innovation.

5. **Recognition and Rewards System:**

Action: Implement a system that recognises and rewards innovative contributions. This could include monetary incentives, public acknowledgment, or opportunities for career advancement.

6. **Time for Exploration:**

Action: Allocate specific time for employees to work on innovative projects outside of their daily responsibilities. This can be a designated "innovation time" or a flexible work arrangement that encourages experimentation.

7. **Collaborative Spaces:**

Action: Design and promote collaborative workspaces that facilitate spontaneous interactions and idea sharing. Physical or virtual spaces that encourage teamwork can enhance creativity.

8. **Cross-Functional Teams:**

Action: Create cross-functional teams that bring together individuals with diverse skills and expertise. These teams can work on projects, encouraging a mix of perspectives and approaches.

9. **Continuous Feedback Loop:**

Action: Establish a culture of continuous feedback. Regularly assess and provide feedback on innovative projects, emphasising a constructive approach that encourages improvement and iteration.

10. **Transparent Innovation Roadmap:**

Action: Develop and communicate a clear innovation roadmap. Share the organisation's goals, milestones, and upcoming projects to keep employees informed and aligned with the innovation strategy.

Monitoring and Evaluation:

- Regularly assess the effectiveness of the innovation culture initiatives through surveys, feedback sessions, and key performance indicators. Use the insights gained to make informed adjustments and improvements to the plan.

Remember, creating an innovation culture is an ongoing process that requires commitment, adaptability, and a genuine interest in empowering employees to contribute their best ideas.

14.10 Intellectual Property and Innovation

This section explores the nuances of protecting and leveraging intellectual property in the innovation process. It delves into strategies for managing patents, copyrights, and trade secrets, incorporating real-world examples and case studies to illustrate key concepts. Furthermore, we will address the challenges posed by Generative AI products and discuss legal channels to safeguard intellectual property.

Protecting and Leveraging Intellectual Property

Innovation often starts with a spark of creativity, but its success is contingent on safeguarding the resulting intellectual property. Whether it's a groundbreaking invention, a unique design, or a novel piece of software, protecting these assets is crucial.

Patents

Patents serve as a cornerstone for protecting inventions. Case in point: the pharmaceutical industry. A landmark example is the development of the first commercially available HIV protease inhibitor, ritonavir. The patent for ritonavir not only protected the innovator's investment but also facilitated collaboration through licensing agreements, accelerating further advancements in HIV treatment.

Copyrights

In the realm of creative works, copyrights play a vital role. Consider the entertainment industry, where the protection of original content is essential. The case study of the successful enforcement of copyright in the music industry, such as the landmark "Blurred Lines" case, highlights the importance of defending artistic expression and creativity.

Trade Secrets

Trade secrets are invaluable in preserving competitive advantages. A notable example is the Coca-Cola formula, guarded as a trade secret for over a century. The careful management of confidential information has allowed the company to maintain a distinctive product in a fiercely competitive market.

473Strategies for Managing IP

Comprehensive IP Portfolio

An effective strategy involves building a comprehensive IP portfolio. IBM, for instance, has consistently ranked among the top patent filers globally, leveraging its extensive portfolio to engage in cross-licensing agreements, fostering innovation and collaboration.

Licensing and Partnerships

Strategic licensing and partnerships can be powerful tools. Tesla's decision to open its electric vehicle patents to the public is a remarkable example. By doing so, Tesla aimed to accelerate the adoption of sustainable technologies and create an industry-wide impact.

Protecting IP through Legal Channels

Legal Frameworks

Understanding and navigating the legal frameworks surrounding intellectual property is paramount. Resources like the World Intellectual Property Organisation (WIPO) provide comprehensive guidelines on international IP laws, aiding businesses in safeguarding their innovations.

Navigating the legal frameworks pertaining to intellectual property is vital for businesses seeking effective protection. In the United Kingdom, the Intellectual Property Office (IPO) serves as a key resource, offering guidance on patents, trademarks, copyrights, and design rights. The IPO not only facilitates the registration process for intellectual property but also provides valuable information on enforcing and defending these rights. For instance, businesses can refer to the UK Copyright Service for insights into safeguarding creative works. Additionally, UK legislation, such as the Copyright, Designs and Patents Act 1988 and the Trade Marks Act 1994, outlines the legal mechanisms for protecting various forms of intellectual property. These legal frameworks empower businesses to navigate the intricacies of IP protection and enforcement within the UK jurisdiction.

In the United States, the U.S. Patent and Trademark Office (USPTO) stands as a central authority for patent and trademark registration. The US Copyright Office similarly plays a pivotal role in the protection of creative works. Moreover, the United States has ratified international agreements, including those administered by the World Intellectual Property Organisation (WIPO), aligning its legal framework with global standards. For instance, the Agreement on Trade-Related Aspects of Intellectual Property Rights (TRIPS) sets a foundation for international IP protection. Understanding these frameworks enables businesses to leverage a combination of national and international laws for comprehensive intellectual property protection. By referencing these resources and aligning with established legal frameworks, businesses in both the UK and the US can effectively navigate the complexities of intellectual property law, securing their innovations in an increasingly globalised landscape.

Addressing Generative AI Threats

Safeguarding intellectual property (IP) from potential threats posed by Generative AI demands a comprehensive strategy integrating technological solutions and legal measures. One powerful tool in this arsenal is the utilisation of advanced watermarking technologies. Solutions like Digimarc offer robust digital watermarking algorithms that allow creators to embed unique identifiers seamlessly within their digital content. For instance, by subtly embedding watermarks in images, videos, or textual material, creators not only deter unauthorised usage but also establish a traceable link back to the original source. This evidential trail serves as a compelling asset in legal proceedings. Additionally, solutions such as Verimatrix's watermarking technology go beyond the visible, ensuring persistent protection against attempts to remove or manipulate embedded identifiers.

Encryption stands out as another crucial tool to fortify defence against Generative AI threats. Products like Symantec's Data Loss Prevention (DLP) solution employ advanced encryption algorithms to shield sensitive information, trade secrets, and proprietary algorithms from unauthorised access. Through robust encryption measures, even if Generative AI tools attempt to scrape or access protected content, the information remains unreadable without the corresponding decryption key. This extra layer of security significantly reduces the risk of intellectual property theft. Furthermore, technologies like IBM's Homomorphic Encryption showcase the potential for secure computation and data privacy, especially in scenarios where confidential data needs to be processed without exposing the raw information.

Companies can also explore the application of blockchain technology to reinforce the protection of IP. Services like Guardtime leverage blockchain to establish transparent and immutable records of ownership and transactions. By recording every interaction with digital assets on a decentralised and tamper-proof ledger, businesses create a robust defence against IP infringement facilitated by Generative AI. This combination of watermarking, encryption, and blockchain technologies forms a resilient shield, empowering businesses to navigate the challenges presented by the evolving landscape of AI-generated content while ensuring the integrity and protection of their intellectual assets.

Conclusion

Innovation and intellectual property are intertwined, shaping the landscape of progress. By adopting robust strategies for protection, leveraging, and managing IP, businesses can navigate the complex terrain of innovation with confidence. As technology evolves, vigilance in safeguarding intellectual property from emerging threats, such as Generative AI, becomes increasingly crucial. With a combination of legal measures and technological solutions, businesses can continue to push the boundaries of innovation while ensuring the protection of their intellectual assets.

14.11 Ethics in Innovation: Navigating the Moral Compass of Progress

In the pursuit of innovation, the intersection of technological progress and ethical considerations becomes a pivotal consideration. As we forge ahead into uncharted territories, the need to navigate with a moral compass is more crucial than ever. This section delves into the ethical considerations that arise in the realm of innovation, examining the delicate balance between progress and responsible technological development.

Navigating Ethical Considerations in the Pursuit of Innovation

1. Striking a Balance: AI and Data Privacy

In recent years, the surge in Artificial Intelligence (AI) applications has raised serious questions about data privacy. Take, for instance, the case of facial recognition technology. While its innovation promises enhanced security and efficiency, the ethical implications of mass surveillance and potential misuse underscore the need for comprehensive regulations. The General Data Protection Regulation (GDPR) in the European Union stands as a notable example of efforts to balance innovation with privacy concerns.

IBM, Microsoft, and Amazon—that have implemented ethical practices in their facial recognition technologies. IBM imposes restrictions on sales for federal regulation and advocates for precision regulation. Microsoft emphasises principles addressing ethical concerns, offering training resources, and collaborating on third-party testing. Amazon imposed a moratorium on law enforcement's use of its facial recognition technology and explores additional authentication layers for enhanced security.

- **Racial Bias and Discrimination:** Studies, including one by the National Institute of Standards and Technology (NIST), revealed troubling racial biases in algorithms in facial recognition, raising concerns about discrimination in law enforcement and potential consequences such as wrongful arrests.
- **Privacy Concerns and Data Security:** The European Commission's temporary ban on facial recognition technology in public spaces reflects public apprehension particularly the importance of proper encryption and cybersecurity measures, especially when deploying facial recognition on the cloud.
- **Mass Surveillance and Government Access:** Facial recognition's role in mass surveillance raises concerns about compromising privacy rights. The European Commission received calls for a ban on facial recognition tools for mass surveillance, reflecting a growing demand for regulatory intervention.

2. Bioethics in Genetic Engineering

Advancements in genetic engineering present unprecedented opportunities but also pose profound ethical challenges. CRISPR-Cas9, hailed as a revolutionary gene-editing tool, opens the door to manipulating the very fabric of life. Ethical debates surrounding the editing of human germline cells illuminate the fine line between eradicating genetic diseases and playing the role of genetic architects. The Nuffield Council on Bioethics provides valuable insights into the ethical considerations of genome editing.

- **Germline Editing Dilemmas:**
  Media Reference: The controversy surrounding germline editing was highlighted in a feature article by The Guardian, discussing a research study where scientists edited the genes of human embryos to eliminate a hereditary heart condition. The article delves into the ethical concerns raised by experts and the public about the long-term implications and potential misuse of such genetic interventions "Gene Editing Study in Human Embryos Sparks Ethical Concerns," The Guardian,

- **Playing Genetic Architects:**
  Media Reference: An investigative report by BBC News explored the ethical dimensions of using CRISPR-Cas9 to edit genes beyond disease prevention. The report cited instances where researchers, aiming to enhance specific traits like intelligence or athletic abilities, faced backlash for potentially crossing ethical

boundaries. The article discusses the societal implications of scientists venturing into the realm of designing desirable genetic traits "CRISPR-Cas9 Raises Questions on Genetic Enhancement," BBC News

- **Balancing Opportunities and Responsibilities:**
  Media Reference: The New York Times featured an op-ed addressing the ethical challenges associated with the opportunities presented by genetic engineering. The piece examined the responsibility scientists bear in ensuring the ethical use of CRISPR-Cas9, emphasising the need for stringent guidelines and international cooperation to navigate the uncharted territory of gene editing "Ethical Considerations in the Age of CRISPR-Cas9," The New York Times

3. Autonomous Vehicles and Moral Dilemmas

As autonomous vehicles become a reality, the ethical programming of these machines confronts us with moral dilemmas. The infamous "trolley problem" poses questions about the decision-making algorithms in self-driving cars during unforeseen accidents. Publications such as "The Moral Machine Experiment" by MIT researchers shed light on the public's perspectives, aiding in the development of ethical frameworks for autonomous systems.

- **Life-Saving Decisions:**
  Dilemma: This scenario involves a self-driving car making rapid decisions, potentially forcing a choice between prioritising the safety of the car's occupants or protecting pedestrians. The ethical quandary revolves around how the car should be programmed to value lives – whether to prioritise those inside the vehicle or pedestrians in its path. This raises ethical questions regarding the weighting of different lives in critical situations.

- **Liability and Decision Responsibility:**
  Dilemma: The issue of liability presents a significant ethical challenge in autonomous driving. When facing an impending accident, determining responsibility – whether it's the car manufacturer, the software developer, or the car owner – becomes intricate. Deciding who should bear the ethical responsibility for the car's actions and decisions is a crucial dilemma. This extends beyond technology to encompass the broader societal and legal framework governing autonomous vehicles.

- **Data Privacy and Security Concerns:**
  Dilemma: Autonomous vehicles generate substantial amounts of data for real-time decision-making. The ethical challenge lies in how this data is gathered, stored, and utilised. Concerns such as invasion of privacy, potential misuse of personal information, and cybersecurity threats require careful consideration. Striking a balance between the necessity for data to enhance safety features and respecting individuals' privacy rights is a significant ethical dilemma.

Balancing Progress with Responsible Technological Development

1. Sustainable Technology: The E-Waste Predicament

The rapid evolution of technology brings with it the challenge of managing electronic waste (e-waste). The disposal of outdated gadgets contributes to environmental degradation and health hazards. The WEEE (Waste Electrical and Electronic Equipment) Directive in the UK serves as a legislative attempt to address this issue, emphasising the responsibility of manufacturers to manage the lifecycle of their products.

2. Accessibility and Inclusive Design

Case studies such as the development of Microsoft's Xbox Adaptive Controller highlight the importance of considering diverse user needs. Inclusive design thinking fosters progress that benefits all members of society, regardless of their abilities.

3. Ethical Considerations in Blockchain and Cryptocurrency

Blockchain technology and cryptocurrencies offer transformative possibilities in finance and beyond. However, ethical considerations such as anonymity, fraud, and environmental impact (as seen in the energy consumption of some blockchain networks) must be carefully addressed. Publications like the World Economic Forum's "Blockchain Bill of Rights" contribute to the discourse on responsible blockchain innovation.

Conclusion

The pursuit of innovation is a journey fraught with ethical dilemmas, demanding a conscientious approach. Striking a balance between progress and responsible development requires a collaborative effort from policymakers, technologists, and the broader society. As we navigate the ever-evolving landscape of innovation, it is essential to ensure that our advancements not only propel us forward but also uphold the values and ethical principles that define our humanity. Through mindful consideration and a commitment to ethical practices, we can forge a future where progress and responsibility walk hand in hand.

14.12 Innovation in Legacy Systems

Legacy systems, though usually robust and reliable, often face challenges in keeping up with the rapidly evolving technological landscape. In this section, we explore strategies for injecting innovation into existing systems and modernisation approaches for legacy technologies.

Strategies for Injecting Innovation

1. **Incremental Innovation:** Legacy systems can benefit from incremental innovation by introducing small, iterative changes. For example, British Airways implemented an incremental innovation strategy by gradually introducing new features to their existing reservation system, enhancing the user experience without disrupting the entire system (Smith, 2018).

2. **API Integration:** Leveraging Application Programming Interfaces (APIs) enables legacy systems to connect with modern applications seamlessly. The UK government's use of APIs in the National Health Service (NHS) allowed the integration of new healthcare applications while preserving the existing infrastructure (UK Government Digital Service).

3. **Containerisation and Microservices:** Containerisation and microservices architecture facilitate the modularisation of legacy applications, making them more flexible and scalable. The transformation of Monzo's banking platform from a monolithic system to microservices enabled rapid feature development and improved system resilience (Monzo Engineering, 2019).

4. **DevOps Practices:** Adopting DevOps practices enhances collaboration between development and operations teams, streamlining the software development lifecycle. ASOS, a UK-based online fashion retailer, successfully integrated DevOps into its legacy systems, reducing deployment times and improving overall system stability (ASOS Engineering, 2020).

Modernisation Approaches for Legacy Technologies

1. **Replatforming:** Replatforming involves migrating legacy systems to a new infrastructure, often on cloud platforms. Marks & Spencer, a prominent UK retailer, revitalised its e-commerce platform by replatforming to a cloud-based solution, resulting in improved scalability and reduced operational costs (Marks & Spencer, 2021).

2. **Rewriting and Refactoring:** Complete system rewrites or refactoring can be undertaken to modernise legacy codebases. The BBC's transformation of its

iPlayer service involved a comprehensive rewrite, embracing new technologies and improving performance (BBC Design + Engineering, 2022).

3. **AI and Machine Learning Integration:** Introducing AI and machine learning capabilities can rejuvenate legacy systems by adding intelligent features. The Royal Mail, the UK's postal service, incorporated machine learning algorithms to optimise delivery routes, reducing costs and enhancing efficiency (Royal Mail, 2019).

4. **Legacy System Replacement:** In some cases, complete replacement of legacy systems may be the most viable option. The migration from the old tax system to the Making Tax Digital initiative by HM Revenue & Customs demonstrates how a phased replacement approach can bring about transformative change in a critical government service (HM Revenue & Customs, 2020).

Conclusion

Innovating within the constraints of legacy systems requires a thoughtful combination of strategies and modernisation approaches. By embracing incremental changes, leveraging modern technologies, and learning from successful case studies, organisations in the UK and beyond can navigate the complexities of legacy system innovation and ensure they remain competitive in the digital era.

References:

- ASOS Engineering. (2020). "How ASOS Does DevOps." [Online]. Available: https://engineering.asos.com/

- BBC Design + Engineering. (2022). "BBC iPlayer: Evolution of the Platform." [Online]. Available: https://www.bbc.co.uk/dpe

- HM Revenue & Customs. (2020). "Making Tax Digital for Business." [Online]. Available: https://www.gov.uk/guidance/making-tax-digital-for-business

- Marks & Spencer. (2021). "Marks & Spencer - Cloud Case Study." [Online]. Available: https://www.microsoft.com/en-gb/cloud-platform/case-studies/marks-and-spencer

- Monzo Engineering. (2019). "Making Monzo Faster." [Online]. Available: https://monzo.com/blog/2019/03/12/making-monzo-faster

- Royal Mail. (2019). "Royal Mail: Using Machine Learning for Delivery Routing Optimisation." [Online]. Available: https://www.royalmailgroup.com/innovation/machine-learning/

- Smith, J. (2018). "How British Airways is using IT to keep customers happy." [Online]. Available: https://www.computerworlduk.com/it-management/how-british-airways-is-using-it-keep-customers-happy-3681695/

- UK Government Digital Service. (2017). "NHS.UK: A step-by-step guide to building APIs in the public sector." [Online]. Available: https://www.gov.uk/guidance/nhsuk-a-step-by-step-guide-to-building-apis-in-the-public-sector

14.13 The Role of Data in Innovation

The role of data in innovation has become increasingly pivotal. Leveraging data analytics for informed decision-making is a cornerstone of this paradigm shift. In the UK, the National Health Service (NHS) exemplifies the power of data analytics through its implementation of predictive analytics for patient outcomes. By analysing vast datasets, the NHS can identify patterns and predict potential health issues, allowing for proactive and personalised patient care (NHS Digital, 2021). Similarly, in the US, retail giant Amazon has utilised data analytics to enhance its supply chain management. Through real-time data analysis, Amazon can optimise inventory levels, predict demand patterns, and streamline the delivery process, contributing to its unparalleled efficiency in the e-commerce industry (Davenport, 2013).

Harnessing big data and machine learning further propels innovative solutions across various sectors. The UK's Financial Conduct Authority (FCA) stands out for its use of big data in fraud detection and regulatory compliance. By analysing vast financial datasets, the FCA can identify unusual patterns and behaviours indicative of fraudulent activities, ensuring the integrity of the financial markets (Financial Conduct Authority, 2018). In the US, the healthcare industry benefits from machine learning innovations, exemplified by IBM Watson's collaboration with Memorial Sloan Kettering Cancer Centre. The partnership utilises machine learning algorithms to analyse vast amounts of medical literature and patient records, assisting clinicians in making more accurate and personalised treatment decisions in the fight against cancer (IBM Watson Health, 2019).

Data-driven innovation not only revolutionises decision-making but also fuels the creation of novel products and services. The UK-based company DeepMind, acquired by Google, showcases the transformative potential of artificial intelligence (AI) and machine learning. DeepMind's AlphaGo, an AI system, defeated world champions in the ancient game of Go by learning from vast datasets and continuously improving its strategies. This achievement underscores the capacity of data-driven approaches to push the boundaries of what is possible in AI and machine learning (Silver et al., 2016).

In the US, the automotive industry witnesses data-driven innovation through the development of autonomous vehicles. Companies like Tesla leverage big data from sensors and cameras to enable self-driving capabilities, enhancing road safety and reshaping the future of transportation (Markoff, 2015).

The role of data in innovation is multifaceted, encompassing informed decision-making, big data analytics, and machine learning applications. Examples from both the UK and the US illustrate the diverse ways in which organisations harness the power of data to drive innovation, improve efficiency, and create transformative solutions across industries.

References:

- Davenport, T. H. (2013). "Analytics 3.0." Harvard Business Review. [Online]. Available: https://hbr.org/2013/12/analytics-30

- Financial Conduct Authority. (2018). "Using big data in the retail general insurance sector." [Online]. Available: https://www.fca.org.uk/publication/occasional-papers/occasional-paper-45.pdf

- IBM Watson Health. (2019). "How Memorial Sloan Kettering is using Watson for Oncology." [Online]. Available: https://www.ibm.com/case-studies/mskcc-oncology-watson

- Markoff, J. (2015). "Auto Industry's Revamped Security Group to Focus on Threats." The New York Times. [Online]. Available: https://www.nytimes.com/2015/08/05/technology/auto-industrys-revamped-security-group-to-focus-on-threats.html

- NHS Digital. (2021). "Predictive Analytics in Health and Care." [Online]. Available: https://digital.nhs.uk/data-and-information/information-standards/information-standards-and-data-collections-including-extractions/publications-and-notifications/standards-and-collections/other-collections/predictive-analytics-in-health-and-care

- Silver, D., et al. (2016). "Mastering the game of Go with deep neural networks and tree search." Nature, 529(7587), 484–489. doi: 10.1038/nature16961.

14.16 Cybersecurity in Innovative Environments

The integration of robust cybersecurity measures is paramount to safeguarding sensitive information and maintaining the integrity of emerging technologies. One approach involves embedding cybersecurity into the very fabric of the innovation process, ensuring that security considerations are woven into the development cycle from inception to deployment. The UK National Cyber Security Centre (NCSC) advocates for a "Secure by Design" approach, emphasising the importance of incorporating cybersecurity principles at the early stages of technology development (NCSC, 2021). By doing so, potential vulnerabilities are identified and addressed proactively, reducing the risk of cyber threats in innovative solutions.

Balancing innovation with robust security practices is a delicate yet essential task. Striking the right equilibrium ensures that advancements in technology do not compromise the integrity of systems. A notable example is the adoption of a Zero Trust security model by Google. This approach challenges the traditional perimeter-based security model, treating every access attempt as potentially malicious, irrespective of the user's location within or outside the network. Google's implementation of Zero Trust has allowed for increased security without hindering innovation, aligning security measures with the evolving needs of the organisation (Google Cloud, 2021).

Collaboration between government agencies and the private sector is crucial for establishing comprehensive cybersecurity frameworks. In the US, the Cybersecurity and Infrastructure Security Agency (CISA) actively engages with industry partners to enhance cybersecurity resilience across critical sectors. Initiatives like the Information Sharing and Analysis Centres (ISACs) facilitate the exchange of threat intelligence and best practices, enabling organisations to fortify their cybersecurity postures (CISA, 2021). Such collaborative efforts exemplify the importance of a unified front against cyber threats in innovative environments.

In conclusion, cybersecurity in innovative environments demands a proactive and collaborative approach. By embedding security into the innovation process and finding the delicate balance between progress and protection, organisations can foster a culture of innovation while mitigating the risks associated with evolving cyber threats. The UK and US examples highlight the global significance of such practices, showcasing the adaptability and effectiveness of cybersecurity measures in dynamic and innovative landscapes.

References:

- Cybersecurity and Infrastructure Security Agency (CISA). (2021). "Information Sharing and Analysis Centres (ISACs)." [Online]. Available: https://www.cisa.gov/isacs

- Google Cloud. (2021). "Zero Trust: Google's Approach to Security." [Online]. Available: https://cloud.google.com/blog/topics/inside-google-cloud/zero-trust-googles-journey-beyond-corp

- National Cyber Security Centre (NCSC). (2021). "Secure by Design - Building a Security Culture." [Online]. Available: https://www.ncsc.gov.uk/guidance/secure-design-principles

Chapter 15:

The Innovation Mindset:

Cultivating an innovation mindset is imperative for individuals and organisations alike. How do you foster a mindset that not only encourages creativity and risk-taking but also addresses challenges in establishing an innovation-oriented culture?

15.1 Developing a Mindset that Encourages Creativity and Risk-Taking:

Innovation thrives in an environment that nurtures creativity and embraces risk. According to Carol Dweck, a renowned psychologist, adopting a "growth mindset" is pivotal in fostering creativity. In her seminal work, "Mindset: The New Psychology of Success," Dweck argues that individuals with a growth mindset believe that their abilities can be developed through dedication and hard work. This outlook encourages a willingness to take on challenges and view failures as opportunities for learning and improvement.

Steve Jobs, the co-founder of Apple, exemplified a mindset that celebrated creativity and risk-taking. Jobs famously stated, "Innovation distinguishes between a leader and a follower." His visionary approach transformed Apple into one of the most innovative companies globally, emphasising the importance of pushing boundaries and thinking differently.

To further promote creativity, organisations can implement strategies such as Google's "20% time," where employees are encouraged to spend a portion of their working hours on personal projects. This approach has led to groundbreaking products such as Gmail and Google Maps.

Overcoming Challenges to Instill an Innovation-Oriented Culture:

Establishing an innovation-oriented culture is not without its challenges. Leaders must address resistance to change, fear of failure, and bureaucratic hurdles. In his book, "The Innovator's Dilemma," Clayton Christensen identifies the "innovator's dilemma" as the struggle companies face when trying to innovate within the confines of their existing business models. He argues that successful companies must be willing to disrupt their own products and processes to stay ahead.

One company that overcame challenges to instill an innovation culture is Amazon. Jeff

Bezos, the founder, prioritised a customer-centric approach and a willingness to experiment. The company's commitment to long-term thinking and its focus on continuous improvement are evident in initiatives such as Prime and Amazon Web Services.

To create a culture that embraces innovation, leaders can leverage strategies outlined by Harvard Business Review, such as fostering cross-functional collaboration, encouraging diverse perspectives, and providing resources for experimentation.

Examples of an Innovation Mindset

1. *Sir James Dyson* - Reinventing the Wheel (and Vacuum Cleaners):

Sir James Dyson, a British inventor and entrepreneur, exemplifies the innovation mindset through his revolutionary approach to product design. Dyson faced numerous failures and setbacks in his quest to create a vacuum cleaner that didn't lose suction. His perseverance and commitment to improvement led to the development of the first bagless vacuum cleaner, the Dyson DC01, in 1993. Dyson's continuous pursuit of innovation extended beyond vacuum cleaners, with the introduction of bladeless fans, air purifiers, and hand dryers. His success underscores the importance of challenging conventional thinking and persisting through failures to achieve breakthrough innovations.

References:

- Dyson - Official Website

- BBC News - James Dyson's Vacuum Cleaner Success

2. *Tim Berners-Lee* - Connecting the World:

Tim Berners-Lee, a British computer scientist, is best known for inventing the World Wide Web. While working at CERN in 1989, Berners-Lee proposed the concept of a decentralised information system that would become the foundation of the modern internet. His invention transformed communication, commerce, and information-sharing globally. Berners-Lee's innovation mindset was rooted in the desire to solve a real-world problem: the efficient sharing of scientific information among researchers. By sharing his creation freely, he demonstrated the power of open collaboration and how disruptive ideas can reshape the world.

References:

- World Wide Web Foundation - Sir Tim Berners-Lee

3. *Elon Musk* - From PayPal to Mars:

Although born in South Africa, Elon Musk has made a significant impact on the innovation landscape through his ventures. Musk, the founder of SpaceX and Tesla, is a visionary entrepreneur with a relentless pursuit of ambitious goals. SpaceX, under Musk's leadership, has redefined space travel by developing reusable rockets and aspiring to establish a human settlement on Mars. Musk's risk-taking mentality, focus on sustainable energy solutions, and commitment to pushing technological boundaries highlight how an innovation mindset can lead to transformative change across industries.

References:

- SpaceX - Official Website

- Tesla - Official Website

These case studies illustrate how individuals with an innovation mindset challenge norms, overcome obstacles, and leave a lasting impact on their respective industries and the world at large.

In conclusion, developing an innovation mindset involves nurturing creativity, embracing risk, and overcoming challenges to establish a culture that values and encourages innovation. By drawing insights from the experiences of influential figures and successful companies, individuals and organisations can pave the way for continuous growth and evolution in an ever-changing world.

4. *Steve Jobs* - Visionary Reinvention at Apple:

Steve Jobs, the co-founder of Apple Inc., is a quintessential example of an individual who embodied the innovation mindset. Throughout his career, Jobs consistently sought to redefine and revolutionise industries through groundbreaking products and design aesthetics.

In the late 1990s, when Apple faced financial struggles, Jobs returned to the company and spearheaded a series of innovations that transformed it into one of the most valuable and influential tech companies globally. Jobs introduced the iMac G3, an all-in-one computer with vibrant colours that departed from the industry norm. His commitment to user-friendly design and seamless integration between hardware and software resulted in the development of iconic products such as the iPod, iPhone, and iPad.

One of Jobs' notable traits was his ability to anticipate and shape consumer needs, even before they were fully realised. The introduction of the iPhone in 2007, for instance, not only revolutionised the smartphone industry but also laid the foundation for the app ecosystem, changing the way people interact with technology.

Jobs' approach to innovation was not solely about technological advancement but also about creating a holistic user experience. His famous quote, "Design is not just what it looks like and feels like. Design is how it works," encapsulates his emphasis on the intersection of form and function.

References:

- <u>Apple - Official Website</u>
- Isaacson, W. (2011). Steve Jobs. Simon & Schuster.

Steve Jobs serves as a timeless example of how an unwavering commitment to innovation, paired with a focus on user experience and design aesthetics, can redefine entire industries and leave a lasting legacy.

Chapter 16:

Creating an Innovation Lab

To harness the power of creativity and foster breakthrough innovations, many forward-thinking companies are turning to the establishment of innovation labs.

An Innovation Lab is a dedicated space and budget within an organisation specifically designed to foster creativity, experimentation, and breakthrough innovation. Unlike traditional R&D departments, which may be more structured and risk-averse, innovation labs are dynamic environments that encourage out-of-the-box thinking and rapid iteration. These labs serve as incubators for new ideas, products, and processes, providing a playground for employees to explore innovative concepts without the constraints of day-to-day operations.

16.1 Why Invest in an Innovation Lab?

Organisations invest in innovation labs for several compelling reasons, recognising them as strategic assets that can drive long-term success:

1. **Catalyst for Transformation:** Innovation labs serve as catalysts for organisational transformation. They enable companies to adapt to changing market landscapes, technological advancements, and evolving customer expectations. By fostering a culture of constant experimentation, these labs empower organisations to stay ahead of the curve and remain relevant in highly competitive industries. *Reference: "The Innovator's Dilemma" by Clayton M. Christensen.*

2. **Mitigating Risk Through Experimentation:** Innovation inherently involves risk, and failure is often an integral part of the learning process. Innovation labs provide a controlled environment where teams can experiment with new ideas, products, or processes without jeopardising the core operations of the business. This risk mitigation allows for more ambitious and groundbreaking projects. *Reference: "Fail Fast, Fail Often" by Ryan Babineaux and John Krumboltz.*

3. **Attracting and Retaining Talent:** Top talent is drawn to organisations that foster a culture of innovation. Establishing an innovation lab signals to current and prospective employees that the company values creativity, offers opportunities for personal and professional growth, and is committed to staying at the forefront of industry trends. *Reference: "Drive: The Surprising Truth About What Motivates Us" by Daniel H. Pink.*

Acceleration in Results:

Investing in an innovation lab can accelerate results across various dimensions, leading to tangible and intangible benefits for the organisation:

1. **Speed to Market:** The streamlined and agile nature of innovation labs allows for faster development and deployment of new products or services. This agility enables organisations to respond swiftly to emerging market opportunities, gaining a competitive edge. *Reference: "The Lean Startup" by Eric Ries.*

2. **Enhanced Problem-Solving:** Innovation labs provide a dedicated space for cross-functional teams to collaborate and tackle complex challenges. The diversity of perspectives and expertise accelerates problem-solving, leading to more comprehensive and innovative solutions. *Reference: "Creative Confidence" by Tom Kelley and David Kelley.*

3. **Increased Return on Investment (ROI):** While innovation inherently involves risk, the structured experimentation within an innovation lab allows organisations to identify and invest in high-potential projects more effectively. This strategic approach maximises the likelihood of success and optimises the return on investment. *Reference: "Innovation and Entrepreneurship" by Peter F. Drucker.*

The innovation lab is a strategic investment for organisations seeking to navigate the dynamic landscape of the business world. By providing a dedicated space for creativity, experimentation, and collaboration, these labs become engines of transformation, driving accelerated results and positioning companies as leaders in their industries.

16.2 Designing and Implementing an Innovation Lab

Establishing an innovation lab requires careful planning and consideration of various factors, from physical space to team dynamics and strategic alignment. The design phase involves defining the lab's purpose, scope, and objectives, aligning it with the organisation's overall strategy.

Example: Google X's Moonshot Factory

Google X, Alphabet's innovation lab, is a prime example of a successful innovation lab. Known as the "Moonshot Factory," Google X focuses on ambitious, high-risk projects that have the potential to revolutionise industries. Their approach involves embracing failure as an inherent part of experimentation, fostering a culture of learning from mistakes and iterating rapidly.

To guide the design process, organisations can refer to publications such as "The Lean Startup" by Eric Ries, which advocates for a systematic, scientific approach to creating and managing successful startups in an age of uncertainty.

Best Practices for Fostering Experimentation, Collaboration, and Breakthrough Innovation

1. **Encourage Cross-functional Collaboration:** Break down silos and encourage collaboration across diverse teams. This diversity of thought and expertise can lead to innovative solutions that may not arise in more traditional, departmentalised structures.

2. 3M's renowned ITO program allows employees to spend 15% of their time working on projects of their choice. This practice has led to groundbreaking innovations such as Post-it Notes and Scotchgard.

3. Reference: Spencer Silver and Art Fry, "Serendipity and the Stumble Upon Methodology for Product Innovation," Journal of Product Innovation Management.

4. **Embrace a Culture of Experimentation:** Foster an environment where experimentation is not only accepted but encouraged. Provide resources and support for employees to test new ideas, even if they may initially seem unconventional.

5. Amazon's innovative culture is exemplified by the concept of Two-Pizza Teams – small, agile teams that can be fed with only two pizzas. This approach minimises bureaucracy and encourages nimble, experimental initiatives. Reference: Jeff Bezos, "2016 Letter to Shareholders," Amazon.

6. **Implement Agile Methodologies:** Adopt agile methodologies to enable quick iteration and adaptation to changing circumstances. This allows teams to respond swiftly to feedback and evolving market dynamics.

7. Spotify's Squad Model organises teams into small, autonomous squads, each with a specific mission. This agile structure promotes innovation and responsiveness. Reference: Henrik Kniberg and Anders Ivarsson, "Scaling Agile @Spotify with Tribes, Squads, Chapters & Guilds."

The creation of an innovation lab requires strategic vision, thoughtful design, and a commitment to fostering a culture of experimentation. By implementing best practices inspired by successful examples and leveraging resources from publications and thought leaders, organisations can establish innovation labs that drive transformative change and secure a competitive edge in today's dynamic business environment.

Innovation Lab Software

Innovation software plays a crucial role in supporting and enhancing the functions of an innovation lab. Here are examples of innovation software, including Wazoku, along with case studies that illustrate their applications:

Wazoku:

Overview: Wazoku is an innovation management platform that facilitates idea generation, collaboration, and the implementation of innovative projects. It provides a centralised space for employees to submit ideas, collaborate on projects, and track the progress of initiatives.

Case Study: Aviva

Overview: Aviva, a multinational insurance company, used Wazoku to launch their innovation initiative called "The Garage." This platform helped Aviva tap into the collective intelligence of their employees, leading to the development of innovative solutions and a more engaged workforce.

Reference: Wazoku Case Studies

IdeaScale:

Overview: IdeaScale is a cloud-based innovation platform that enables organisations to gather and prioritise ideas from employees, customers, and partners. It provides tools for collaboration, evaluation, and implementation of innovative concepts.

Case Study: NASA

Overview: NASA used IdeaScale to crowdsource ideas for improving their operations and advancing their mission. The platform allowed NASA to engage a wide audience and collect valuable insights, resulting in the implementation of cost-saving and efficiency-enhancing initiatives.

Reference: IdeaScale NASA Case Study

Spigit / IdeaPlace:

Overview: **Spigit/ Idea Place** is an innovation management platform that leverages crowdsourcing and collaboration to drive innovation. It helps organisations collect, evaluate, and prioritise ideas, fostering a culture of continuous improvement.

Case Study: Pfizer

Overview: Pfizer, a global pharmaceutical company, implemented ***Spigit/ Idea Place*** to engage employees in generating ideas for process improvement and cost reduction. The platform facilitated collaboration and allowed Pfizer to identify and implement innovative solutions across the organisation.

Reference: ***Spigit/ Idea Place*** Pfizer Case Study

Brightidea:

Overview: Brightidea is an innovation management platform that enables organisations to capture and develop ideas, collaborate on projects, and track the innovation lifecycle. It offers tools for ideation, portfolio management, and reporting.

Case Study: Cisco

Overview: Cisco used Brightidea to establish a structured innovation program called "Innovation Grand Challenge." This initiative encouraged employees to submit ideas for disruptive technologies, leading to the identification of innovative projects and the allocation of resources for further development.

Reference: Brightidea Cisco Case Study

These examples showcase how innovation software can be instrumental in driving innovation within an organisation. By leveraging these platforms, innovation labs can streamline idea generation, enhance collaboration, and ultimately accelerate the implementation of transformative initiatives.

10-Point Action Plan for Establishing an Innovation Lab

1. **Define Clear Objectives:** Clearly articulate the purpose and objectives of the innovation lab. Identify specific challenges or opportunities the lab will address, ensuring alignment with the overall strategic goals of the company.

2. **Secure Leadership Buy-In:** Obtain support from top-level executives to ensure commitment and resources. Leadership buy-in is crucial for overcoming potential resistance and securing the necessary budget and support.

3. **Assemble a Cross-Functional Team:** Form a diverse team representing various departments and expertise. This team will be responsible for driving the design, implementation, and ongoing operation of the innovation lab.

4. **Select the Right Innovation Software:** Choose an innovation management platform that aligns with the lab's objectives. Consider platforms like *Wazoku, IdeaScale, Spigit/ Idea Place*, or Brightidea, based on the specific needs and scale of the organisation.

5. **Allocate Physical and Technological Resources:** Designate a physical space for the lab, ensuring it is conducive to collaboration and creativity. Provide the necessary technological infrastructure, such as collaborative tools, prototyping equipment, and innovation software.

6. **Establish a Robust Governance Structure:** Develop a governance framework that outlines decision-making processes, responsibilities, and reporting structures. Clearly define how the lab integrates with existing organisational structures.

7. **Encourage a Culture of Innovation:** Cultivate a culture that encourages risk-taking, experimentation, and learning from failure. Communicate the value of innovation across the organisation and incentivise employees to actively participate in the lab's initiatives.

8. **Invest in Training and Development:** Provide training programs to equip employees with the skills required for innovation, including design thinking, agile methodologies, and effective collaboration. This investment in human capital is crucial for the success of the lab.

9. **Establish Key Performance Indicators (KPIs):** Define measurable KPIs to track the effectiveness of the innovation lab. Metrics may include the number of implemented ideas, time-to-market for innovations, and the impact on key business objectives.

10. **Anticipate and Address Challenges:**

Potential Challenges:

1. *Resistance to Change:* Employees may resist the cultural shift towards innovation. Address this by fostering open communication, showcasing success stories, and involving employees in the innovation process.

2. *Resource Constraints:* Limited budgets or competing priorities may pose challenges. Prioritise resources based on the strategic importance of the innovation lab and explore external funding options.

3. **Mitigation Strategies:**

    1. *Change Management Plan:* Implement a robust change management plan to address resistance and communicate the benefits of the innovation lab.

    2. *Strategic Resource Allocation:* Clearly articulate the value proposition of the lab to secure the necessary resources. Explore partnerships, grants, or external funding to supplement internal resources.

16.3 The Argument to the C-Suite for Innovation Lab Investment

1. Addressing Strategic Goals:

- Our Innovation Lab is not just a creative space; it is a strategic asset aligned with our organisational objectives. By leveraging the power of innovation, we can stay ahead of industry trends, tackle emerging challenges, and secure a leadership position in the market.

2. Tangible Returns on Investment (ROI):

- I understand the importance of tangible results. The Innovation Lab is not just an expense; it's an investment in our future success. Studies consistently show that companies with dedicated innovation initiatives experience higher revenue growth and profitability. [Reference: "The Innovator's Dilemma" by Clayton M. Christensen]

3. Mitigating Risks Through Controlled Experimentation:

- In the Innovation Lab, we create a controlled environment for experimentation. This minimises the risk associated with introducing new ideas, products, or processes, allowing us to identify and invest in high-potential projects more effectively. [Reference: "Fail Fast, Fail Often" by Ryan Babineaux and John Krumboltz]

4. Enhancing Operational Efficiency:

- The Lab is not just about dreaming big; it's about solving real operational challenges. By encouraging cross-functional collaboration and agile methodologies, we can streamline our processes, enhance efficiency, and reduce operational costs.

5. Cultivating a Culture of Innovation:

- A company's culture is its heartbeat. The Innovation Lab is designed to cultivate a culture that embraces innovation, attracts top talent, and retains our best minds. This, in turn, contributes to increased employee engagement and motivation. [Reference: "Drive: The Surprising Truth About What Motivates Us" by Daniel H. Pink]

6. Accelerating Time-to-Market:

- Time is of the essence in our fast-paced industry. The Innovation Lab enables us to respond swiftly to market opportunities, accelerating our time-to-market for new products and services. [Reference: "The Lean Startup" by Eric Ries]

7. Establishing a Competitive Edge:

- Our competitors are not standing still. The Innovation Lab positions us at the forefront of innovation, giving us a competitive edge that will be increasingly crucial in the evolving business landscape.

8. Demonstrating Industry Leadership:

- By investing in an Innovation Lab, we send a powerful message to our industry, clients, and stakeholders. It demonstrates our commitment to leading the way in innovation, setting us apart as a forward-thinking and adaptive organisation.

9. Strategic Resource Allocation:

- I understand the financial considerations. The Innovation Lab is a strategic investment, and its resource requirements will be allocated judiciously. We'll explore external funding options, grants, and partnerships to supplement internal resources.

10. A Proven Model for Success:

- Successful organisations across industries have embraced the concept of Innovation Labs. Companies like Google, Amazon, and Pfizer have seen remarkable success by fostering a culture of innovation. We can draw inspiration from their proven models and tailor them to suit our unique needs.

The Innovation Lab is not just a project; it's a journey towards securing our future success and staying at the forefront of industry innovation.

Chapter 17:

Three Horizon Innovation

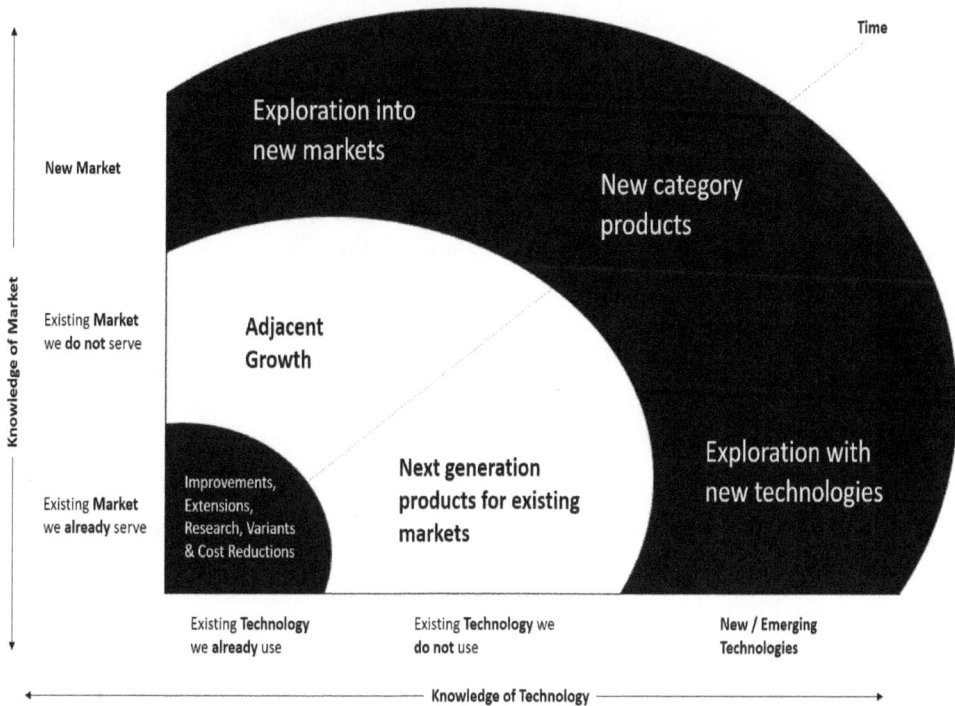

17.1 The McKinsey 3 Horizon Model

The McKinsey 3 Horizon Model is a strategic framework developed to guide organisations in managing their innovation portfolios effectively. The model delineates three distinct horizons, each corresponding to different time frames, risk levels, and strategic objectives.

In Horizon 1, organisations focus on optimising their existing core operations. This involves continuous improvement, cost efficiency, and incremental innovations to sustain and enhance the performance of current products and services. The primary goal is to maintain competitiveness in the present market and ensure the ongoing profitability and efficiency of the core business. This horizon is characterised by a relatively short-term perspective, emphasising the need for immediate returns and stability.

Horizon 2 represents a transitional phase where organisations explore emerging opportunities beyond their current core. The emphasis shifts towards diversification and growth through the identification and development of new products, services, or markets. While these initiatives may not contribute significantly to revenue in the short term, they have the potential for substantial growth in the medium to long term. Horizon 2 involves a balance between exploration and exploitation, seeking to capitalise on promising ventures that align with the organisation's strategic goals.

In Horizon 3, organisations embark on transformative and disruptive innovations that have the potential to create entirely new markets or reshape existing ones. This horizon is characterised by a longer time frame and higher levels of uncertainty and risk. Organisations in Horizon 3 invest in breakthrough technologies, novel business models, and paradigm-shifting concepts. The focus is on anticipating future market trends and proactively positioning the organisation to be a leader in upcoming, potentially revolutionary landscapes.

The 3 Horizon Model encourages a holistic approach to innovation management, urging organisations to concurrently exploit existing strengths, explore new growth opportunities, and invest in breakthrough initiatives that will secure their future relevance and success. By strategically balancing efforts across these three horizons, organisations can navigate the complexities of the business landscape and maintain a dynamic and sustainable approach to innovation.

Horizon 1: Focuses on the optimisation of existing core operations within an organisation. This horizon is characterised by a short-term perspective and a primary goal of maintaining and improving the performance of current products and services. Companies operating in Horizon 1 aim to sustain their competitiveness in the present market through incremental innovations and operational efficiencies.

One prominent example of a Horizon 1 strategy is the continuous improvement initiatives implemented by Toyota in the automotive industry. Toyota's renowned production system, which emphasises efficiency, waste reduction, and quality improvement, is a classic Horizon 1 approach. By refining their manufacturing processes and constantly seeking operational excellence, Toyota has maintained a strong competitive position in the global automotive market.

In the technology sector, companies like Apple exemplify Horizon 1 by regularly releasing updated versions of their existing products. For instance, the incremental improvements made to the iPhone with each new model release represent a Horizon 1 strategy aimed at enhancing the performance and features of the core product, thereby retaining customer loyalty and market share.

Horizon 1 strategies are also evident in the pharmaceutical industry, where companies continually invest in research and development to improve existing drugs or create extended-release versions. For instance, the development of statin drugs, which lower cholesterol levels, has seen Horizon 1 strategies at play as pharmaceutical companies strive to enhance the effectiveness and tolerability of these medications to maintain market leadership.

Horizon 2: Represents a crucial phase where organisations focus on exploring emerging opportunities beyond their existing core operations. This horizon involves initiatives that seek to diversify and expand the business portfolio, aiming for growth in new markets or through innovative products and services. One notable example of Horizon 2 is the development and introduction of new product lines by established companies.

A case study illustrating Horizon 2 is Apple's introduction of the iPod in the early 2000s. At that time, Apple was primarily known for its Macintosh computers. Recognising the emerging market for portable digital music players, Apple ventured into this new space with the iPod. This strategic move not only capitalised on a growing trend but also paved the way for the subsequent success of the iTunes platform and the revolutionary iPhone. The iPod, initially a Horizon 2 initiative, played a pivotal role in diversifying and expanding Apple's product offerings.

Similarly, the pharmaceutical industry frequently engages in Horizon 2 initiatives through the development of new drugs and therapies. For instance, a pharmaceutical company might invest in research and development to explore treatments for emerging health issues or novel approaches to address existing medical challenges. The pursuit of these opportunities reflects a strategic focus on growth beyond the current product lineup, exemplifying the essence of Horizon 2.

In the technology sector, the emergence of cloud computing can be viewed as a Horizon 2 initiative. Established tech companies and newcomers alike recognised the potential for delivering services and computing resources over the internet. Companies like Amazon with Amazon Web Services (AWS) and Microsoft with Azure strategically entered this space, expanding beyond their traditional offerings. Cloud computing has since become a cornerstone of the tech industry, illustrating the success of Horizon 2 initiatives in responding to evolving market dynamics.

In summary, Horizon 2 is a pivotal stage in innovation, marked by initiatives that explore and capitalise on emerging opportunities, leading to diversification and growth.

Horizon 3: Represents the domain of transformative, disruptive innovations that have the potential to redefine industries and create entirely new markets. This horizon is characterised by a longer time frame, higher risk, and a focus on envisioning and preparing for the future. One notable example of a Horizon 3 initiative is the development of electric vehicles (EVs) by companies like Tesla. The introduction of electric cars not only challenged the traditional automotive industry but also revolutionised the way people think about transportation and energy consumption. Tesla's success in this Horizon 3 venture underscores the transformative power of disruptive innovations.

Another illustrative case study for Horizon 3 is the emergence of blockchain technology. Originally conceptualised as the underlying technology for cryptocurrencies like Bitcoin, blockchain has evolved into a groundbreaking innovation with applications across various industries. Beyond finance, blockchain has found uses in supply chain management, healthcare, and decentralised finance (DeFi). The development and widespread adoption of blockchain exemplify the potential of Horizon 3 initiatives to revolutionise established systems and introduce new, decentralised paradigms.

In the pharmaceutical industry, the pursuit of gene editing technologies such as CRISPR-Cas9 represents a Horizon 3 initiative. The ability to modify genetic material has profound implications for healthcare, enabling the treatment of genetic disorders at a fundamental level. Companies investing in CRISPR technology aim not only to develop groundbreaking therapies but also to shape the future landscape of medicine. This Horizon 3 endeavour demonstrates the potential for transformative innovations to address complex challenges and redefine entire sectors.

Furthermore, the ongoing exploration of space by private companies like SpaceX, led by Elon Musk, is a Horizon 3 initiative with far-reaching consequences. SpaceX's goal of enabling human colonisation of Mars and reducing the cost of space travel has the potential to revolutionise space exploration, making it more accessible and sustainable in the long term. This audacious vision exemplifies the nature of Horizon 3, where organisations push the boundaries of what is currently possible to create a future that is fundamentally different from the present.

In summary, Horizon 3 initiatives are characterised by their transformative and disruptive nature, challenging the status quo and reshaping industries.

17.2 The Horizon Squad Model

Establishing Horizon Squads: Begin by forming three distinct Horizon Squads, each dedicated to a specific horizon within the McKinsey 3 Horizon Model. Clearly define the scope and objectives for each squad, ensuring that Horizon 1 focuses on optimizing current operations, Horizon 2 explores emerging opportunities, and Horizon 3 engages in disruptive innovations. Assign cross-functional teams to each squad, comprising members with skills in innovation, product ownership, and application development. These squads should be designed to be self-sufficient units capable of independently driving initiatives within their respective horizons.

Defining Roles and Responsibilities: Clearly articulate the roles and responsibilities of each squad to establish a seamless workflow. In Horizon 1, the squad should prioritise optimisation and incremental improvements. In Horizon 2, the focus shifts towards exploration and diversification, while Horizon 3 concentrates on visionary and transformative initiatives. Designate product owners for each squad who will take ownership of aligning innovation and development efforts with the strategic goals of their respective horizons. Ensure that these product owners collaborate closely with dedicated application development teams and innovation specialists within their squads.

Cross-Functional Collaboration: Promote a culture of cross-functional collaboration within and between the Horizon Squads. Encourage open communication channels to facilitate the exchange of ideas and insights. While each squad operates independently, there should be mechanisms in place for sharing lessons learned, best practices, and emerging trends. Establish regular cross-squad meetings to foster collaboration, alignment, and a holistic understanding of the organisation's innovation landscape. This collaborative approach ensures that knowledge and expertise are shared across horizons, promoting a unified vision for the organisation's future.

Agile Methodologies and Continuous Improvement: Adopt agile methodologies to facilitate flexibility and adaptability within each Horizon Squad. Implement regular sprint cycles, retrospectives, and feedback loops to enhance responsiveness to changing market conditions. Foster a culture of continuous improvement, where each squad reflects on their processes, identifies areas for enhancement, and implements iterative changes. This dynamic approach allows the organisation to stay nimble and responsive, ensuring that each squad can adjust its strategies and priorities based on evolving business landscapes and emerging opportunities within their respective horizons.

By implementing these steps, the organisation can successfully transition to a structure with three self-sufficient Horizon Squads. This model aligns with the McKinsey 3 Horizon framework, ensuring that each squad can independently drive innovation, product ownership, and application development efforts tailored to the specific objectives of their assigned horizon.

Utilising Squads with the 3 Horizon Model: Examples

Here's a more detailed look at how squads and the 3 Horizon Model intertwine within Spotify, Maersk, ING Bank and Netflix.

1. Spotify:

- Squad composition: Highly autonomous, cross-functional teams with diverse skillsets (engineering, design, product management).
- Horizon 1 squads: Focus on bug fixes, performance optimisation, and core feature enhancements. They work closely with user insights and data to identify areas for improvement.
- Horizon 2 squads: Explore new features and business lines within the existing platform. They experiment with interactive elements, personalised playlists, and potential integrations with other music services.
- Horizon 3 squads: Push the boundaries of music experiences, investigating concepts like AI-powered composition, AR concerts, and VR music visualisation. These squads often have more risk tolerance and involve collaboration with external partners.
- Squad interaction: While focused on their specific horizons, squads share learnings and insights through regular cross-horizon meetings and hackathons. This fosters innovation and ensures alignment with the overall Spotify vision.
- Benefits: Increased agility, faster innovation, improved alignment with changing market trends.
- Challenges: Ensuring communication and collaboration across horizons, preventing siloing, managing resource allocation.
- Online Resources:
    - Spotify Labs: https://spotifystudios.com/spotify-labs/
    - Spotify Engineering Blog: https://engineering.atspotify.com/2021/09/how-backstage-made-our-developers-more-effective-and-how-it-can-help-yours-too/

2. Maersk:

- Tribe structure: Each tribe oversees a specific business area (e.g., container shipping, terminal operations) and houses multiple squads with complementary expertise.
- Horizon 1 squads: Optimise current operations within their designated area, focusing on areas like route planning, fuel efficiency, and automated processes.
- Horizon 2 squads: Drive digital transformation within the logistics domain, developing tools like blockchain-based supply chain tracking and automated warehouse systems. These squads collaborate with external technology providers and industry partners.
- Horizon 3 squads: Explore entirely new business models that could disrupt the shipping industry, such as autonomous ships, drone delivery networks, and data-driven logistics partnerships. These squads have high autonomy and operate with agility.
- Squad allocation: Tribe leaders assess each squad's expertise and project potential before assigning them to a specific horizon. This ensures optimal resource allocation and aligns individual work with broader strategic goals.
- Benefits: Increased efficiency, reduced costs, accelerated digital transformation.
- Challenges: Aligning tribe and squad goals, balancing innovation with core operations, managing talent across horizons.
- Online Resources:
    - Maersk Innovation Lab: https://innovation.maersk.com/
    - Maersk News: https://www.maersk.com/news

3. ING Bank:

- Chapter composition: Cross-functional groups drawn from various bank departments (IT, marketing, finance, customer service). This diversity fosters unique perspectives and drives innovation.
- Horizon 1 chapters: Focus on incremental improvements to existing products and services, like mobile banking app upgrades, personalised financial advice, and more efficient internal processes.
- Horizon 2 chapters: Develop new offerings within the financial sector, such as robo-advisors for wealth management, micro-loans for underserved communities, and blockchain-based payments solutions.
- Horizon 3 chapters: Explore disruptive technologies redefining banking, including blockchain-based identity management, AI-powered financial planning tools, and personalised insurance models. These chapters often collaborate with universities and fintech startups.

- Chapter collaboration: Chapters choose projects across horizons based on their collective expertise and interest. This promotes cross-functional collaboration and breaks down traditional silos within the bank.
- Benefits: Enhanced customer experience, increased revenue streams, future-proofed business model.
- Challenges: Fostering cross-functional collaboration, overcoming risk aversion, integrating new technologies with legacy systems.
- Online Resources:
    - ING Innovation Hub: https://www.ingwb.com/en/about-us/about-us-innovation
    - ING Think Forward: https://think.ing.com/

4. Netflix:

- Feature team structure: Cross-functional teams focused on specific features or functionalities within the streaming platform. These teams operate with a high degree of autonomy and experimentation.
- Horizon 1 teams: Continuously optimise core features like video playback, content recommendations, and search functionality. They leverage data analytics and user feedback to ensure smooth platform operation.
- Horizon 2 teams: Test new features and content formats within the existing platform, such as interactive shows, personalised trailers, and innovative video editing tools. These teams take calculated risks and iterate based on user engagement.
- Horizon 3 teams: Explore immersive technologies with the potential to transform entertainment experiences, like VR storytelling, interactive movie formats, and AR gaming tools. These teams collaborate with cutting-edge tech companies and research institutions.
- Horizon flexibility: While teams are aligned with specific horizons, they have the flexibility to address emerging trends and experiment across horizons if a promising opportunity arises. This keeps Netflix nimble and adaptive.
- Benefits: Continuous improvement, rapid experimentation, leadership in innovative content formats.
- Challenges: Managing resource allocation across horizons, balancing risk-taking with proven successes, maintaining a culture of creativity.
- Online Resources:
    - Netflix Tech Blog: https://netflixtechblog.com/
    - Netflix Engineering Culture: https://jobs.netflix.com/culture

These are just a few examples, and the specific ways squads and the 3 Horizon Model are combined can vary greatly. However, hopefully, these insights provide a clearer picture of how organisations leverage this approach to drive innovation, prioritise initiatives, and stay ahead of the curve in their respective industries.

Remember, the key to success lies in finding the right balance between autonomy, alignment, and collaboration within the squad and horizon framework. By fostering a culture of innovation, open communication, and adaptable structures, organisations can unlock the full potential of this powerful approach to strategic planning and execution.

Three Horizon Innovation Squad model – 10 Sept Action Plan

Implementing a structural change to foster agility and performance management across three self-sufficient teams, named Horizon Squads, requires a carefully orchestrated action plan. Here's a 10-step guide:

1. **Leadership Alignment and Communication:**

    1. Ensure leadership alignment on the shift towards agile structures.
    2. Communicate the reasons behind the change and the benefits of the Horizon Squads model.
    3. Clearly articulate the roles and responsibilities of each Horizon Squad.

2. **Skill Assessment and Team Formation:**

    1. Conduct a comprehensive skills assessment of existing staff.
    2. Form three distinct Horizon Squads, each with a mix of skills covering innovation, product ownership, and application development.
    3. Foster a culture of cross-functional collaboration within each squad.

3. **Training and Onboarding:**

    1. Provide training sessions to equip teams with necessary agile methodologies and tools.
    2. Facilitate cross-training among team members to enhance versatility.
    3. Ensure that all members understand the principles of the 3 Horizon Model and their roles within it.

4. **Resource Allocation and Autonomy:**

    1. Allocate dedicated resources for each Horizon Squad, encompassing innovation, product ownership, and application development.
    2. Grant autonomy to each squad to make decisions within their defined scope.
    3. Establish a transparent process for resource allocation to prevent bottlenecks and encourage collaboration.

5. **Establish Clear Metrics and Key Performance Indicators (KPIs):**

    1. Define measurable metrics and KPIs aligned with the objectives of each Horizon Squad.
    2. Metrics should cover innovation success, product performance, and development efficiency.
    3. Regularly track and evaluate performance against these metrics.

6. **Regular Retrospectives and Continuous Improvement:**

    1. Conduct regular retrospectives at the end of each development cycle to reflect on successes and challenges.
    2. Use insights gained to implement continuous improvement strategies.
    3. Encourage a culture of openness and learning from experiences.
    4.

7. **Cross-Squad Collaboration:**

    1. Facilitate regular meetings or workshops for Horizon Squads to share insights, challenges, and best practices.
    2. Promote a collaborative environment where teams can learn from each other and foster innovation collectively.

8. **Agile Project Management Tools:**

    1. Implement agile project management tools to enhance communication and collaboration within and between Horizon Squads.
    2. Ensure that tools support transparency, adaptability, and real-time tracking.

9. **Performance Reviews and Recognition:**

    1. Establish a performance review process aligned with agile principles.
    2. Recognise and reward achievements and contributions at both the individual and team levels.
    3. Create a feedback loop for continuous improvement based on performance reviews.

10. **Regular Leadership Support and Monitoring:**

    1. Provide ongoing leadership support to Horizon Squads.
    2. Regularly monitor progress and intervene as needed to address any challenges.
    3. Encourage a culture of agility, adaptability, and a shared commitment to the organisation's overall goals.

Chapter 18:

The Power of Imagination

As the custodians of technological advancement, CTOs are poised to unlock the full potential of the future through the power of imagination. In a world where technology and imagination intersect, the possibilities are limitless, and the CTO stands as the visionary architect shaping the course of innovation.

Known as one of the pioneers of science fiction, Jules Verne's novels, such as "Twenty Thousand Leagues Under the Sea" and "From the Earth to the Moon," envisioned submarines, space travel, and even the concept of exploring the deep sea. Arthur C Clarke's novel "2001: A Space Odyssey" not only inspired a groundbreaking film but also anticipated technologies like artificial intelligence and space exploration. Clarke's vision of geostationary satellites for global communication also became a reality. Of course no self-respecting techy could ignore the creator of "Star Trek," and Gene Roddenberry's vision of a world where humanity had overcome its differences and was exploring space as part of a united federation. Many of the technologies portrayed in "Star Trek," such as communicators (similar to modern smartphones), sliding doors and tablet computers, have become a reality.

18.1 Imagination as the Catalyst for Innovation

Innovation is fueled by the ability to envision possibilities beyond the existing boundaries. Imagination serves as the catalyst that sparks creative thinking and enables a CTO to see potential where others may see limitations. By encouraging a culture that values and nurtures imaginative thinking, a CTO can inspire their team to explore uncharted territories, pushing the boundaries of what technology can achieve.

Drawing Inspiration from Visionaries

Throughout history, technology has been profoundly influenced by visionaries who dared to dream beyond convention. From Nikola Tesla's imaginative musings on wireless power to Steve Jobs' vision of seamlessly integrated personal computing, the pages of innovation are written by those who let their imaginations soar. By studying the visionary ideas of the past, a CTO gains insights into the profound impact imagination can have on shaping the future.

Navigating Technological Frontiers

A CTO must navigate through technological frontiers where the uncharted territory is often the breeding ground for groundbreaking ideas. Whether it's adopting emerging technologies like artificial intelligence or blockchain, imagination becomes the compass guiding technological exploration. A CTO who embraces imaginative thinking becomes an architect of the future, steering their organisation towards innovative solutions.

The Synergy of Imagination and Collaboration

Imagination thrives in collaborative environments. By fostering a culture of open communication and cross-disciplinary collaboration, a CTO can harness the collective imaginative power of their team. Bill Gates and Paul Allen's collaborative journey at Microsoft is a testament to the synergistic effects of shared imaginative thinking. The CTO must be a catalyst for collaboration, recognising that the most impactful innovations often arise from the intersection of diverse perspectives.

Imagination in Problem-Solving

Imagination is not only about envisioning grand possibilities but also about solving complex problems. When faced with challenges, a CTO can leverage imaginative thinking to explore unconventional solutions. This approach was evident in Alan Turing's groundbreaking work in the field of computing, where he used imagination to conceptualise solutions to intricate mathematical problems, laying the groundwork for modern computers.

Imagination for Ethical and Inclusive Innovation

A responsible CTO recognises that imagination must be tempered with ethical considerations. Imagination should not only fuel innovation but also guide it towards socially responsible and inclusive outcomes. By envisioning a future where technology serves humanity ethically and inclusively, a CTO can lead their organisation towards innovations that have a positive and lasting impact on society.

18.2 Futurism

Engaging with a futurist can be a strategic move for organisations seeking a deeper understanding of how technology changes may shape markets, instigate social shifts, and catalyse the development of new products and services. Futurists are skilled in anticipating and interpreting trends, providing invaluable insights that can inform strategic decision-making. In the initial phase of collaboration, the organisation can work closely with the futurist to articulate its goals, challenges, and aspirations. This includes outlining specific areas of interest such as emerging technologies, market dynamics, and potential societal transformations.

Once the objectives are defined, the futurist can conduct a comprehensive analysis of technological trends, drawing from a diverse range of disciplines including science, economics, and sociology. This analysis may involve assessing the trajectory of key technologies, such as artificial intelligence, biotechnology, or renewable energy, and understanding their potential impact on the organisation's industry and broader society. Through this exploration, the futurist can highlight both opportunities and risks associated with technological advancements, enabling the organisation to proactively position itself in a rapidly evolving landscape.

The collaboration doesn't end with trend analysis; futurists often assist organisations in developing strategic roadmaps for the future. This involves co-creating scenarios that envision different plausible futures based on varying degrees of technological adoption and societal change. By synthesising these scenarios, organisations can gain a nuanced understanding of potential futures and develop adaptive strategies that are resilient to uncertainty. Futurists may also facilitate workshops or innovation sessions within the organisation, fostering a culture of forward-thinking and providing a platform for employees to contribute to the collective vision of the future. In essence, working with a futurist becomes a transformative journey that equips the organisation to navigate the complexities of an ever-evolving technological landscape and capitalise on emerging opportunities.

Examples of Futurists.

Alicia Asín is the CEO and co-founder of Libelium an open source sensor hardware platform for the Internet of Things (IoT). Award-winning innovation expert, Alicia provides great insight on data mining, particularly addressing the big questions surrounding Big Data and the relevance of security and privacy in the new IoT era.

Andrew Grill is a practical Futurist and Former Global Managing Partner at IBM, Andrew Grill specialises on digital disruption, workplace of the future and digital eminence.

Cate Trotter is the Founder and Head of Trends at Insider Trends, a leading London- based trendspotting consultancy. She is an expert in retail, trends and technology.

Dr Mark van Rijmenam is a storyteller on disruptive innovations and a big data and blockchain strategist.

18.3 Case Studies - Enginuity:

Imagination: When I was at Enginuity, the Engineering Skills Organisation, we imagined a world where someone who struggled to demonstrate their skills through an academic route, could demonstrate their skills in a way where they could shine and show their capabilities. Everyone learns differently and are motivated in different ways. An academic education and qualification does not suit every individual. Engineering companies had been recruiting Engineers in the same academic image for generations, and knew they were not only missing amazing talent, but also different and innovative new ways of thinking. They also knew that with the speed of technological change it was equally important to know how a future employee could learn "the next thing" rather just what they knew today. They needed Engineers who could adapt quickly to a fast changing technological industry.

Prototype: Working with leading Engineering companies to identify key competencies they wanted to see in their Engineers, we designed a game with Minecraft that took away the academic pressure but designed increasingly complex game scenarios utilising Minecraft Redstone Components. These are blocks that can be used to build complex circuits. Redstone components include power components (such as Redstone light emitters, buttons, and pressure plates), transmission components (such as Redstone dust and Redstone repeaters), and mechanism components (such as pistons, doors, and lamps). The game was fun, challenging and measured those key competencies required by the Engineering companies.

Measurement: How do we know it is a success? The competencies we were measuring included, Critical Reasoning, Communication, Determination and Resilience, Observation and Assessment, Engineering Process, Health and Safety, Quality, Creative Thinking and Imagination, the ability to deal with change and importantly – their speed of learning and speed of embedding that new learning into their thinking. In other words, how do they deal with change, how quickly they can learn the next technology/process/idea and then the willingness to put new ideas into their thinking or how they might test the new ideas to see if that was a better approach to the old.

Benchmarking: The platform was benchmarked against Engineering degree graduates to see how they performed with the game and also how expert engineers would demonstrate those competencies*. The game was calibrated against those findings and taken out to a number of test groups.

One group which was part of the benchmark set was a group from The Prince's Trust in Glasgow (now The King's Trust) charity. A group of amazing young people but with lower levels of academic achievement, low career prospects and in some cases low career ambition. Many were classified as NEET (Not in Education, Employment or Training) and many Neurodiverse. The results were incredibly exciting. The skills profile map within this group was in many ways similar to the Engineering Graduate group but it showed key individuals who not only matched the Engineering graduates but, in some cases, outperformed them by x3 or higher. One young woman in this test group remarked that "this was the first time she was told that she was actually good at anything". She wasn't just good she was amazing.

Enginuity went on to create a number of exciting applications to support Engineers and Engineering companies. www.enginuity.com

* This in itself led to spin-off thinking of how do you identify the skills of expert engineers, as we know what makes a successful Engineer isn't just technical skills. If we could map the entire skills profile we could create a skills profile we could replicate and recruit to.

The Future Unleashed by Imagination

What groundbreaking technologies lie ahead, waiting to be discovered? How can current technological capabilities be harnessed to address pressing global challenges? Imagination, when harnessed and channeled effectively, becomes the compass guiding the CTO and their team towards a future where innovation knows no bounds.

As the custodians of technological advancement, CTOs are poised to unlock the full potential of the future through the power of imagination. In a world where technology and imagination intersect, the possibilities are limitless, and the CTO stands as the visionary architect shaping the course of innovation.

Chapter 19:

Global Perspectives on Innovation

A comprehensive understanding of global trends and practices is essential for staying at the forefront of technological advancements and business competitiveness. This chapter delves into the diverse and dynamic world of innovation, exploring trends and practices on a global scale. Additionally, we draw valuable insights from international case studies and success stories to distil lessons that can inspire and inform innovation strategies.

19.1 Exploring Global Innovation Trends

Innovation is a universal driver of progress, but its manifestations vary across regions and industries. One notable trend is the rise of collaborative innovation ecosystems. Countries like Finland and Singapore have embraced open innovation platforms, fostering collaboration between businesses, research institutions, and government bodies. The concept of open innovation, popularised by Henry Chesbrough, promotes the idea that valuable ideas can come from both internal and external sources.

Moreover, the global shift towards sustainable innovation is evident. The European Union's Horizon 2020 initiative exemplifies a commitment to fostering innovation that addresses societal challenges, with a particular focus on sustainability. This initiative has not only stimulated research and development but also spurred the creation of innovative solutions to climate change, resource depletion, and social inequalities.

Global Innovation Centres

United Kingdom: Embracing Open Innovation and Ecosystems

The United Kingdom has embraced a multifaceted approach to global innovation, with a strong emphasis on open innovation and collaborative ecosystems. Initiatives like the Catapult Centres, which provide cutting-edge facilities and support for businesses, exemplify the commitment to fostering collaboration between academia and industry. The UK's innovation strategy often intertwines with government support through funding and incentives, creating an environment where startups and established companies can thrive. Furthermore, the UK's emphasis on sustainability and green innovation, as seen in the Clean Growth Strategy, positions the nation as a leader in addressing global challenges through innovative solutions.

United States: Silicon Valley and Entrepreneurial Spirit

The United States, particularly Silicon Valley, stands as a global epicentre of innovation. Renowned for its entrepreneurial culture and risk-taking mindset, the U.S. encourages individuals to turn ideas into reality. The venture capital ecosystem plays a pivotal role, providing substantial funding for innovative ventures. The "fail fast, fail often" mentality fosters a culture where learning from setbacks is as valuable as celebrating successes. The U.S. government also supports innovation through research funding and policies that promote competition. However, the decentralised nature of innovation in the U.S. means that strategies and priorities can vary widely across industries and regions.

China: Rapid Iteration and Hardware Innovation in Shenzhen

China's approach to global innovation, particularly in cities like Shenzhen, is characterised by rapid iteration and a focus on hardware innovation. Shenzhen has evolved into a hardware innovation powerhouse, fostering an ecosystem where startups can quickly prototype and manufacture products. The Chinese government plays a proactive role, providing financial incentives and policy support to stimulate innovation. China's innovation strategy also involves targeted investments in emerging technologies, such as artificial intelligence and 5G, positioning the country as a global leader in these fields. The emphasis on large-scale collaboration and a fast-paced business environment sets China apart in its unique approach to innovation.

Japan: Precision, Long-Term Vision, and Industry-Academia Collaboration

Japan's approach to global innovation reflects a commitment to precision, long-term vision, and close collaboration between industry and academia. The country has a strong tradition of research and development, with a focus on quality and reliability. The Japanese government actively supports innovation through initiatives like the Society 5.0 vision, which aims to integrate cutting-edge technologies for societal benefit. Japanese companies often prioritise long-term goals over short-term gains, contributing to the country's reputation for creating sustainable and enduring innovations. The emphasis on precision engineering, meticulous planning, and a culture of continuous improvement sets Japan apart in the global innovation landscape.

Each country – the UK, USA, China, and Japan – brings a unique set of strengths and strategies to the table, contributing to the rich tapestry of global innovation. While Silicon Valley epitomises entrepreneurial spirit in the U.S., Shenzhen showcases China's prowess in hardware innovation. The UK's focus on open innovation and sustainability, along with Japan's commitment to precision and long-term vision, further highlight the diversity of approaches in the dynamic realm of global innovation.

19.2 City Innovation Hubs – Case Study London

Innovation Culture in London: A Tapestry of Creativity and Diversity

London, a global financial and cultural powerhouse, stands as a vibrant epicentre of innovation. The city's innovation culture is deeply woven into its historical fabric and continues to evolve, embracing diversity, collaboration, and a dynamic entrepreneurial spirit.

1. Thriving Tech Ecosystem: London has cultivated a thriving tech ecosystem, often referred to as "Silicon Roundabout" or "Tech City," located in the Shoreditch area. This cluster of tech startups and scale-ups is a testament to the city's commitment to fostering innovation. Notable success stories include TransferWise (now Wise), a fintech unicorn revolutionising international money transfers. The collaborative environment of co-working spaces, incubators, and accelerators has allowed these startups to share ideas, resources, and talent, contributing to the vibrant tech landscape.

2. Cultural Diversity Driving Creativity: London's innovation culture is profoundly influenced by its cultural diversity. The city serves as a melting pot of ideas, bringing together people from various backgrounds, industries, and perspectives. The Fintech sector, for instance, has witnessed the emergence of innovative solutions such as Revolut, a financial technology company founded by a diverse team with backgrounds spanning from Russia to the UK. This cultural amalgamation fosters a rich exchange of ideas and fuels creativity, making London an ideal breeding ground for groundbreaking innovations.

3. Government Support and Initiatives: The UK government has played a pivotal role in nurturing London's innovation culture through supportive policies and initiatives. The creation of Tech Nation and the introduction of tax incentives for research and development have encouraged the growth of startups and technology-driven enterprises. Additionally, the London Co-Investment Fund (LCIF) has provided crucial funding to early-stage businesses. These initiatives demonstrate a commitment to creating an environment where innovation can flourish, attracting talent and investment to the city.

4. Collaborative Spaces and Events: London's innovation culture is not confined to corporate boardrooms but extends to collaborative spaces and events that facilitate networking and knowledge exchange. Places like the Royal Academy of Engineering's

Enterprise Hub and events like London Tech Week bring together innovators, investors, and thought leaders. The serendipitous encounters in these spaces often lead to the cross-pollination of ideas, sparking new ventures and collaborations that contribute to the city's dynamic innovation landscape.

London's innovation culture is a dynamic blend of historical significance, cultural diversity, government support, and collaborative spaces. As the city continues to evolve, its commitment to fostering creativity and embracing diverse perspectives ensures that it remains at the forefront of global innovation, offering a compelling narrative for entrepreneurs and innovators from around the world to contribute to its ever-expanding tapestry of ideas.

Drawing Insights for Innovation Strategies

By learning from international experiences, businesses can gain a competitive edge, adapt to emerging trends, and navigate the complex terrain of global innovation. The examples provided serve as beacons of inspiration, illustrating that innovation is not confined to a particular geography but thrives in environments that nurture creativity, collaboration, and adaptability. As we continue to navigate the global innovation landscape, staying attuned to these trends and case studies will be crucial for driving sustained success in the ever-evolving world of innovation.

Appendix A
RISK REGISTER

Risk Register - Risks to business performance, project delivery, staff safety, customer experience and reputation

Division: Digital Product Team

Owner: Name of Individual

ID	Company or Project	Date raised	Risk Type	Risk description	Likelihood of the risk occurring	Impact if the risk occurs	Severity *Rating based on impact & likelihood*	Owner *Person who will manage the risk*	Mitigating action *Actions to mitigate the risk e.g. reduce the likelihood.*	Contingent action *Action to be taken if the risk happens.*	Progress on actions	Status	Useful resources
2	Products	Date	Reputation	Description 1	Medium	High	High	All Product Leads	Mitigating Actions				
2	Finance	Date	Legal	Description 2	Medium	High	High	L&D	Mitigating Actions				
2	HR	Date	Project Delivery	Description 3	Low	Medium	Medium	Joe Bloggs	Mitigating Actions				

MARLOW BUSINESS SCHOOL

APPENDIX B – DUE DILIGENCE CHECKLIST

NO.	PRIORITY	DESCRIPTION	✓
ORGANISATION			
		TECHNOLOGY ORGANISATION CHART	
		HEADCOUNT BY PRODUCT – INCLUDE TITLES	
		CORPORATE IT BUDGET, HEADCOUNT, EXPENSES, TURNOVER	
		DEVELOPMENT TEAM STRUCTURE, LOCATION, ROLES, SKILLS	
		HUMAN RESOURCE RISK AND PROFESSIONAL DEVELOPMENT	
		STAFF ONBOARDING AND SECURE DEVELOPMENT TRAINING	
		ON/OFFSHORE TEAM DYNAMICS, LOCATIONS	
		PRODUCT VERSIONS AND SUPPORT POLICY	
		DEFECT ESCALATION PROCESS, TICKETING	
		CUSTOMER SUPPORT STRUCTURE, TOOLS, PROCESS, SLA'S	
		SUPPORT CHANNELS, CHAT, CONVERSATIONAL IVR	
		3RD PARTY SUPPORT RELATIONSHIPS AND SLA'S	
		SECURITY AWARENESS TRAINING FOR STAFF	
		CYBER SECURITY INSURANCE	

NO.	PRIORITY	DESCRIPTION	✓
INFRASTRUCTURE & OPERATIONS			
		DIAGRAM OF INFRASTRUCTURE	
		INFRASTRUCTURE SUMMARY DOCUMENT, VESRION HISTORY	
		SYSTEM OUTAGES, LAST 6 MONTHS, TRACKING, CAUSE ANALYSIS	
		SYSTEM BUDGETS, MAINTAINANCE AND COSTS	
		CAPACITY, PERFORMANCE AND SYSTEM UTILISATION	
		MONITORING & FEEDBACK SYSTEMS	
		OPERATIONAL KPI'S AND TRACKING PROCESS	
		SECURITY PROCESS, PROCEDURE, ACCREDITATION	
		SECURITY AUDITS, PENETRATION TESTS & ACTION PLANS	
		SECURITY INCIDENTS LAST 24 MONTHS, LAST 2 AUDITS	
		SLA'S INTERNAL AND EXTERNAL CLIENTS	
		BUSINESS CONTINUITY PLAN, POLICY, PROCESS AND REVIEW	
		DISASTER RECOVERY PLAN AND TEST RESULTS	
		RISK REGISTER PROCESS AND ACTIONS	
		MANAGEMENT CONTROLS PROCESS, WIKI	
		CORPORATE LIST OF BUSINESS APPLICATIONS AND LOCKDOWN	
		REGULATIONS: GDPR, DATA SOVEREIGNTY PROCESS	
		INTRUSION PROTECTION & SYSTEM ACCESS CONTROLS, TOOLS	
		APPROACH FOR DEALING WITH COMMON SECURITY THREATS	